T0323927

Arousal in Neurological and Psychiatric Diseases

Arousal in Neurological and Psychiatric Diseases

Edited by

Edgar Garcia-Rill

Academic Press is an imprint of Elsevier
125 London Wall, London EC2Y 5AS, United Kingdom
525 B Street, Suite 1650, San Diego, CA 92101, United States
50 Hampshire Street, 5th Floor, Cambridge, MA 02139, United States
The Boulevard, Langford Lane, Kidlington, Oxford OX5 1GB, United Kingdom

Library of Congress Cataloging-in-Publication Data
A catalog record for this book is available from the Library of Congress

British Library Cataloguing-in-Publication Data
A catalogue record for this book is available from the British Library

ISBN 978-0-12-817992-5

For information on all Academic Press publications
visit our website at https://www.elsevier.com/books-and-journals

Publisher: Nikki Levy
Acquisition Editor: Melanie Tucker
Editorial Project Manager: Samantha Allard
Production Project Manager: Bharatwaj Varatharajan
Cover Designer: Miles Hitchen

Typeset by SPi Global, India

Dedication

This book is dedicated to Erica Beecher-Monas, a colleague with whom I published a half-dozen law reviews. Often times posthumously rendered dedications are exaggerated. This is not true for Erica Beecher-Monas. In her case the perseverance, honesty, and grace with which she lived her personal life were applied in equal measure to her work.

Contents

Contributors

Numbers in paraentheses indicate the pages on which the authors' contributions begin.

Veronica Bisagno (115), Department of Physiology, Molecular and Cellular Biology "Dr. Héctor Maldonado"; CONICET, Institute of Physiology, Molecular Biology and Neurosciences (IFIBYNE) and CONICET, Institute of Pharmacological Research (ININFA), University of Buenos Aires, Buenos Aires, Argentina

Santiago Castro-Zaballa (1), Laboratory of Sleep Neurobiology, Department of Physiology, School of Medicine, Republic University, Montevideo, Uruguay

Matías Cavelli (1), Laboratory of Sleep Neurobiology, Department of Physiology, School of Medicine, Republic University, Montevideo, Uruguay

Ethan M. Clement (195), Department of Neurology, University of Arkansas for Medical Sciences, Little Rock, AR, United States

Stasia D'Onofrio (55), Center for Translational Neuroscience, Department of Neurobiology and Developmental Sciences, University of Arkansas for Medical Sciences, Little Rock, AR, United States

Edgar Garcia-Rill (25,43,55,67,83,131,161,179,221,235), Center for Translational Neuroscience, Department of Neurobiology and Developmental Sciences, University of Arkansas for Medical Sciences, Little Rock, AR, United States

Joaquín Gonzalez (1), Laboratory of Sleep Neurobiology, Department of Physiology, School of Medicine, Republic University, Montevideo, Uruguay

James Hyde (83), Southern Arkansas University, Magnolia, AR, United States

Muna Irfan (161), Minnesota Regional Sleep Disorders Center, Department of Neurology, Hennepin County Medical Center, Veterans Affair Medical Center, Minneapolis, University of Minnesota Medical School, Minneapolis, MN, United States

Young-Jin Kang (195), Department of Neurology, University of Arkansas for Medical Sciences, Little Rock, AR, United States

Sang-Hun Lee (195), Department of Neurology, University of Arkansas for Medical Sciences, Little Rock, AR, United States

Giacomo Della Marca (143), Institute of Neurology, Catholic University, Rome, Italy

Paolo Mazzone (143), Stereotactic and Functional Neurosurgery, Centro Chirurgico Toscano, Arezzo, Italy

Erick Messias (43), Department of Psychiatry, University of Arkansas for Medical Sciences, Little Rock, AR, United States

Eugenio Scarnati (143), Department of Biotechnological and Applied Clinical Sciences, University of L'Aquila, L'Aquila, Italy

Carlos H. Schenck (161), Minnesota Regional Sleep Disorders Center, Department of Psychiatry, Hennepin County Medical Center, University of Minnesota Medical School, Minneapolis, MN, United States

Pablo Torterolo (1), Laboratory of Sleep Neurobiology, Department of Physiology, School of Medicine, Republic University, Montevideo, Uruguay

Francisco J. Urbano (115), Department of Physiology, Molecular and Cellular Biology "Dr. Héctor Maldonado"; CONICET, Institute of Physiology, Molecular Biology and Neurosciences (IFIBYNE) and CONICET, Institute of Pharmacological Research (ININFA), University of Buenos Aires, Buenos Aires, Argentina

Preface

Virtually every neurological and psychiatric disorder manifests sleep-wake dysregulation. In many cases the sleep disturbances begin years before the presentation of the specific neurological or psychiatric disorder. Disorders in which sleep dysregulation presages the clinical signs and symptoms are detailed herein. The reticular activating system (RAS), in addition to controlling sleep-wake states, modulates arousal and posture and gait. That is, when the RAS is engaged during waking, there is simultaneous promotion of arousal by ascending projections that determine whether to fight or flee and also the setting of muscle tone to carry out fight or flight. If these functions are disturbed, how would these be manifested in specific neurological and psychiatric disorders?

(1) Since the RAS participates in fight-or-flight responses, we would expect that responses to sudden alerting stimuli will be abnormal. For disorders in which the RAS is overactive, this would mean that such stimuli will produce exaggerated responses that would be manifested as exaggerated startle responses or hyperactive reflexes such as the blink reflex.
(2) Another property of the RAS is its rapid habituation to repetitive stimuli. This is reflected in its lack of responsiveness to rapidly repeating stimuli, that is, its habituation. This endows the RAS with its capacity for sensory gating, the property of decreasing responsiveness of repetitive events in favor of novel or different stimuli. For disorders in which this property is affected, we expect a decrease in habituation or a sensory gating deficit.
(3) The RAS controls waking and sleep, so that sleep patterns would be dysregulated. If the RAS is downregulated by a disorder, we expect an inability to remain awake; the presence of excessive daytime sleepiness; and an excess of total sleep time, especially an increase in slow-wave sleep (SWS). If, on the other hand, the RAS is upregulated, we expect difficulty in getting to sleep and maintaining sleep. This would be reflected in decreased SWS, insomnia, and fragmented sleep during the night, as well as increased rapid eye movement (REM) sleep drive, which is characterized by nightmares and frequent awakenings. In development, what would happen if the developmental decrease in REM sleep drive did not occur? Such a condition would lead to lifelong increased REM sleep drive during sleep (resulting in intense dreaming) but perhaps also during waking, that is, resulting in dreaming while awake or hallucinations, along with hypervigilance during waking.

(4) The RAS also modulates the *maintenance* of waking, a property ignored by many but one that affects a host of cognitive functions. The inability to maintain a steady waking state, in the form of maintained gamma-band activity, will interfere with attention, learning, and memory, to name a few processes.
(5) Another factor in all of these disorders is the level of frontal lobe blood flow. Decreased frontal lobe blood flow, or hypofrontality, is present in many neurological and psychiatric disorders to some extent. This state is normally present during REM sleep and may account for the lack of critical judgment present in our dreams; however, such a state during waking would lead to impulsive "knee-jerk" or reflexive reactions with the lack of consideration for the consequences. This condition is probably related to the lack of habituation to repetitive stimuli or sensory gating deficit. Under the condition of decreased cortical modulation, fight-or-flight responses and reflexes would be exaggerated. Whether hypofrontality is a cause of RAS dysregulation or RAS dysregulation leads to hypofrontality remains to be determined. However, successful treatment of any of these disorders should be marked by normalization of wake-sleep rhythms and reflexes, as well as frontal lobe blood flow.

Therefore, in the assessment of the role of arousal in neurological and psychiatric disorders, we need to consider, (a) responses to reflexes such as the startle response and the blink reflex using EMG, (b) sensory gating using paired stimuli or prepulse inhibition paradigms of sensory responses such as the P50 potential, (c) sleep patterns or somnography using nighttime EEG, (d) the maintenance or interruption of gamma-band activity during waking using EEG or MEG and finally, (e) the level of frontal lobe blood flow. These measures will allow us to determine in detail the role of arousal in specific neurological and psychiatric disorders and provide quantifiable measures for monitoring progress and assessing treatment.

In this book, we will deal with the role of arousal in normal cognitive function and then consider **three of the most important discoveries in the RAS in the last 10 years**, and throughout the book, we consider ways of clinically modulating arousal and vigilance in neurological and psychiatric disorders. The first discovery was the presence of **electrical coupling in the RAS**, the second was the presence of **intrinsic gamma-band oscillations** in the RAS, and the most recent is the presence of **gene transcription that may begin upon waking**. We address treatments that affect electrical coupling and intrinsic gamma-band activity within the reticular activating system (RAS). Modafinil has been found to increase electrical coupling in the RAS, thus increasing coherence and allowing ensembles of cells to fire together. This will increase the power of activity at whatever frequency the system is firing, whether during slow-wave sleep or waking. That is, sleep and waking states will be more coherent and thus more easily maintained. On the other hand, by increasing gamma-band activity using electrical stimulation, such as deep brain stimulation (DBS), at gamma frequencies, such activity in the circuit will result in waking and arousal. We see these as potential therapies for a number of disorders that include dysregulation of arousal.

The last discovery promises to open a host of new therapeutic avenues based on modulation of gene transcription but may also lead to a better understanding of the symptoms generated in these devastating disorders. At the end of the book, we deal with the potential functions of gene transcription taking place immediately upon waking, one of which may be in the formulation of the self.

Acknowledgments

I am grateful to my collaborators, students, and trainees. In particular, Francisco Urbano has been a driving force behind many of our studies. They have performed the studies that allowed us to understand arousal and how it affects brain function in health and disease. I am forever grateful and wish all of them long and productive careers. Rebeka Henry and Melanie Tucker at Elsevier were extremely helpful and supportive in the publication of this book, my sincere thanks to them. Susan Mahaffey was stellar in drafting and editing many of the figures for this book, and I am extremely grateful for her help.

Chapter 1

Arousal and normal conscious cognition

Pablo Torterolo, Santiago Castro-Zaballa, Matías Cavelli, Joaquín Gonzalez
Laboratory of Sleep Neurobiology, Department of Physiology, School of Medicine, Republic University, Montevideo, Uruguay

Introduction

The knowledge of the neurophysiological processes that generates and maintains consciousness provides the clinician the foundations to understand its absence and alterations. Consciousness is probably the main feature of human wakefulness (W), which is lost in the falling asleep process. However, we have a hint that during our night sleep, our mind is very active and fly without control during our dreams. Dreams are a different type of cognitive state, with its own rules. Neurological syndromes such as comma or vegetative state suppress consciousness. Psychiatric conditions such as psychosis generate an alteration of consciousness. Furthermore, while general anesthetic drugs suppress consciousness, several drugs, such as hallucinogens, alter it.

Which are the neuronal networks involved in the generation consciousness? How do they work? What are the adjustments in these networks that determine that consciousness is not supported during sleep? In the present chapter, focusing on the information provided by the electroencephalogram (EEG), we reviewed the most relevant concepts of the electrocortical correlates of normal conscious cognition and the physiological and drug-induced network modification that are involved in its absence or alteration.

Arousal and consciousness

Arousal is the physiological and psychological state of being awoken from sleep and the increase in vigilance or alertness during W. It involves the function of the activating system (AS) in the brain; one of its main components is the reticular activating system (RAS) whose soma is located in the mesopontine brain stem. The RAS is a phylogenetically conserved system that modulates fight-or-flight responses (Yates and Garcia-Rill, 2015). An increase in the firing

Arousal in Neurological and Psychiatric Diseases. https://doi.org/10.1016/B978-0-12-817992-5.00001-5

rate of the RAS neurons mediates the activation of the thalamocortical system (i.e., the main neuroanatomical structure associated with consciousness), the sympathetic autonomic nervous system, and the motor and the endocrine systems (Yates and Garcia-Rill, 2015). This increase in the firing rate of RAS neurons generates sensory alertness, mobility, and readiness to respond, that is, accompanied by an increase in heart rate and blood pressure, respiratory activity, and other phenomena related with fight-or-flight responses. Hence, during W, there are periods with low level of arousal (quiet or relaxed W) and periods with high level of arousal. A novel, painful, or motivational stimuli can induce high level of arousal; in any case, the result is alertness or full attention status. In humans (and supposedly in animals with high cognitive abilities), arousal is accompanied by consciousness.

"There is nothing we know more intimately than consciousness, but there is nothing harder to explain," stated the mind philosopher David Chalmers (Chalmers, 2005). Dictionaries usually define consciousness as the ability to be aware of surroundings and ourselves. Although this is a circular definition (because awareness and consciousness are synonyms), it captures the essence: consciousness allows us to know about ourselves and the existence of objects and events (Damasio and Meyer, 2009). In the present work, following the directives of Edelman and Tononi, we define consciousness in practical terms: "Everyone knows what consciousness is: it is what abandons you every evening when you fall asleep and reappears the next morning when you wake up" (Edelman and Tononi, 2000). This definition suggests that for normal W, consciousness is a sine qua non condition. However, we must keep in mind that dreams are considered a special (or altered) type of consciousness (see succeeding text).

The concept of "neural correlates of consciousness" (NCC) represents the smallest set of neural events and structures sufficient for a given conscious percept, explicit memory, or cognitive function. Where is the structural (neural) basis of consciousness? The thalamocortical system is the ultimate responsible for the generation of consciousness, and the associative cortices play a major role (Llinas and Pare, 1991; Tononi and Laureys, 2009).

Due to fact that the thalamocortical system is also the main responsible for the electric activity recorded in the EEG, in this chapter, we will focus in the EEG phenomena related to arousal and normal conscious condition and in its physiological and nonphysiological suppression or alteration.

Electroencephalogram

The EEG is produced by the summed electric activities of populations of neurons, with a modest contribution from glial cells (Lopes da Silva, 2010). Pyramidal neurons of the cortex are the main contributor of the EEG signal, since they are arranged in palisades with the apical dendrites aligned perpendicularly to the cortical surface. The electric fields generated by these neurons can be recorded by means of electrodes located at a short distance from the source (local field potentials, LFPs), from the cortical surface (electrocorticogram or ECoG), or at longer distances such as from the scalp (standard EEG). In the standard EEG, oscillations higher than 30 Hz are difficult to observe because they are filtered out by the skull and scalp and there is more distance from the source and worse spatial resolution. On the contrary, oscillations up to 200 Hz can be recorded with LFPs or ECoG.

Several oscillatory rhythms can be observed in the EEG. These rhythms are generated in the thalamus and/or at cortical levels and are modified according to the behavioral state (W and sleep).

Wakefulness

In humans (and mammals in general), three behavioral states can be distinguished: W, nonrapid eye movement (NREM) sleep (also called slow-wave sleep), and rapid eye movement (REM) sleep. These behavioral states can be recognized by means of polysomnography, which consists of the simultaneous recording of various physiological parameters such as EEG, electromyogram (EMG), and electrooculogram.

The EEG recording during W is characterized by the presence of high-frequency and low-voltage oscillations (cortical activation). The EEG (ECoG in sensu stricto) during W (alert wakefulness, AW; quite wakefulness, QW) of a cat is shown in Fig. 1.1. EEG recordings during W show relatively low-amplitude and high-frequency oscillations (active EEG). As it is shown in Fig. 1.2, the analysis of the frequency content of the EEG signal (i.e., the power spectrum) shows that in comparison with other behavioral states the power of the low-frequency bands (delta, theta, and sigma bands) during AW is low, while there is an increment in high-frequency bands, especially the low gamma band (30–45 Hz).

High EEG gamma activity during W has been described in several species, including humans (Maloney et al., 1997; Cantero et al., 2004; Cavelli et al., 2017a). In the ECoG of the cat, gamma activity is readily observed in the raw recordings during W (indicated with "a" in Fig. 1.1). As is displayed in Fig. 1.3, low gamma (30–45 Hz) oscillations take place as "bursts" of approximately 25 μV of amplitude and 200–500 ms of duration; these "bursts" are enhanced (in frequency of appearance, amplitude, and duration) during arousal produced by a stimulus that produces alertness (sound and light) or motivation (smell of food). In Figs. 1.3 and 1.4A, AW was produced with random sound stimulation, and gamma bursts seem to be coupled among several cortices. As it is shown in Fig. 1.5, when this intercortical gamma coupling is analyzed by the magnitude square coherence function, gamma coherence increases during AW in comparison with QW and sleep (see Castro et al., 2013, 2014). Another clear example of gamma coherence increment during AW is exhibited in Fig. 1.6. In this case, a person unknown to the animal entered in the recording room, and there was a large increase in gamma power (Fig. 1.6A) and coherence (Fig. 1.6B). A large gamma coherence between two cortical areas strongly suggests that there is a

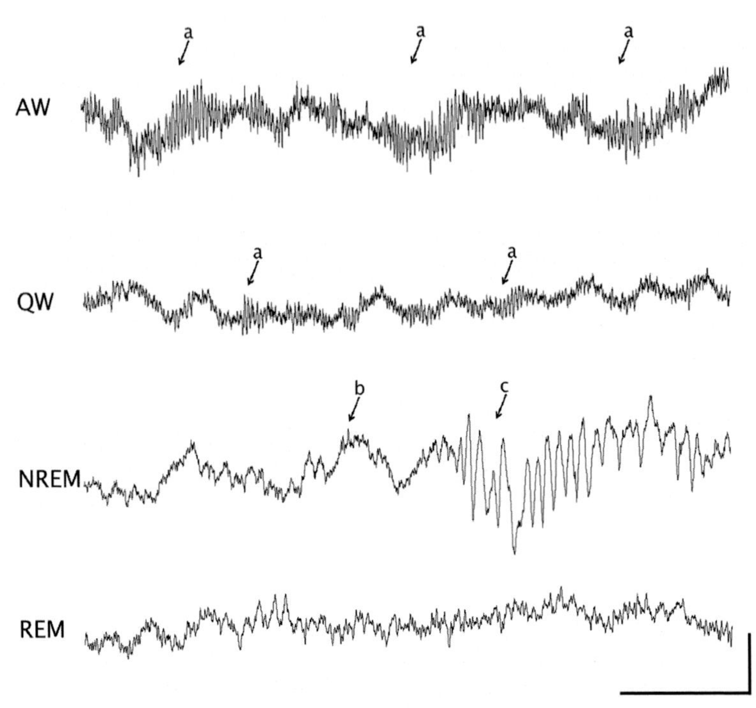

FIG. 1.1 EEG raw recordings of the dorsolateral prefrontal cortex of the cat during alert wakefulness (AW), quiet wakefulness (QW), and NREM and REM sleep. *a*, gamma (30–45 Hz) oscillations; *b*, slow waves; *c*, sleep spindles. Calibration bars, 1 s and 200 μV.

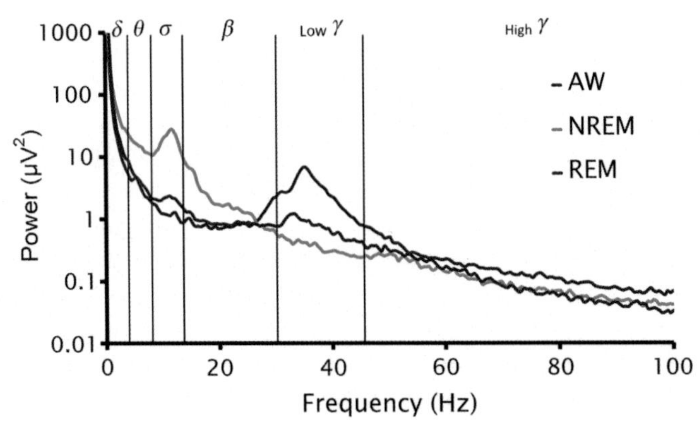

FIG. 1.2 Power spectrum (0.5–100 Hz) during wakefulness and sleep. The figure shows the average profile of 10,100 s' windows from the prefrontal cortex EEG of a cat during alert wakefulness (AW), NREM sleep, and REM sleep. Delta (0.5–4 Hz), theta (5–9 Hz), sigma (10–15 Hz), beta (16–30 Hz), and low (31–45 Hz) and high gamma (46–100 Hz) bands are shown between vertical lines.

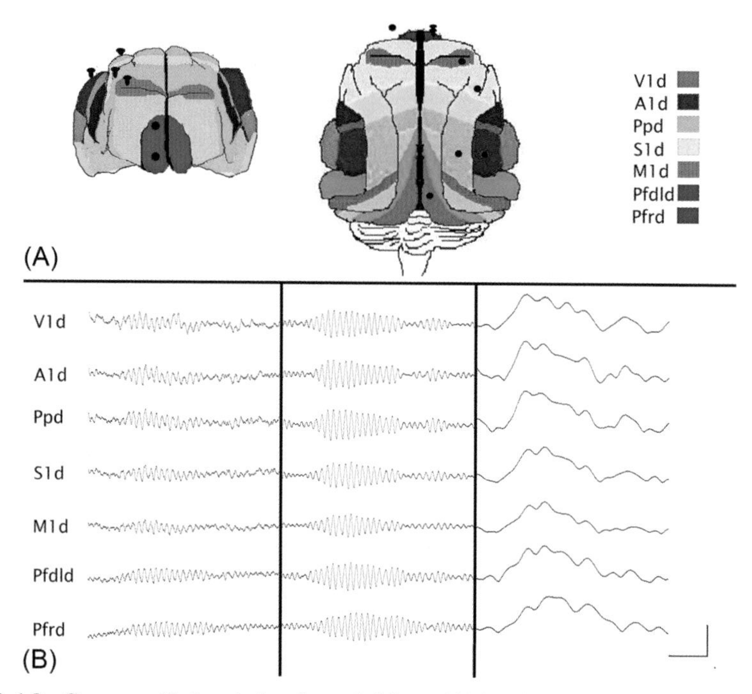

FIG. 1.3 Gamma oscillations during alert wakefulness. (A) Anterior and top view of the cat brain. The position of the cortical recording electrodes on the right cerebral hemisphere is displayed. The recordings were monopolar and referenced to an electrode located in the left frontal sinus. (B) The simultaneous recordings of the cortical gamma oscillation are exhibited. Raw recordings are shown on the left, filtered recordings (band pass 30–45 Hz) are on the middle, and the envelope of the gamma oscillations is displayed on the right. There is a large coupling in the gamma oscillations among cortical areas. Horizontal calibration bar, 200 ms. Vertical calibration bar: raw recording, 200 μV; filtered recording, 100 μV; envelopes, 50 μV. *Pfrd*, right rostral prefrontal cortex; *Pfdld*, right dorsolateral prefrontal cortex; *M1d*, right primary motor cortex; *S1d*, right primary somatosensory cortex; *Ppd*, right posterior parietal cortex; *A1d*, right auditory cortex; *V1d*, right visual cortex.

high degree of communication between these areas at the gamma band. This gamma coupling during aroused W has been also observed in rodents and humans (Llinas and Ribary, 1993; Cantero et al., 2004; Voss et al., 2009; Cavelli et al., 2015, 2017a).

During relaxed or QW, gamma activity decrease, and oscillations at lower frequencies begin to appear. This fact is readily observed in humans; during relaxed W with eyes closed, a high-amplitude alpha (8–12 Hz) oscillation appears mainly in the occipital (visual) cortex. The frequency of these oscillations is considered the basic idle (resting) speed of the brain during W (Garcia-Rill, 2015a).

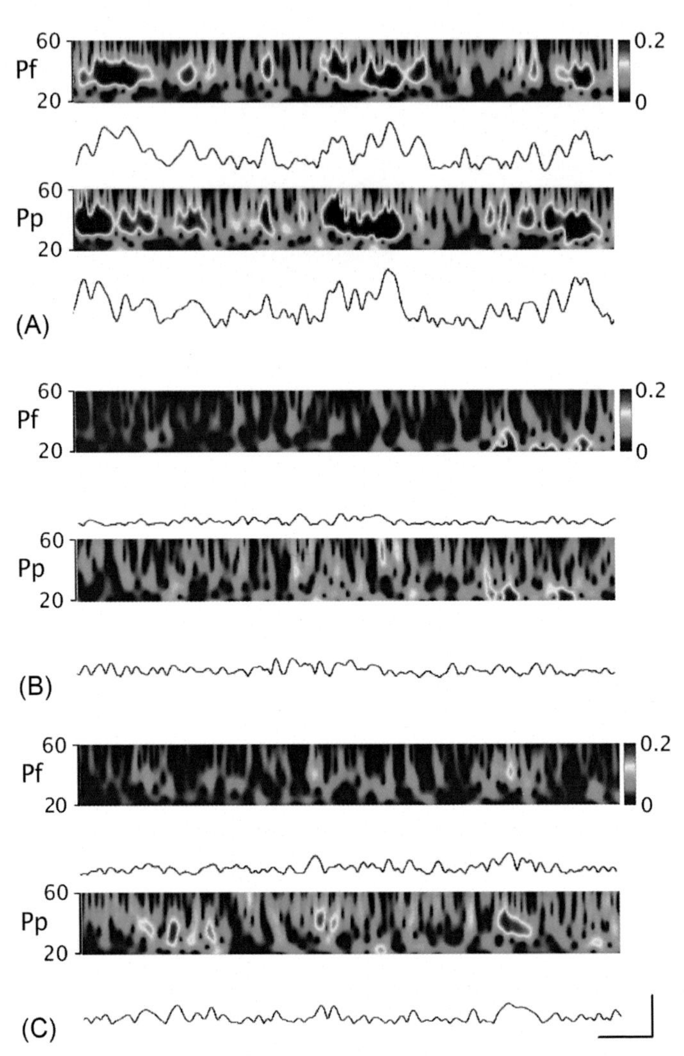

FIG. 1.4 Spectrograms (by means of wavelet function) and rectified gamma band (30–45 Hz) or gamma envelopes, during alert (A) wakefulness (AW) and (B and C) NREM and REM sleep. Calibration bars, 30 μV and 400 ms. The color code of the spectrograms shows a wavelet coefficient that represents in relative units the energy of the signal.

In summary, during W, the EEG activity transits from slower EEG rhythms such as alpha during relaxed W to higher-frequency rhythms during aroused W (especially at frequencies around 40 Hz in humans and cats). Both cortical gamma power (associated with synchronized neuronal oscillations within a cortical area) and long-range gamma coherence (associated with gamma coupling between distant cortical areas) tend to increase in correlation with the level of arousal.

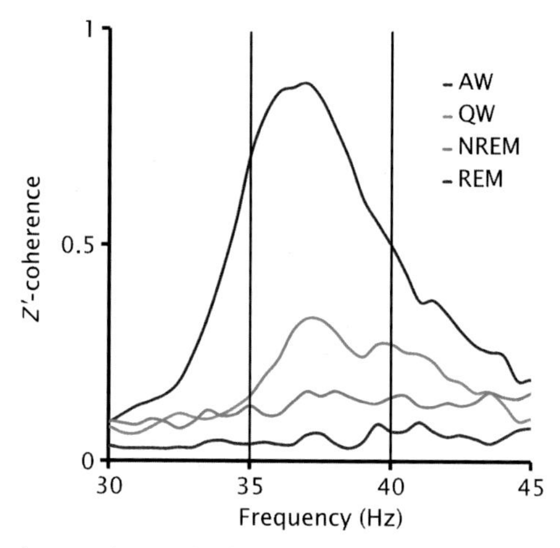

FIG. 1.5 Gamma coherence. Average EEG gamma z'-coherence profiles (between prefrontal and posterior parietal cortices) of 12,100 s' windows during alert (AW) and quiet wakefulness (QW) and NREM and REM sleep. The gamma coherence peak is between 35 and 40 Hz and is shown between vertical lines.

EEG correlates of wakefulness and arousal

As previously mentioned, consciousness (awareness) is the cognitive counterpart of normal W. It is considered that two "components" are needed to support consciousness (Posner et al., 2007; Garcia-Rill, 2015b). One is the "content" of consciousness. In spite of the fact that several neural networks contribute to the cognitive well-being (such as the basal ganglia, neocerebellum, hippocampus, and reticular formation), the thalamocortical system constitutes its main anatomical site where the "content" is processed; the associative cortical areas and related thalamic nuclei are considered to play the major role. These areas are fed with information provided by sensory pathways. The other component that supports consciousness is activation or arousal, which is also supposed to provide the "context" of sensory experience. This function is supported by the AS, in which the RAS and nonspecific thalamic nuclei play a critical role. A disturbance in the "content" of W is characteristic of diffuse cortical lesions and metabolic or toxic disorders that affect the cortex or thalamic nuclei; these injuries may produce what it is known as vegetative state. On the other hand, subtle injuries or deficits of the AS may produce comma, usually accompanied by an increase in the EEG slow activity (Posner et al., 2007).

Which are the electrocortical correlates of waking consciousness? In a very schematic way, the main EEG correlates of W consciousness are listed in Table 1.1. As commented before, an active EEG is needed to support W. In other words, widespread slow waves (delta waves) and sleep spindles that are features of NREM sleep do not support W.

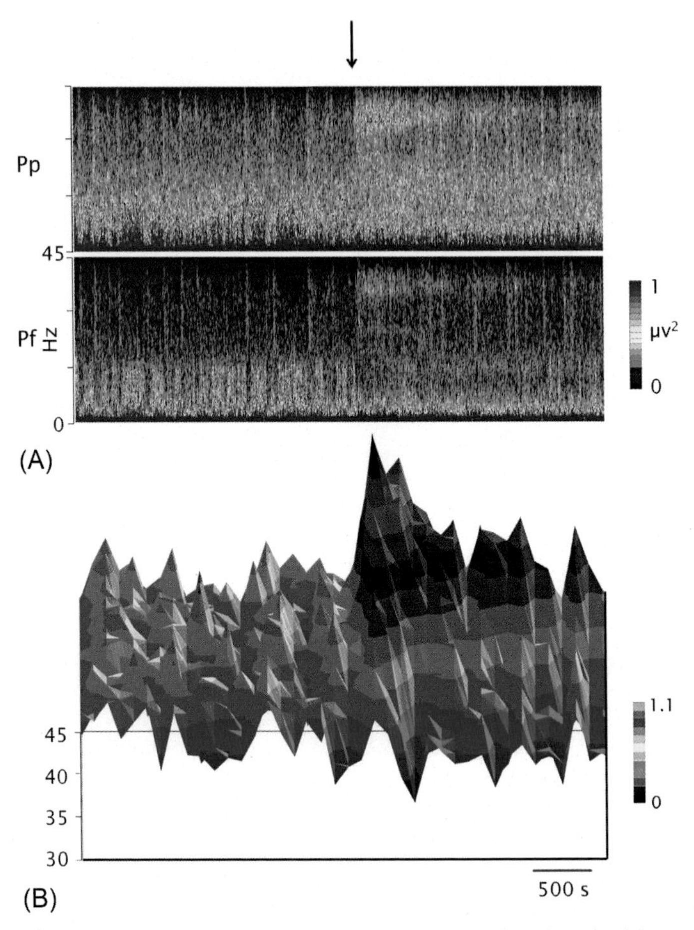

FIG. 1.6 Dynamic evolution of the EEG gamma z'-coherence when the animal is aroused. (A) Gamma power spectrograms of prefrontal (Pf) and posterior parietal (Pp) cortices when the animal is alerted by a stimulus that consisted on unknown people entering to the recording room *(arrow)*. (B) Three-dimensional spectrogram of the gamma z'-coherence between Pf and Pp cortices (same recordings as in A). Time and frequency are displayed on the horizontal and vertical axes (depth), respectively; the z'-coherence is represented in a color code.

For unified perceptual experiences, the brain integrates fragmentary neural events that occur at different times and locations. Synchronization of neuronal activity by phase locking of network oscillations has been proposed for integration or binding mechanism ("binding by synchrony") (Singer, 1999). Gamma activity, especially gamma coherence, has been involved in the explanation of this "binding problem" (Varela et al., 2001; Uhlhaas et al., 2009; Buzsaki et al., 2013; Buzsaki and Schomburg, 2015) and is one of the most studied neural correlates of consciousness (Noreika, 2015). In this regard, gamma coherence is lost during general anesthesia (see succeeding text). Higher-frequency

TABLE 1.1 Main EEG features across physiological and nonphysiological states

	W	NREM	REM	Isoflurane	Ketamine	S/A
Slow waves	−	+	−	+	−	+
Sleep spindles	−	+	−	+	−	+
Gamma power	+	−	+	−	+	+
Gamma coherence	+	−	−	−	−	+

Main electrographic features during wakefulness (W), NREM sleep, REM sleep, isoflurane general anesthesia, ketamine (subanesthetic dose), and scopolamine or atropine treatment (S/A). These profiles could explain the cognitive differences between these physiological and pharmacological conditions. Positive symbols indicate the presence of these features in the EEG; negative signs indicate absence.

oscillations, known as HFO (up to 160 Hz), may also play a role in this function (Cavelli et al., 2017b).

Wakefulness-promoting neuronal networks

Thalamocortical, premotor/motor, autonomic, and hypothalamic neuroendocrine neuronal networks modify their function during the waking-sleep cycle. However, the "primary engine" that determines changes in these neuronal networks during W is the AS. This system is composed of neurons that utilize different neurotransmitters (such as acetylcholine, noradrenaline, serotonin, dopamine, histamine, and hypocretins) and have widespread projections (Torterolo and Vanini, 2010; Torterolo et al., 2016b). The firing rate of the W-promoting neurons and the release of their neurotransmitters into the synaptic cleft tend to be maximal during W and decrease during NREM sleep.

NREM sleep

In the falling asleep process, adults enter into NREM sleep. In addition to the quiescent behavior and deep modification of autonomic and endocrine activity that regulate visceral functions, NREM sleep is associated with impressive cognitive alterations. The manifestation of the changes in thalamocortical activity on passing from W to NREM sleep can be partially appreciated in the EEG.

In humans, three NREM sleep phases are recognized: N1, N2, and N3, according to the depth of the state. N1 is the transitional stage from W, where hypnagogic imaginary (dreamlike activity) is common. This transition into NREM sleep is complex and heterogeneous from the EEG point of view. In fact, Tanaka et al. (1996) divided the transition in nine "hypnagogic states" (from relaxed W with alpha activity to N2). N2 is characterized by the presence of sleep spindles (11–15 Hz oscillatory events with a duration of 0.5–2 s) and K-complexes. K-complex, which is often associated with sleep spindles, consists of a brief negative sharp high-voltage peak (usually greater than 100 μV), followed by a slower positive complex and a final negative peak.

The presence of high-amplitude (approximately 70 μV), low-frequency (0.5–4 Hz, delta) oscillations characterizes N3 (Carskadon and Dement, 2011). Fig. 1.1 shows the EEG activity during NREM sleep in the cat; slow-wave oscillations and sleep spindles are indicated (indicated with "b" and "c," respectively). Fig. 1.2 depicts the power spectrum during NREM sleep. Large values of delta and sigma power produced by slow waves and spindles, respectively, are distinctive features of NREM sleep. Also, the decrease in the gamma band power and coherence is another remarkable feature of NREM sleep (Figs. 1.2, 1.4B, and 1.5).

Somnambulism or sleepwalking is an NREM sleep parasomnia that can be explained as a dissociated state, with both waking and NREM sleep features (Mahowald and Schneck, 2011; Canclini et al., 2018). In other words, part of the brain is active (i.e., as in waking state, with probable activation of motor cortical and subcortical regions), while other cortical regions present slow waves (as in NREM sleep) in the EEG. As a result, the individual is awake enough to carry out complex motor acts but is unconscious and irresponsible for these actions (because is partially asleep). It is likely that slow waves during these events are mainly present in associative cortical areas that are critical for awareness (Tononi and Laureys, 2009). The slow cortical activity during somnambulism is a pathological manifestation of what is known as local sleep. Nowadays, it is accepted that during W, part of the cortical columns behaves as they were asleep, especially when there is high sleep pressure, that is, during sleep deprivation or prolonged W (Vyazovskiy et al., 2011).

The presence of slow waves and/or sleep spindles during NREM sleep is against the generation of consciousness (Table 1.1). The high-amplitude slow waves of NREM sleep are widespread throughout the cortex and are produced by the synchronization of a large number of pyramidal neurons. This electrocortical condition suggests that large groups of neurons are doing quite the same at the same time; this reduction in the degree of freedom would wane consciousness. On the contrary, a feature of cortical activity during W is regional "functional differentiation," and according to Tononi's information integration theory, functional differentiation between different areas is critical for consciousness (Tononi, 2010). This feature is lost during NREM sleep. A similar circumstance occurs in generalized seizures such as "petit mal," where unconsciousness of the event is associated with widespread stereotyped slow waves. In addition, functions such as cortical lateral inhibition that is critical for perception are lost when pyramidal neurons behave in an homogeneous manner such as in deep NREM sleep (Garcia-Rill, 2015c).

Oneiric activity is scarce or absent during deep NREM sleep (N3) (Dement and Kleitman, 1957; Pace-Schott, 2011; Siclari et al., 2017). As previously mentioned, widespread slow waves and spindles, as well as low gamma power and coherence in the EEG, do not support cognitive activity (neither W consciousness nor dreams) (Table 1.1). However, oneiric activity may appear by local REM sleep–like activation of critical areas in a background of light NREM sleep (N1 and N2 at the end of a nocturnal sleep period). In fact, both during NREM and REM sleep, dream reports were associated with local decrease in low-frequency activity in posterior cortical regions, while an increase in high-frequency activity within these regions is correlated with specific dream contents (Siclari et al., 2017).

NREM sleep-promoting system

Cognitive activity (waking consciousness and dreams) and the different EEG rhythms that support these functions are mainly generated by the activity of cortical and thalamic neuronal networks, which are mutually interconnected. Thalamic neurons have a complex electrophysiology that allows them to operate differently according to their level of polarization (Steriade et al., 1993). When hyperpolarized, the thalamic neurons that project to the cortex (thalamocortical neurons) oscillate at low frequency (0.5–4 Hz) and tend to block the sensory information that travels toward the cortex. This "oscillatory mode" of function synchronizes cortical neurons and, accompanied by other phenomena of cortical origin, generates the slow waves of NREM sleep (Huguenard and McCormick, 2007; Crunelli et al., 2015). Moreover, the reticular nucleus of the thalamus is the site of generation of the sleep spindles that characterize N2 (Fuentealba and Steriade, 2005; Huguenard and McCormick, 2007). On the contrary, when thalamic neurons are relatively depolarized, they enter in the "tonic mode" of function. In this condition, the thalamocortical neurons transmit sensory information toward the cortex in a reliable way. This mode of function occurs during W and REM sleep because the AS maintains a depolarized membrane potential in these neurons.

Neurons from the preoptic area (POA) of the hypothalamus are critical in the generation and maintenance of NREM sleep (Torterolo and Vanini, 2010; Torterolo et al., 2016b). Most of these neurons are GABAergic; these inhibitory neurons project in monosynaptic form toward the activating nuclei. On the other hand, experimental evidence suggests that W-promoting neurons inhibit NREM sleep–promoting POA neurons (Gallopin et al., 2000; Williams et al., 2014). This reciprocal inhibition between activating and hypnogenic neurons is critical for the transition between sleep and W and the basis of the flip-flop state switch model (Saper et al., 2010).

Other neuronal networks also seem to play a role in NREM sleep generation, such as neurons located in the medullary reticular formation (Anaclet et al., 2012) and the melanin-concentrating hormone (MCH)-containing neurons of the lateral hypothalamus and incertohypothalamic area (Torterolo et al., 2011; Monti et al., 2013).

REM sleep

REM sleep (also called stage R) is a deep sleep stage even though it exhibits similar electrographic characteristics to that of W, that is, has an active EEG. Hence it is also called "paradoxical" sleep. REM sleep is also characterized by REM, muscle atony, and phasic changes in autonomic activity.

REM sleep EEG in rodents and cats is similar to W (Fig. 1.1) (Torterolo et al., 2016b). However, the EEG during REM sleep in humans has more similarities to N1; in both states, the EEG is described as low-voltage, mixed-frequency activity (Keenan and Hirshkowitz, 2011).

In humans, during nighttime sleep, REM sleep episodes occur with a period of approximately 90 min; in fact, there are four to five "sleep cycles" per night. The "sleep cycles" are the period between the onset of sleep until the end of the first episode of REM sleep or the period from the end of an episode of REM sleep to the end of the subsequent REM sleep episode (Carskadon and Dement, 2011).

Dreams occur mainly during REM sleep and are considered a special kind of cognitive activity or protoconsciousness (Hobson, 2009). REM sleep dreams are characterized by their vividness, single-mindedness, bizarreness, and the loss of voluntary control over the plot. Attention is unstable and rigidly focused, facts and reality are not checked, violation of physical laws and bizarreness are passively accepted, contextual congruence is distorted, time is altered, and memories become labile (Rechtschaffen, 1978; Hobson, 2009; Nir and Tononi, 2010). Interestingly, some authors have suggested that cognition during REM sleep resembles psychosis (Gottesmann and Gottesman, 2007). In fact, Hobson stands that "dreaming is, by definition, a psychosis" (Hobson, 1997).

High local cortical gamma activity (and hence relatively large gamma power) is present during REM sleep (Fig. 1.2), both in humans and animals (Maloney et al., 1997; Cantero et al., 2004; Cavelli et al., 2017a). However, long-range gamma coherence is almost absent during REM sleep (Figs. 1.4C and 1.5). High gamma power accompanied by minimal gamma coherence is a trait that characterizes REM sleep (Fig. 1.7 and Table 1.1), which is conserved in rodents, felines, and humans (Cantero et al., 2004; Voss et al., 2009; Castro et al., 2013, 2014; Cavelli et al., 2015, 2017a; Torterolo et al., 2016a).

Coherent EEG gamma activity has been observed during REM sleep solely during lucid REM sleep dreaming; the level of gamma coherence during lucid dreaming is intermediate between W and nonlucid REM sleep (Voss et al., 2009). Lucid dreams are a relatively infrequent phenomenon, whereby the "sleeper" reports being aware that he/she is dreaming and, in some cases, is able to deliberately modify the events of the ongoing dream. Interestingly, externally imposed resonance at 40 Hz by means of electric stimulation produces self-awareness (lucidity) during REM sleep (Voss et al., 2014). From the electrocortical point of view, lucid dreams share features for both W and REM sleep; hence it could be considered a type of dissociate state.

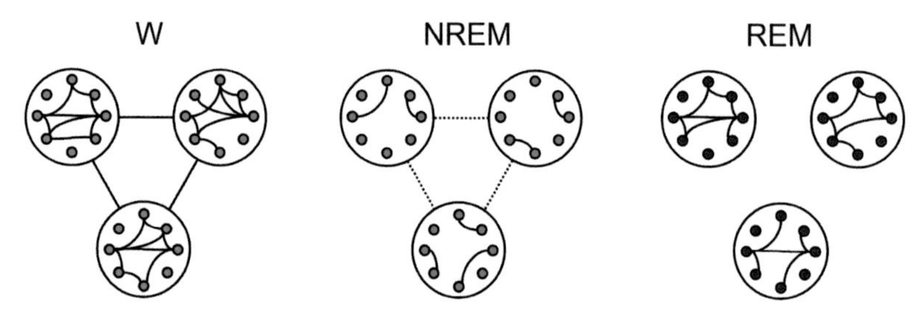

FIG. 1.7 Schematic representation of the short- and long-range gamma synchronization during wakefulness (W), NREM sleep, and REM sleep. The *small circles* represent neurons, while *large circles* represent the areas of the cortex where these neurons are located. Colors in neurons represent the behavioral states (*blue*, W; *green*, NREM sleep; *red*, REM sleep) and connecting lines between the circles represent the gamma synchronization between distant cortical areas. Short-range (local) and long-range (distant) gamma synchronization occurs during W. During NREM sleep, both short- and long-range gamma synchronization decrease. During REM sleep, while gamma synchronization is present at local level, distant gamma coupling is absent.

Neural systems that promote the generation of REM sleep

The neural networks necessary and sufficient for the generation and maintenance of REM sleep are found in the mesopontine reticular formation (Siegel, 2011). In fact, most of the mesopontine neurons that play a role in the maintenance of W coincide with the neurons that are responsible for the generation of REM sleep.

Within these areas, monoaminergic (noradrenergic and serotonergic) neurons that are active during W turn off during REM sleep (REM-off neurons). On the other hand, cholinergic neurons increase their firing rate both during W and REM sleep (REM-ON neurons) (McCarley, 2007). Mesopontine GABAergic and glutamatergic neurons (Luppi et al., 2007) and hypothalamic MCH-containing neurons also play a critical role in REM sleep generation (Torterolo et al., 2011, 2015; Monti et al., 2013).

Drug-induced loss of consciousness: General anesthesia

Comma, vegetative state, and seizures are the more salient conditions related to pathological loss of consciousness. Since the variety and complexity of these conditions, they will not be analyzed in the present report. However, we will focus in drug-induced loss and alteration of consciousness.

General anesthesia is a drug-induced state, characterized by a relatively safe and reversible loss of consciousness. The ability to render a patient unconscious (hypnosis) and insensible to pain made modern surgery possible, and general anesthetics have become one of the most widely used class of drug (Franks, 2006). Sleep and anesthesia share many behavioral and electroencephalographic characteristics (Vanini et al., 2011). In addition, several authors suggest that sleep and anesthesia (induced by most anesthetic) share an underlying mechanism. In fact, several studies showed that most anesthetics suppress consciousness by recruiting or inhibiting regions that regulate sleep and W (Vanini et al., 2011).

When the anesthetic plane is reached, NREM sleep–like slow waves are present during this drug-induced state (Lydic and Baghdoyan, 2005). In addition, as it is shown in Fig. 1.8, isoflurane decreases EEG gamma power and coherence activity in the cat. Similar effects were observed for long-range gamma coherence in humans and rats (John, 2002; Mashour, 2006; Pal et al., 2016). Hence both the presence of slow waves and the decrease in gamma power and coherence are associated with the absence of consciousness induced by anesthesia (Table 1.1).

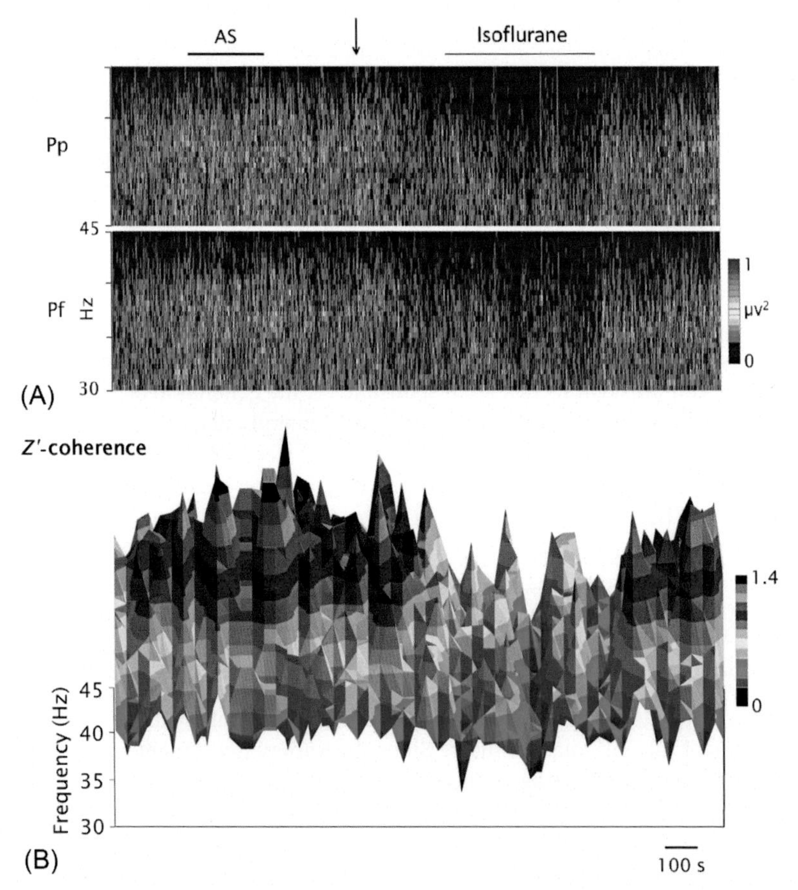

FIG. 1.8 Dynamic evolution of the EEG gamma z'-coherence following isoflurane administration (bar). (A) Gamma power spectrograms of prefrontal (Pf) and posterior parietal (Pp) cortices. The first bar (AS) shows when the animal is alerted by a stimulus that consisted of unknown people entering the recording room during wakefulness. The *arrow* indicates an auditory stimulus. The time during the administration of an anesthetic dose of isoflurane is also displayed. (B) Three-dimensional spectrogram of the gamma z'-coherence between Pf and Pp cortices (same recordings as in A). Time and frequency are displayed on the horizontal and vertical axes (depth), respectively; the z'-coherence is represented in a color code. The decrease in gamma power and coherence is readily observed.

Drug-induced alteration of consciousness

Ketamine, a pharmacological model of psychosis

The word psychosis (from Greek "disorder of the mind") is used in psychiatry to define a mental state in which there is a loss of contact with reality. The *Diagnostic and Statistical Manual of Mental Disorders (DSM), 5th Edition (2013)* classifies psychotic disorders in a chapter entitled "Schizophrenia Spectrum and Other Psychotic Disorders," highlighting among them schizophrenia (Bhati, 2013). This pathology is characterized by the presence of positive or psychotic (visual and auditory hallucinations, delusions, and paranoia) and negatives symptoms (apathy, the loss of motivation, and serious social isolation) and memory and executive function disorders.

Several hypotheses that attempt to explain the pathophysiology of psychotic disorders have been postulated. Among them, it is widely accepted that glutamatergic hypofunction mediated by the *N*-methyl-D-aspartate receptor (NMDA-R) is a key mechanism contributing to the positive, negative, and cognitive symptoms observed in this condition (Krystal et al., 1994; Pomarol-Clotet et al., 2006; Javitt, 2010). This is based on clinical reports showing that the consumption of noncompetitive antagonists of NMDA-R, such as ketamine, induces in healthy individuals the characteristic alterations of the psychotic disorders and exacerbates the symptoms in schizophrenic patients (Krystal et al., 2003; Pomarol-Clotet et al., 2006). Therefore models involving NMDA-R hypofunction is considered a valid pharmacological approach for the study of the psychotic disorders (Corlett et al., 2007; Scorza et al., 2008; Javitt, 2010).

Ketamine in subanesthetic doses produces an activated state, with relatively high gamma power (Fig. 1.9); however, it also produces a deep decrease in gamma (30–45 Hz) band coherence (Fig. 1.10). This decrease was similar to that occurring during REM sleep, which is considered a natural model of psychosis (Hobson, 1997; Gottesmann, 2006; Gottesmann and Gottesman, 2007). Furthermore, under ketamine, the gamma coherence was not affected by novel stimuli (Fig. 1.10), which in basal conditions alert the animal causing a large increase in gamma coherence (Castro-Zaballa et al., 2019b).

Disruptions in gamma activity similar to the induced by ketamine, have been described in psychosis (Lee et al., 2003; Light et al., 2006; Yeragani et al., 2006; Uhlhaas and Singer, 2010; Sun et al., 2011; White and Siegel, 2016).

In summary, it is possible that an active state with high local gamma band synchronization (i.e., high gamma power), accompanied with low long-range gamma coherence, is associated to the cognitive features shared by REM sleep and psychosis (Table 1.1).

Dissociative state induced by atropine and scopolamine

Mesopontine and basal forebrain cholinergic neurons are critically involved in the EEG activation during W and REM sleep (Torterolo and Vanini, 2010;

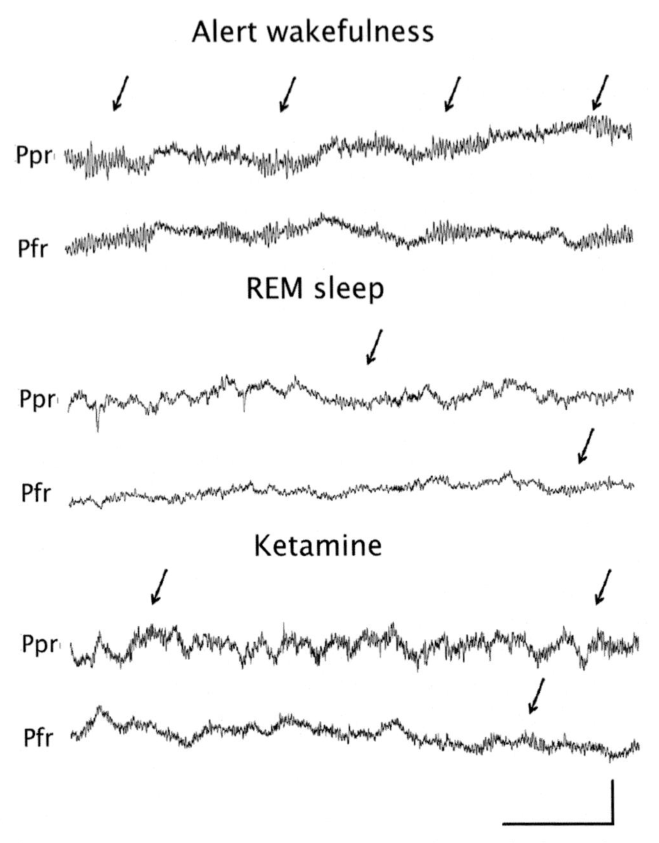

FIG. 1.9 Simultaneous raw recordings of the prefrontal (Pf) and parietal posterior (Pp) cortices during alert wakefulness and REM sleep and under the administration of ketamine (15 mg/kg). The *arrows* indicate the gamma "bursts." Calibration bars, 1 s and 200 μV. Gamma activity is present under ketamine; however, this gamma activity is not coupled between cortical areas.

Torterolo et al., 2016b). In this regard, animals treated with muscarinic antagonists (atropine or scopolamine) display high-voltage slow waves and spindles in EEG that resembles NREM sleep; however, they remain behaviorally awake and active (Wikler, 1952). Furthermore, these drugs decrease the electrocortical arousal response elicited by either sensory or midbrain reticular formation stimulation, but the gross behavior in response to such stimuli is not affected (Rinaldi and Himwich, 1955; Bradley and Key, 1958). This "dissociation" in which waking behavior coexists with NREM sleep–like EEG was observed in different animals and humans (Wikler, 1952; Longo, 1956; Chow and John, 1959; Lindsley et al., 1968; Yamamoto, 1988).

Classic pharmacological studies have also shown cognitive disturbances in humans treated with muscarinic antagonists (Ostfeld et al., 1960). They produce

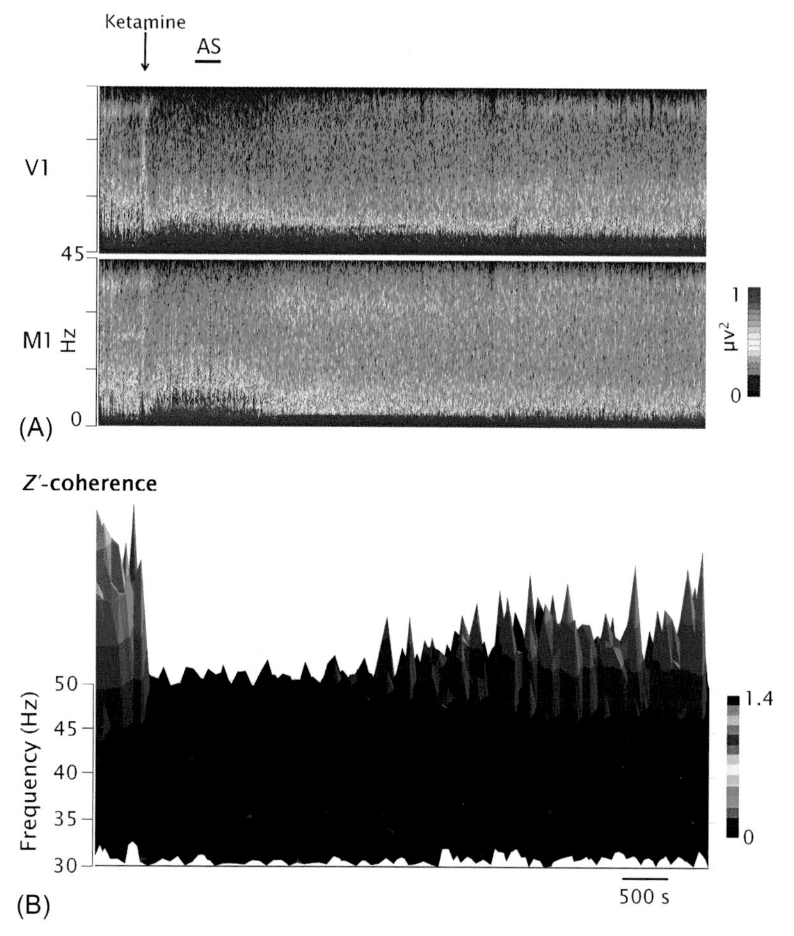

FIG. 1.10 Dynamic evolution of EEG gamma power and z'-coherence following the administration of subanesthetic dose of ketamine. (A) Gamma power spectrograms of primary motor (M1) and primary visual (V1) cortices following ketamine administration (*arrow*). The horizontal bar represents at random sound stimulation (AS) in order to arouse the animal. (B) Three-dimensional spectrogram of the gamma z'-coherence between M1 and V1 cortices (same recordings as in A). Time and frequency are displayed on the horizontal and vertical axes (depth), respectively; the z'-coherence is represented in a color code. Ketamine produced a large decrement in gamma coherence that was not affected by sensory stimulation.

a decrease in spontaneous speech and movement, impairment in memory and attention, and drowsiness. However, subjects are able to answer simple questions and perform tasks without the requirement of prolonged attention or memory. They can sit, stand, open or close their eyes, or extend their extremities on request, although they move more slowly than in the predrug period. In addition, muscarinic antagonists produce sedation, impairment of coordinative and reactive skills, visual disturbances, and diminution of short-term memory (Nuotto, 1983).

Furthermore, these drugs affect simple and choice reaction time, number matching, and memory scanning tasks (Ebert et al., 1998). It is considered that the most prominent effects of scopolamine involve discrimination processes, vigilance, selective attention, and consolidation and retrieval of memories (Sahakian, 1988).

Anticholinergic drugs have been used for recreational or ritualistic purposes. One of the most widely described religious or magical experiences dating back to ancient times is the alteration of consciousness with the induction of hallucinations by a member of the Solanaceae family of plants (belladonna, henbane, or datura), which contain scopolamine, atropine, and other closely related alkaloids (Perry and Perry, 1995). Perhaps the most extraordinary example of cognitive dysfunction is the criminal use of anticholinergic substances that are present in datura extracts (called "burundanga"), to induce amnesia and submissive behavior or "obedience" in victims (Ardila and Moreno, 1991; Ardila-Ardila et al., 2006).

We recently demonstrated that under the effect of atropine or scopolamine, coherent gamma (≈40 Hz) oscillations are conspicuous (Castro-Zaballa et al., 2019a) (Fig. 1.11). This "dissociated" EEG not only with slow waves and sleep

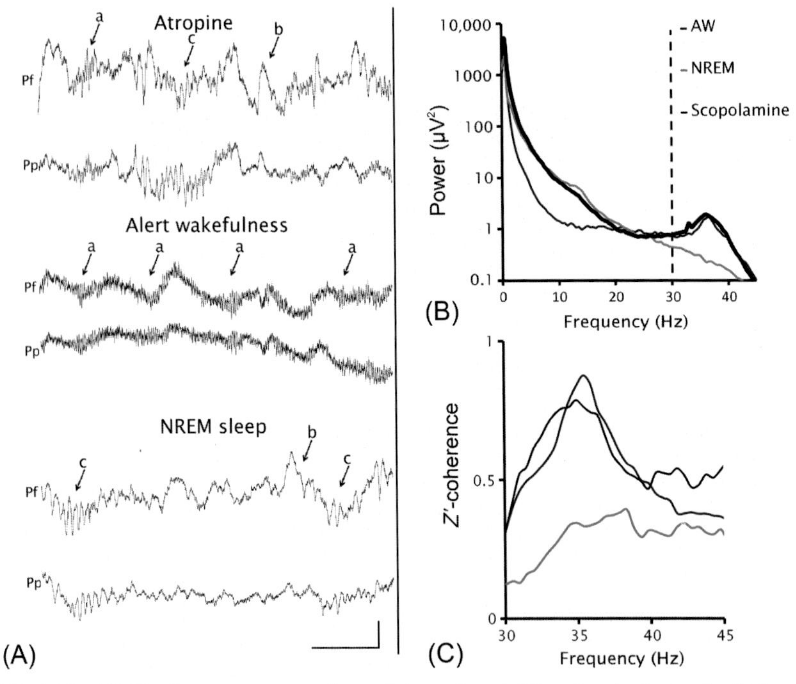

FIG. 1.11 (A) Simultaneous raw recordings of the prefrontal (Pf) and parietal posterior (Pp) cortices. Raw recordings during alert wakefulness and REM sleep and following the administration of atropine. *a*, gamma (30–45 Hz) oscillations; *b*, slow waves; *c*, sleep spindles. Calibration bars, 1 s and 200 μV. (B) Power spectrum (0–45 Hz) of the posterior parietal cortex during alert wakefulness (AW), NREM sleep, and scopolamine administration. (C) Average gamma z'-coherence profiles (30–45 Hz) in the same conditions as in B. The coherence profile during AW and scopolamine is similar.

spindles but also with coherent 40 Hz oscillations (a trait of AW) may be the neurophysiological basis of the "classic" EEG and behavior dissociation that is produced by these drugs. Hence the alteration of consciousness produced by antimuscarinic drugs is associated with slow waves and spindles (NREM sleep feature), combined with high gamma power and coherence, a trait of AW (Table 1.1).

Conclusions

Consciousness is the cognitive counterpart of normal W, at least for animals with higher cognitive abilities. W is characterized by an EEG with the absence of slow-wave (delta) activity and the presence of coherent high-frequency waves, mainly at about 40 Hz. These coherent gamma oscillations are highly dependent of the level of arousal. On the other hand, important adjustments of the delta waves and/or gamma activity (power and coherence) are associated with physiological absence or alteration of consciousness (NREM and REM sleep, respectively). An important modulation of gamma and delta activity is also associated with either the loss or alteration of consciousness induced by drugs.

Acknowledgments

This study was supported by the "Agencia Nacional de Investigación e Innovación, Fondo Clemente Estable FCE-1-2017-1-136550" grant, the "Comisión Sectorial de Investigación Científica I+D-2016-589" grant, and the "Programa de Desarrollo de las Ciencias Básicas, PEDECIBA" from Uruguay.

References

Anaclet, C., Lin, J.S., Vetrivelan, R., Krenzer, M., Vong, L., Fuller, P.M., Lu, J., 2012. Identification and characterization of a sleep-active cell group in the rostral medullary brainstem. J. Neurosci. 32, 17970–17976.

Ardila, A., Moreno, C., 1991. Scopolamine intoxication as a model of transient global amnesia. Brain Cogn. 15, 236–245.

Ardila-Ardila, A., Moreno, C.B., Ardila-Gomez, S.E., 2006. Scopolamine poisoning ('burundanga'): loss of the ability to make decisions. Rev. Neurol. 42, 125–128.

Bhati, M.T., 2013. Defining psychosis: the evolution of DSM-5 schizophrenia spectrum disorders. Curr. Psychiatry Rep. 15, 409.

Bradley, P.B., Key, B.J., 1958. The effect of drugs on arousal responses produced by electrical stimulation of the reticular formation of the brain. Electroencephalogr. Clin. Neurophysiol. 10, 97–110.

Buzsaki, G., Schomburg, E.W., 2015. What does gamma coherence tell us about inter-regional neural communication? Nat. Neurosci. 18, 484–489.

Buzsaki, G., Logothetis, N., Singer, W., 2013. Scaling brain size, keeping timing: evolutionary preservation of brain rhythms. Neuron 80, 751–764.

Canclini, M.R., Canzani, M.B., Luna, M.J., Royol, M.J., Rusiñol, M.C., Vega, M., Torterolo, P., 2018. Parasomnias: state of the art. An. Facultad Med. (Univ. Repúb. Urug.) 5, 29–43.

Cantero, J.L., Atienza, M., Madsen, J.R., Stickgold, R., 2004. Gamma EEG dynamics in neocortex and hippocampus during human wakefulness and sleep. Neuroimage 22, 1271–1280.

Carskadon, M.A., Dement, W., 2011. Normal human sleep: an overview. In: Kryger, M.H., Roth, T., Dement, W. (Eds.), Principles and Practices of Sleep Medicine. Elsevier-Saunders, Philadelphia, pp. 16–26.

Castro, S., Falconi, A., Chase, M.H., Torterolo, P., 2013. Coherent neocortical 40-Hz oscillations are not present during REM sleep. Eur. J. Neurosci. 37, 1330–1339.

Castro, S., Cavelli, M., Vollono, P., Chase, M.H., Falconi, A., Torterolo, P., 2014. Inter-hemispheric coherence of neocortical gamma oscillations during sleep and wakefulness. Neurosci. Lett. 578, 197–202.

Castro-Zaballa, S., Cavelli, M., Gonzalez, J., Monti, J., Falconi, A., Torterolo, P., 2019a. EEG dissociation induced by muscarinic receptor antagonists: coherent 40 Hz oscillations in a background of slow waves and spindles. Behav. Brain Res. 359, 27–38.

Castro-Zaballa, S., Cavelli, M., Gonzalez, J., Nardi, A.E., Machado, S., Scorza, C., Torterolo, P., 2019b. EEG 40 Hz coherence decreases in REM sleep and ketamine models of psychosis. *Front.Psychiatr*. 766, https://doi.org/10.3389/fpsyt.2018.00766.

Cavelli, M., Castro, S., Schwarzkopf, N., Chase, M.H., Falconi, A., Torterolo, P., 2015. Coherent neocortical gamma oscillations decrease during REM sleep in the rat. Behav. Brain Res. 281, 318–325.

Cavelli, M., Castro-Zaballa, S., Mondino, A., Gonzalez, J., Falconi, A., Torterolo, P., 2017a. Absence of EEG gamma coherence in a local activated neocortical state: a conserved trait of REM sleep. Transl. **Brain** Rhythm. 2, 1–13.

Cavelli, M., Rojas-Libano, D., Schwarzkopf, N., Castro-Zaballa, S., Gonzalez, J., Mondino, A., Santana, N., Benedetto, L., Falconi, A., Torterolo, P., 2017b. Power and coherence of cortical high-frequency oscillations during wakefulness and sleep. Eur. J. Neurosci. 48, 272–273, https://doi.org/10.1111/ejn.13718.

Chalmers, D.J., 2005. Facing up to the problem of consciousness. J. Conscious. Stud. 2, 200–219.

Chow, K.L., John, E.R., 1959. Acetylcholine metabolism and behavior of rats. Science 129, 64.

Corlett, P.R., Honey, G.D., Fletcher, P.C., 2007. From prediction error to psychosis: ketamine as a pharmacological model of delusions. J. Psychopharmacol. 21, 238–252.

Crunelli, V., David, F., Lorincz, M.L., Hughes, S.W., 2015. The thalamocortical network as a single slow wave-generating unit. Curr. Opin. Neurobiol. 31, 72–80.

Damasio, A., Meyer, K., 2009. Consciousness:an overview of the phenomenon and of its possible neural basis. In: Laureys, S., Tononi, G. (Eds.), The Neurology of Consciousness. Academic Press, London, pp. 3–14.

Dement, W., Kleitman, N., 1957. The relation of eye movements during sleep to dream activity: an objective method for the study of dreaming. J. Exp. Psychol. 53, 339–346.

Ebert, U., Siepmann, M., Oertel, R., Wesnes, K.A., Kirch, W., 1998. Pharmacokinetics and pharmacodynamics of scopolamine after subcutaneous administration. J. Clin. Pharmacol. 38, 720–726.

Edelman, G.M., Tononi, G., 2000. A Universe of Consciousness. Basic Books, New York.

Franks, N.P., 2006. Molecular targets underlying general anaesthesia. Br. J. Pharmacol. 147 (Suppl. 1), S72–S81.

Fuentealba, P., Steriade, M., 2005. The reticular nucleus revisited: intrinsic and network properties of a thalamic pacemaker. Prog. Neurobiol. 75, 125–141.

Gallopin, T., Fort, P., Eggermann, E., Cauli, B., Luppi, P.H., Rossier, J., Audinat, E., Muhlethaler, M., Serafin, M., 2000. Identification of sleep-promoting neurons in vitro. Nature 404, 992–995.

Garcia-Rill, E., 2015a. The 10 Hz fulcrum. In: Garcia-Rill, E. (Ed.), Waking and the Reticular Activating System in Health and Disease. Elsevier, London, pp. 157–170.

Garcia-Rill, E., 2015b. Governing principles of brain activity. In: Garcia-Rill, E. (Ed.), Waking and the Reticular Activating System in Health and Disease. Elsevier, pp. 1–16.

Garcia-Rill, E., 2015c. The science of waking and public policy. In: Garcia-Rill, E. (Ed.), Waking and the Reticular Activating System in Health and Disease. Elsevier, London, pp. 292–306.

Gottesmann, C., 2006. The dreaming sleep stage: a new neurobiological model of schizophrenia? Neuroscience 140, 1105–1115.

Gottesmann, C., Gottesman, I., 2007. The neurobiological characteristics of rapid eye movement (REM) sleep are candidate endophenotypes of depression, schizophrenia, mental retardation and dementia. Prog. Neurobiol. 81, 237–250.

Hobson, J.A., 1997. Dreaming as delirium: a mental status analysis of our nightly madness. Semin. Neurol. 17, 121–128.

Hobson, J.A., 2009. REM sleep and dreaming: towards a theory of protoconsciousness. Nat. Rev. Neurosci. 10, 803–813.

Huguenard, J.R., McCormick, D.A., 2007. Thalamic synchrony and dynamic regulation of global forebrain oscillations. Trends Neurosci. 30, 350–356.

Javitt, D.C., 2010. Glutamatergic theories of schizophrenia. Isr. J. Psychiatry Relat. Sci. 47, 4–16.

John, E.R., 2002. The neurophysics of consciousness. Brain Res. Brain Res. Rev. 39, 1–28.

Keenan, S., Hirshkowitz, M., 2011. Monitoring and staging human sleep. In: Kryger, M.H., Roth, T., Dement, W.C. (Eds.), Principles and Practices of Sleep Medicine. Elsevier-Saunders, Philadelphia, pp. 1602–1609.

Krystal, J.H., Karper, L.P., Seibyl, J.P., Freeman, G.K., Delaney, R., Bremner, J.D., Heninger, G.R., Bowers Jr., M.B., Charney, D.S., 1994. Subanesthetic effects of the noncompetitive NMDA antagonist, ketamine, in humans. Psychotomimetic, perceptual, cognitive, and neuroendocrine responses. Arch. Gen. Psychiatry 51, 199–214.

Krystal, J.H., D'Souza, D.C., Mathalon, D., Perry, E., Belger, A., Hoffman, R., 2003. NMDA receptor antagonist effects, cortical glutamatergic function, and schizophrenia: toward a paradigm shift in medication development. Psychopharmacology (Berl) 169, 215–233.

Lee, K.H., Williams, L.M., Haig, A., Gordon, E., 2003. "Gamma (40 Hz) phase synchronicity" and symptom dimensions in schizophrenia. Cogn. Neuropsychiatry 8, 57–71.

Light, G.A., Hsu, J.L., Hsieh, M.H., Meyer-Gomes, K., Sprock, J., Swerdlow, N.R., Braff, D.L., 2006. Gamma band oscillations reveal neural network cortical coherence dysfunction in schizophrenia patients. Biol. Psychiatry 60, 1231–1240.

Lindsley, D.F., Carpenter, R.S., Killam, E.K., Killam, K.F., 1968. EEG correlates of behavior in the cat. I. Pattern discrimination and its alteration by atropine and LSD-25. Electroencephalogr. Clin. Neurophysiol. 24, 497–513.

Llinas, R.R., Pare, D., 1991. Of dreaming and wakefulness. Neuroscience 44, 521–535.

Llinas, R., Ribary, U., 1993. Coherent 40-Hz oscillation characterizes dream state in humans. Proc. Natl. Acad. Sci. U. S. A. 90, 2078–2081.

Longo, V.G., 1956. Effects of scopolamine and atropine electroencephalographic and behavioral reactions due to hypothalamic stimulation. J. Pharmacol. Exp. Ther. 116, 198–208.

Lopes da Silva, F., 2010. EEG: origin and measurement. In: Mulert, C., Lemieuz, L. (Eds.), EEG-fMRI. Springer-Verlag, Berlin, pp. 19–38.

Luppi, P.H., Gervasoni, D., Verret, L., Goutagny, R., Peyron, C., Salvert, D., Leger, L., Fort, P., 2007. Paradoxical (REM) sleep genesis: the switch from an aminergic-cholinergic to a GABAergic-glutamatergic hypothesis. J. Physiol. Paris 100, 271–283.

Lydic, R., Baghdoyan, H.A., 2005. Sleep, anesthesiology, and the neurobiology of arousal state control. Anesthesiology 103, 1268–1295.

Mahowald, M.W., Schneck, C., 2011. Non-REM arousal parasomnias. In: Kryger, M.H., Roth, T., Dement, W.C. (Eds.), Principles and Practices of Sleep Medicine. Elsevier-Saunders, Philadelphia, pp. 1075–1082.

Maloney, K.J., Cape, E.G., Gotman, J., Jones, B.E., 1997. High-frequency gamma electroencephalogram activity in association with sleep-wake states and spontaneous behaviors in the rat. Neuroscience 76, 541–555.

Mashour, G.A., 2006. Integrating the science of consciousness and anesthesia. Anesth. Analg. 103, 975–982.

McCarley, R.W., 2007. Neurobiology of REM and NREM sleep. Sleep Med. 8, 302–330.

Monti, J.M., Torterolo, P., Lagos, P., 2013. Melanin-concentrating hormone control of sleep-wake behavior. Sleep Med. Rev. 17, 293–298.

Nir, Y., Tononi, G., 2010. Dreaming and the brain: from phenomenology to neurophysiology. Trends Cogn. Sci. 14, 88–100.

Noreika, V., 2015. It's not just about the contents: searching for a neural correlate of consciousness. In: Metzinger, T., Windt, J.M. (Eds.), Open Mind, pp. 1–12.doi: 10.15502/9783958570504.

Nuotto, E., 1983. Psychomotor, physiological and cognitive effects of scopolamine and ephedrine in healthy man. Eur. J. Clin. Pharmacol. 24, 603–609.

Ostfeld, A.M., Machne, X., Unna, K.R., 1960. The effects of atropine on the electroencephalogram and behavior in man. J. Pharmacol. Exp. Ther. 128, 265–272.

Pace-Schott, E., 2011. The neurobiology of dreaming. In: Kryger, M.H., Roth, T., Dement, W.C. (Eds.), Principles and Practices of Sleep Medicine. Elsevier-Saunders, Philadelphia, pp. 563–575.

Pal, D., Silverstein, B.H., Lee, H., Mashour, G.A., 2016. Neural correlates of wakefulness, sleep, and general anesthesia: an experimental study in rat. Anesthesiology 125, 929–942.

Perry, E.K., Perry, R.H., 1995. Acetylcholine and hallucinations: disease-related compared to drug-induced alterations in human consciousness. Brain Cogn. 28, 240–258.

Pomarol-Clotet, E., Honey, G.D., Murray, G.K., Corlett, P.R., Absalom, A.R., Lee, M., McKenna, P.J., Bullmore, E.T., Fletcher, P.C., 2006. Psychological effects of ketamine in healthy volunteers. Phenomenological study. Br. J. Psychiatry 189, 173–179.

Posner, J., Saper, C.B., Schiff, N.D., Plum, F., 2007. The Diagnosis of Stupor and Coma. Oxford University Press, New York.

Rechtschaffen, A., 1978. The single-mindedness and isolation of dreams. Sleep 1, 97–109.

Rinaldi, F., Himwich, H.E., 1955. Alerting responses and actions of atropine and cholinergic drugs. A.M.A. Arch. Neurol. Psychiatry 73, 387–395.

Sahakian, B., 1988. Cholinergic drugs and human cognitive performance. In: Handbook of Psychopharmcology, pp. 393–424. https://doi.org/10.1007/978-1-4613-0933-8_9.

Saper, C.B., Fuller, P.M., Pedersen, N.P., Lu, J., Scammell, T.E., 2010. Sleep state switching. Neuron 68, 1023–1042.

Scorza, M.C., Meikle, M.N., Hill, X.L., Richeri, A., Lorenzo, D., Artigas, F., 2008. Prefrontal cortex lesions cause only minor effects on the hyperlocomotion induced by MK-801 and its reversal by clozapine. Int. J. Neuropsychopharmacol. 11, 519–532.

Siclari, F., Baird, B., Perogamvros, L., Bernardi, G., LaRocque, J.J., Riedner, B., Boly, M., Postle, B.R., Tononi, G., 2017. The neural correlates of dreaming. Nat. Neurosci. 20, 872–878.

Siegel, J.M., 2011. REM sleep. In: Kryger, M.H., Roth, T., Dement, W.C. (Eds.), Principles and Practices of Sleep Medicine. Elsevier-Saunders, Philadelphia, pp. 92–111.

Singer, W., 1999. Neuronal synchrony: a versatile code for the definition of relations? Neuron 24, 49–65. 111–125.

Steriade, M., McCormick, D.A., Sejnowski, T.J., 1993. Thalamocortical oscillations in the sleeping and aroused brain. Science 262, 679–685.

Sun, Y., Farzan, F., Barr, M.S., Kirihara, K., Fitzgerald, P.B., Light, G.A., Daskalakis, Z.J., 2011. Gamma oscillations in schizophrenia: mechanisms and clinical significance. Brain Res. 1413, 98–114.

Tanaka, H., Hayashi, M., Hori, T., 1996. Statistical features of hypnagogic EEG measured by a new scoring system. Sleep 19, 731–738.

Tononi, G., 2010. Information integration: its relevance to brain function and consciousness. Arch. Ital. Biol. 148, 299–322.

Tononi, G., Laureys, S., 2009. The neurology of consciousness: an overview. In: Laureys, S., Tononi, G. (Eds.), The Neurology of Consciousness: Cognitive Neuroscience and Neuropathology. Elsevier, San Diego, pp. 375–412.

Torterolo, P., Vanini, G., 2010. New concepts in relation to generating and maintaining arousal. Rev. Neurol. 50, 747–758.

Torterolo, P., Lagos, P., Monti, J.M., 2011. Melanin-concentrating hormone (MCH): a new sleep factor? Front. Neurol. 2, 1–12.

Torterolo, P., Scorza, C., Lagos, P., Urbanavicius, J., Benedetto, L., Pascovich, C., Lopez-Hill, X., Chase, M.H., Monti, J.M., 2015. Melanin-concentrating hormone (MCH): role in REM sleep and depression. Front. Neurosci. 9, 475.

Torterolo, P., Castro-Zaballa, S., Cavelli, M., Chase, M.H., Falconi, A., 2016a. Neocortical 40 Hz oscillations during carbachol-induced rapid eye movement sleep and cataplexy. Eur. J. Neurosci. 43, 580–589.

Torterolo, P., Monti, J.M., Pandi-Perumal, S.R., 2016b. Neuroanatomy and neuropharmacology of sleep and wakefulness. In: Pandi-Perumal, S.R. (Ed.), Synopsis of Sleep Medicine. Apple Academic Press, Oakville, Canada.

Uhlhaas, P.J., Singer, W., 2010. Abnormal neural oscillations and synchrony in schizophrenia. Nat. Rev. Neurosci. 11, 100–113.

Uhlhaas, P.J., Pipa, G., Lima, B., Melloni, L., Neuenschwander, S., Nikolic, D., Singer, W., 2009. Neural synchrony in cortical networks: history, concept and current status. Front. Integr. Neurosci. 3, 17.

Vanini, G., Torterolo, P., Baghdoyan, H., Lydic, R., 2011. Effects of general anesthetics on sleep-wake centers. In: Mashour, G.A., Lydic, R. (Eds.), The Neuroscientific Foundations of Anesthesiology. Oxford University Press, Oxford.

Varela, F., Lachaux, J.P., Rodriguez, E., Martinerie, J., 2001. The brainweb: phase synchronization and large-scale integration. Nat. Rev. Neurosci. 2, 229–239.

Voss, U., Holzmann, R., Tuin, I., Hobson, J.A., 2009. Lucid dreaming: a state of consciousness with features of both waking and non-lucid dreaming. Sleep 32, 1191–1200.

Voss, U., Holzmann, R., Hobson, A., Paulus, W., Koppehele-Gossel, J., Klimke, A., Nitsche, M.A., 2014. Induction of self awareness in dreams through frontal low current stimulation of gamma activity. Nat. Neurosci. 17, 810–812.

Vyazovskiy, V.V., Olcese, U., Hanlon, E.C., Nir, Y., Cirelli, C., Tononi, G., 2011. Local sleep in awake rats. Nature 472, 443–447.

White, R.S., Siegel, S.J., 2016. Cellular and circuit models of increased resting-state network gamma activity in schizophrenia. Neuroscience 321, 66–76.

Wikler, A., 1952. Pharmacologic dissociation of behavior and EEG "sleep patterns" in dogs; morphine, n-allylnormorphine, and atropine. Proc. Soc. Exp. Biol. Med. 79, 261–265.

Williams, R.H., Chee, M.J., Kroeger, D., Ferrari, L.L., Maratos-Flier, E., Scammell, T.E., Arrigoni, E., 2014. Optogenetic-mediated release of histamine reveals distal and autoregulatory mechanisms for controlling arousal. J. Neurosci. 34, 6023–6029.

Yamamoto, J., 1988. Roles of cholinergic, dopaminergic, noradrenergic, serotonergic and GABAergic systems in changes of the EEG power spectra and behavioral states in rabbits. Jpn. J. Pharmacol. 47, 123–134.

Yates, C., Garcia-Rill, E., 2015. Descending projections of the RAS. In: Garcia-Rill, E. (Ed.), Waking and the Reticular Activating System in Health and Disease. Elsevier, London, pp. 129–156.

Yeragani, V.K., Cashmere, D., Miewald, J., Tancer, M., Keshavan, M.S., 2006. Decreased coherence in higher frequency ranges (beta and gamma) between central and frontal EEG in patients with schizophrenia: a preliminary report. Psychiatry Res. 141, 53–60.

Chapter 2

Physiology of arousal

Edgar Garcia-Rill
Center for Translational Neuroscience, Department of Neurobiology and Developmental Sciences, University of Arkansas for Medical Sciences, Little Rock, AR, United States

Introduction

Two of the most important discoveries in sleep-wake control over the last 10 years are the description of electric coupling in some neurons of the reticular activating system (RAS) and the presence of intrinsic membrane gamma frequency oscillations in three arousal-related nuclei (Garcia-Rill et al., 2013). Coherence, most commonly mediated by electric coupling through gap junctions, is the process whereby groups of neurons fire in coordination to create a signal that is mirrored precisely by other groups of neurons (Thiagarajan et al., 2010). We demonstrated the presence of electric coupling in the pedunculopontine nucleus (PPN) (Garcia-Rill et al., 2007), in one of its ascending targets in the intralaminar thalamus and the parafascicular nucleus (Pf) (Garcia-Rill et al., 2008) and in one of its descending targets, the subcoeruleus nucleus dorsalis (SubCD) (Heister et al., 2007), a mechanism that allows these groups of neurons to fire coherently. This mechanism has significant clinical implications. For example, the mechanism of action of the stimulant modafinil, which is used for the treatment of narcolepsy and excessive daytime sleepiness, is to increase coupling via gap junctions, promoting coherence (Urbano et al., 2007). Most neurons in the RAS that are electrically coupled via gap junctions are GABAergic, so that modafinil-induced coupling decreases input resistance and decreases GABA release, thus disinhibiting a host of other systems, which leads to increased overall activity and higher coherence in the RAS, thalamocortical systems, and elsewhere. On the other hand, blocking gap junctions leads to rapid loss of consciousness, such as when using the fast-acting anesthetics halothane and propofol, both of which efficiently block gap junctions (Evans and Boitano, 2001; He and Burt, 2000). Oleamide and anandamide both promote sleep by blocking gap junctions (Boer et al., 1998; Murillo-Rodriguez et al., 2003; Gigout et al., 2006). The antiulcer agent carbenoxolone and the antimalarial agents quinine and mefloquine are also efficient gap junction blockers and share pronounced soporific side effects (Rozental et al., 2001). Thus drugs that

Arousal in Neurological and Psychiatric Diseases. https://doi.org/10.1016/B978-0-12-817992-5.00002-7

affect gap junctions also affect sleep-wake control by helping synchronize or desynchronize ensemble activity (Leznik and Llinas, 2005).

The other aspect of large-scale activity that is essential to neural encoding is firing frequency. Another major discovery in the RAS is the presence of intrinsic gamma band activity in the same three RAS nuclei named earlier (Kezunovic et al., 2011a,b; Simon et al., 2010). As we will see in the succeeding text, gamma band activity has been associated with attentional mechanisms, sensory perception, and memory. Conditions that block gamma band activity such as anesthetics and disease processes, schizophrenia, and the Parkinson's disease (PD) all lead to interrupted gamma band activity, with serious consequences on complex brain processes. Electric coupling and intrinsic oscillations at dedicated frequencies work together to allow oscillations to persist over a wide range of membrane potentials, making their firing frequency independent of membrane potential (Leznik and Llinas, 2005; Llinas and Yarom, 1986) and thus maintaining their activity. This chapter will stress the role of gamma band activity in the RAS since it is directly related to the control of arousal.

Arousal and the maintenance of arousal

The organizational features of brain activity for generating conscious awareness were long ago proposed to begin through the activation of the RAS by incoming and continuously changing sensory input and by intrinsically generated activity (Sperry, 1969). Most neuroscientists have never appreciated the prescience nature of Sperry's conclusions. More recent evidence suggests that, when we awaken, blood flow first increases in the upper brain stem and thalamus and only later increases in the frontal lobes (Balkin et al., 2002). Thus awakening entails a rapid reestablishment of consciousness within seconds followed by a relatively slow reestablishment of full awareness. Using positron-emission tomography, cerebral blood flow upon waking was found to increase soonest in the brain stem and thalamus, suggesting that the reactivation of these regions underlies the reestablishment of basic consciousness. Over the next 15–20 min, blood flow increased primarily in anterior cortical regions. These results question ideas that the cortex is solely responsible for achieving conscious awareness.

Moreover, sleep and waking do not appear to be a continuous, but rather a stepwise process. There are slow transitions in the order of minutes between waking and sleep and in the transition between slow-wave sleep (SWS) and rapid eye movement (REM) sleep (Steriade, 1999). In the absence of strong sensory stimuli to induce waking, waking and sleep appear to be "recruited" and may be part of a stepwise process, in which frequencies decrease from gamma, to beta, to alpha and lower, leading to SWS, or increase from delta, to theta, to alpha and higher, leading to waking (Garcia-Rill, 2017). This stepwise process also suggests that the modulation of waking and sleep is progressive but piecemeal and that it is the coherence at particular frequencies that determine the particular new state. That is, the new state may be "recruited" to another step or level rather than gradually formed. Clinically, this may be reflected in varying levels of consciousness from coma (complete absence of wakefulness), to a vegetative state (achieves partial arousal), to a minimally conscious state (partial preservation of consciousness), to locked-in syndrome (awareness without movement). These may all represent stages of waking based on the degree of damage to gamma band generating processes (Steriade, 1999).

The original description of the RAS proposed that it participates in tonic or continuous arousal (Moruzzi and Magoun, 1949). Moreover, lesions of this region were found to eliminate tonic arousal (Watson et al., 1974). Thus both sensory input and intrinsic activity help maintain gamma band activity for prolonged periods and are critically important for such functions as perception, learning, and memory, not to mention consciousness. It has been suggested that consciousness is associated with continuous gamma band activity, as opposed to an interrupted pattern of activity (Vanderwolf, 2000a,b). Clinically, the lack of consistent gamma band activity would alter waking and REM sleep drive, decrease attention, disrupt the binding of perceptions, and the like, accounting for a number of symptoms in neurological and psychiatric disorders. Dysregulation of this system thus can elicit a host of abnormal sensory and motor responses, both tonic and phasic, that can account for a number of symptoms expressed in schizophrenia, bipolar disorder, posttraumatic stress disorder, PD, the Alzheimer's disease (AD), and certain strokes (Garcia-Rill, 2015). It should be noted that gamma band activity is present throughout various brain regions including the cortex (Eckhorn et al., 1988; Gray and Singer, 1989; Philips and Takeda, 2009), thalamus (Llinas et al., 2002; Steriade and Llinás, 1988), hippocampus (Colgin and Moser, 2010), basal ganglia (Jenkinson et al., 2013), cerebellum (Middleton et al., 2008), and, of course, the RAS (Garcia-Rill, 2015). But how is gamma band activity generated in the RAS?

Gamma activity in the RAS

Two calcium channels

The PPN is the only nucleus in the RAS that is active during both states of high-frequency (beta/gamma) EEG activity as in waking and REM sleep. Paradoxically, the EEG pattern during these two markedly different states is very similar, thus the name "paradoxical sleep" for REM sleep because of its paradoxically similar EEG pattern to waking. However, beta/gamma band activity may not be the same during waking as during REM sleep. Classic studies of recordings of PPN neurons during sleep-wake states revealed neurons that were active during waking and REM sleep, referred to as "wake-REM on" cells; others were active only during REM sleep, referred to as "REM on" cells, and yet others only during waking, referred to as "wake on" cells, and none were more active during SWS (Sakai et al., 1990; Steriade et al., 1990a,b; Kayama et al., 1992; Datta and Siwek, 2002; Datta et al., 2009; Boucetta et al., 2014). Gamma band activity has been observed in the cortical EEG of the cat in vivo and in PPN neurons when the animal is active (Steriade et al., 1990a,b) and in the region of the PPN in humans during stepping, but not at rest (Fraix et al., 2008), and firing at low frequencies ~10 Hz at rest in the primate but firing at gamma band frequencies when the animal woke up or when the animal began walking on a treadmill (Goetz et al., 2016). Thus the same cells were involved in both arousal and motor control in the PPN in vitro, in vivo, and across species, including man.

Our findings established that every PPN neuron fired maximally at beta/gamma frequencies (Simon et al., 2010), that all PPN neurons manifested beta/gamma frequency intrinsic membrane oscillations (Urbano et al., 2012), and that these oscillations were mediated by high-threshold, voltage-dependent N- and/or P/Q-type calcium channels (Kezunovic et al., 2011a,b). We found that these channels were distributed along the dendrites of PPN neurons (Hyde et al., 2013a,b) and that some cells exhibited both N- and P/Q-type calcium channels, some had only N-type channels, and some had only P/Q-type channels (Luster et al., 2015, 2016). We suggested that these three cell types represented neurons that were active in relation to waking and REM sleep (N + P/Q cells = "wake-REM on" cells), only during REM sleep (N only = "REM on" cells), or only during waking (P/Q only = "wake on" cells), respectively (Garcia-Rill et al., 2014b; Luster et al., 2015, 2016). Fig. 2.1 shows the protocol used to identify these neurons. If the cell was a P/Q-only neuron (~25%), ramp-induced oscillations were not affected by the N-type channel blocker conotoxin, but were entirely blocked by agatoxin, a specific P/Q-type channel blocker. If the cell was an N-only neuron (~25%), ramp-induced oscillations were not affected by agatoxin but were completely blocked by conotoxin. If the cell was an N + P/Q neuron (~50%), conotoxin only partially reduced oscillations, and agatoxin blocked the remainder of the oscillations.

Two intracellular pathways

In freely moving animals, injections of glutamate into the PPN increased both waking and REM sleep (Datta et al., 2001), but injections of *N*-methyl-D-aspartate (NMDA), one glutamatergic receptor agonist, increased only waking (Datta et al., 2002), while injections of kainic acid (KA), another glutamatergic receptor agonist, increased only REM sleep (Datta, 2002). These data indicated a differential control of waking versus REM sleep by different glutamatergic receptor subtypes. Protein kinase C (PKC), which modulates KA receptors, enhances N-type channel activity and has no effect on P/Q-type channel function (Stea et al., 1995); however, CaMKII, which modulates NMDA receptors, was shown to modulate P/Q-type channel function (Jiang et al., 2008). These results suggest that the two calcium channel subtypes are modulated by different intracellular pathways, N-type by the cAMP/PK pathway and P/Q-type via the CaMKII pathway. The implications from all of these results are that there is a "waking" pathway mediated by CaMKII and P/Q-type channels and a

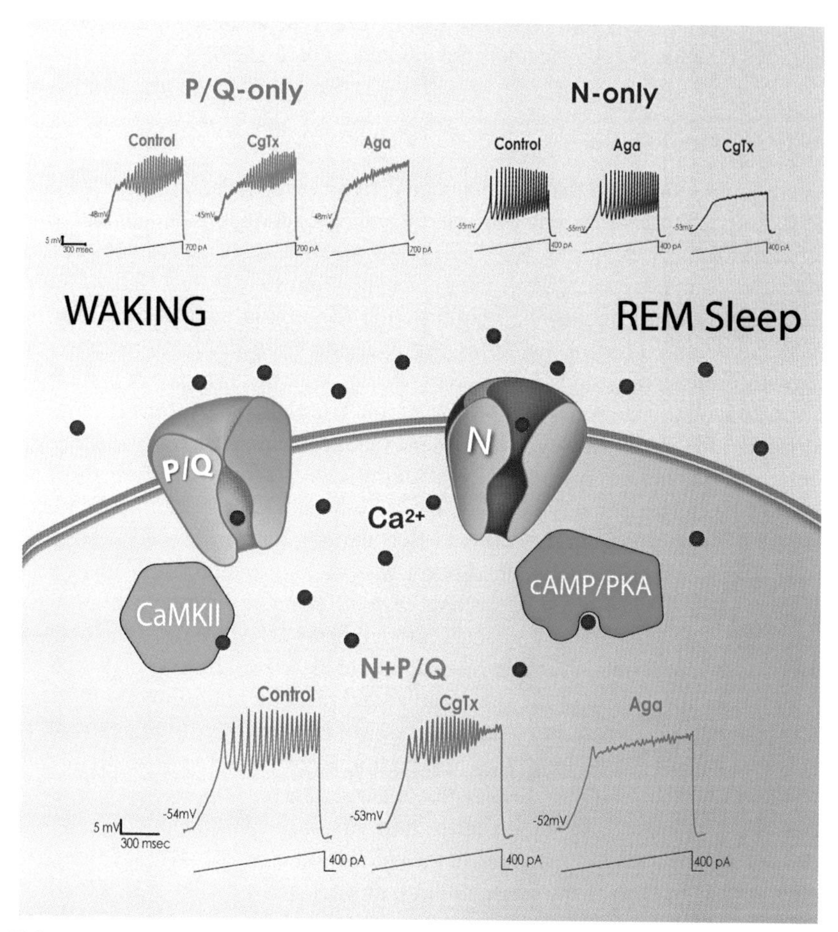

FIG. 2.1 Two types of calcium channels modulate waking versus REM sleep. P/Q-type calcium (Ca^{2+}) channels *(green channel)* are modulated by CaMKII *(green sextagon)* intracellularly, while N-type channels *(blue channel)* are modulated by cAMP/PKA *(blue structure)* intracellularly. If the cell was a P/Q-only neuron (~25%), ramp-induced oscillations were not affected by the N-type channel blocker ω-conotoxin-GVIA (CgTx), but were entirely blocked by ω-agatoxin-IVA (Aga), a specific P/Q-type channel blocker (*green* records top left). If the cell was an N-only neuron (~25%), ramp-induced oscillations were not affected by Aga but were completely blocked by CgTx (*purple* records top right). If the cell was an N + P/Q neuron (~50%), CgTx only partially reduced oscillations, and Aga blocked the remainder of the oscillations (*blue-green* records bottom). Calcium (Ca^{2+}), *red dots*.

"REM sleep" pathway mediated by cAMP/PK and N-type channels; moreover, different PPN cells fire during waking (those with N + P/Q and only P/Q type) or REM sleep (those with N + P/Q and only N type) or both (N + P/Q) (Garcia-Rill, 2015, 2017; Garcia-Rill et al., 2014b; Urbano et al., 2012). Fig. 2.1 also shows the intracellular mechanisms modulating the two types of high-threshold, voltage-dependent calcium channels, that is, P/Q-type channels are modulated by CaMKII while N-type channels are modulated by cAMP/PKA.

Cortical manifestations

Recent findings showed that gamma band activity at the level of the cortex during waking is characterized by coherence across distant regions but gamma band activity in the cortex during REM sleep has an absence of coherence (Castro et al., 2013; Cavelli et al., 2015). Thus since the brain stem is the origin of REM sleep drive, it is clear that the manifestation of gamma band activity during REM sleep at the level of the cortex begins in the brain stem. We assume that the manifestation of gamma band activity during waking is at least in part originating in the brain stem as well. Moreover, injections of the cholinergic agonist carbachol induced REM sleep with cataplexy that is characterized by decreased gamma band coherence in the cortex (Torterolo et al., 2015). Therefore, this line of evidence suggests that not only do brain stem centers drive gamma band activity that is manifested in the cortical EEG but also, during waking, brain stem-thalamic projections include coherence across regions and, during REM sleep, which is controlled by the subcoeruleus region (lesion of this region eliminates REM sleep, and injection of carbachol into it induces REM sleep), drive cortical EEG rhythms without coherence (Garcia-Rill et al., 2016a,b).

Fig. 2.2 shows more of the intracellular and receptor control of the two calcium channels described previously. Not shown on this figure are the effects of activation of, for example, a muscarinic cholinergic receptor typically acting through G protein coupling to phospholipase C that in turn cleaves phospholipid phosphatidylinositol biphosphate into inositol triphosphate (IP3). IP3 is released and binds to IP3 receptors in the endoplasmic reticulum (ER) to release calcium (Ca^{2+}). One of the intracellular pathways activated involves CaMKII, which modulates P/Q-type calcium channels, and the other pathway involves cAMP/PKA, which modulates N-type calcium channels. The CaMKII/P/Q-type pathway mediates beta/gamma band activity during waking, while the cAMP/PKA/N-type pathway mediates beta/gamma band activity during REM sleep. Moreover, activation of NMDA receptors modulates P/Q-type channel activity, while activation of kainic acid receptors modulates N-type channel activity. Thus gamma band oscillations generated by the PPN during waking ultimately lead to gamma activity in the cortical EEG with coherence across distant sites, while gamma activity during REM sleep leads to a similar fast cortical EEG but without coherence across distant sites.

Bottom-up gamma

Fig. 2.3 depicts one origin of bottom-up gamma via sensory afferents that induce gamma oscillations in PPN neurons. These cells project to the parafascicular nucleus in the ILT that also manifests N- and P/Q-type calcium

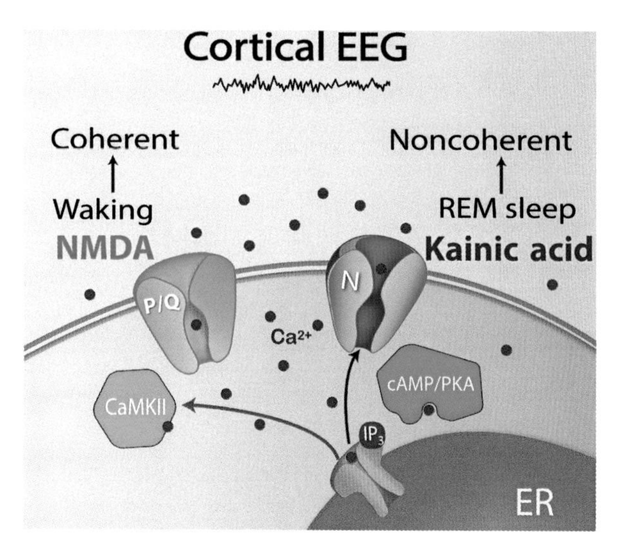

FIG. 2.2 Differential effects on cortical EEG of bottom-up gamma during waking versus REM sleep. Upon receptor activation, IP_3 is released and binds to IP_3 receptors in the endoplasmic reticulum (ER) to release calcium (Ca^{2+}, *red dots*). One of the intracellular pathways activated involves CaMKII *(green sextagon)*, which modulates P/Q-type calcium channels, and the other pathway involves cAMP/PKA *(blue structure)*, which modulates N-type calcium channels. The CaMKII/P/Q-type pathway mediates beta/gamma band activity during waking, while the cAMP/PKA/N-type pathway mediates beta/gamma band activity during REM sleep. Moreover, activation of the NMDA receptor modulates P/Q-type channel activity, while activation of the kainic acid receptor modulates N-type channel activity. The gamma band oscillations generated by the PPN during waking ultimately lead to gamma activity in the cortical EEG with coherence across distant sites, while gamma activity during REM sleep leads to a similar fast cortical EEG but without coherence across distant sites (see text for relevant citations).

channel-mediated gamma oscillations (Kezunovic et al., 2011a,b), and these channels are located all along the dendrites of Pf neurons (Hyde et al., 2013a,b). The ILT sends arousal-related input to the upper layers of the cortex, modulating low-amplitude, high-frequency EEG and providing the context of sensory perception (Llinas et al., 2002). These suggestions point to a new direction for determining, for example, the role of N-type and P/Q-type calcium channel expression in various neurological and psychiatric disorders, which has not been well explored. The role of intracellular pathways through CaMKII and waking and cAMP/PKA and REM sleep also requires further attention since it is at intracellular sites that pathological states may more easily be treated (since most drugs act on intracellular G-coupled proteins). These and additional questions remain, but the discoveries described herein provide a physiological mechanism that accounts for many of the arousal symptoms of a number of diseases and multiple novel directions for developing more effective treatments for these disorders.

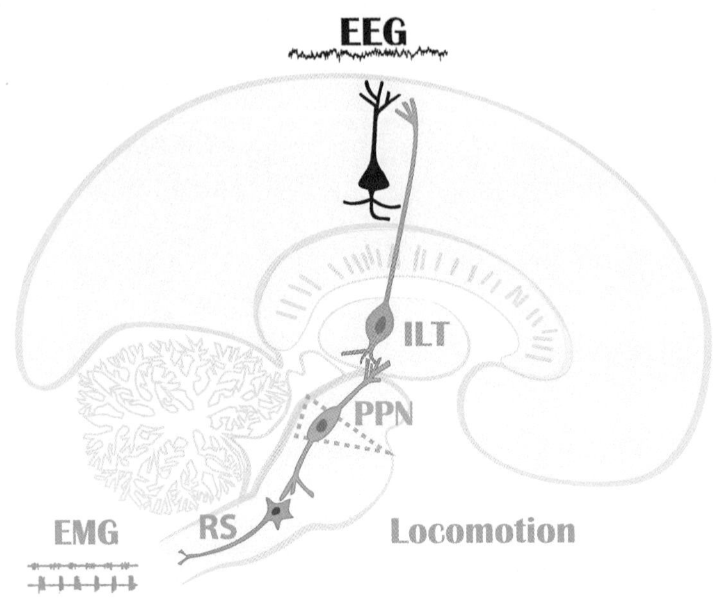

FIG. 2.3 Role of the PPN in cortical arousal and motor control. The PPN sends ascending projections to the intralaminar thalamus (ILT) that in turn project to the upper layers of the cortex to help modulate low-amplitude, high-frequency EEG and cortical arousal. The PPN also sends descending projections to reticulospinal (RS) pathways to modulate the startle response (through inhibition of giant reticulospinal projections, not shown) and locomotion through excitation of medium reticulospinal neurons that in turn trigger spinal pattern generators to induce locomotion in fight-or-flight responses reflected in alternating electromyograms (EMG).

Such efforts are important because disturbances in gamma band maintenance at the level of the PPN will disrupt a basic survival process. The ubiquitous generation of gamma band activity by all PPN neurons has been proposed, not to participate in the temporal process of sensory binding as in the cortex, but, as mentioned earlier, to underlie the more basic process of preconscious awareness (Garcia-Rill, 2015; Urbano et al., 2014). That is, PPN gamma activity may participate in a process that provides the essential stream of information necessary for the formulation of many of our actions. Preconscious awareness is the state that allows us to reliably assess the world around us on a continuous basis, providing a unifying picture and promoting survival. Although this process does not involve selective attention to specific stimuli, it provides the background upon which we assess our environment and act upon it. Without this function, our very survival is at risk. Interruptions in this process, therefore, will markedly disrupt perception and planning of motor strategies, adding uncertainty to the wildly shifting world of the patient.

In addition, Fig. 2.3 shows the simultaneous activation of descending projections to reticulospinal pathways that influence the spinal cord in the form of postural changes in muscle tone resulting from the startle response and trigger locomotor events in fight-or-flight responses (Garcia-Rill, 2015). That is, the RAS is inextricably entwined with motor and postural control. Not only does the RAS modulate locomotion in fight-or-flight responses but also muscle tone in the atony of REM sleep and the startle response. However, the involvement in motor control of the RAS, and of the PPN in particular, has been misunderstood.

Posture and locomotion

The mesencephalic locomotor region

The mesencephalic locomotor region (MLR) was originally described in very specific terms. In the precollicular-postmamillary transected cat, stimulation using long-duration (0.5–1.0 ms), low-amplitude (10–100 uA) pulses delivered at specific frequencies (40–60 Hz) could elicit locomotion on a treadmill when the body was suspended on a sling (Shik et al., 1966). Moreover, increasing stimulus amplitude would change stepping from a walk to a trot to a gallop, otherwise known as "controlled" locomotion on a treadmill. This is markedly different from the "unspecific" progression elicited by stimulation of multiple brain regions in an intact animal (Shik et al., 1966; Garcia-Rill, 1986). Such studies in decerebrate animals were designed to identify a locus independent of higher structures that could drive stepping, in contrast to studies on intact animals in which stimulation of this region could activate multiple systems to drive unknown and uncontrolled behavioral approach and escape, appetitive response, etc. We investigated the neuronal substrates mediating this very specific effect in decerebrate animals and found that at least some low-threshold sites in the rat and cat brain were located within the PPN when labeled for cholinergic neurons (Garcia-Rill et al., 1987, 1985). We further showed that the region could be chemically activated, eliminating the potential for electric stimulation of fibers of passage (Garcia-Rill et al., 1985). However, we pointed out that there were other nearby sites that would lead to locomotion, especially when stimulated using different parameters (Garcia-Rill, 1986; Garcia-Rill and Skinner, 1988). We also pointed out that locomotion had to be "recruited" rather than induced, since it did not immediately follow the onset of stimulation but required at least 1 s of stimulation, and that the region overlapped with elements of the RAS, especially the PPN (Garcia-Rill and Skinner, 1988; Skinner and Garcia-Rill, 1990).

We repeatedly cast doubt on the specificity of this region, concluding that the MLR was not a "locomotion-specific" area, but rather a region that could be stimulated to induce rhythmic activity, that is, a "rhythmogenic" area (Garcia-Rill et al., 1987; Garcia-Rill and Skinner, 1988; Skinner and Garcia-Rill, 1990). In other words, stimulating at the natural frequency of this region would activate all of its efferent projections, including ascending ones projecting to the thalamus and basal ganglia modulating arousal and descending ones to reticulospinal systems modulating posture and gait (Garcia-Rill, 1991; Reese et al., 1995). These projections explain why PPN stimulation can simultaneously elicit cortical arousal and postural and locomotor changes (Garcia-Rill et al., 1996). It was not until much later that we discovered the reason for the need to stimulate at 40–60 Hz, this natural frequency or unique intrinsic oscillation beta/gamma frequency in these cells, as described earlier, that could be optimally activated by stimulation at 40–60 Hz. That is, the fact that all PPN cells, regardless of transmitter type, manifest intrinsic membrane oscillations in the beta/gamma range mediated by high-threshold calcium channels imparts on them a natural frequency of firing (Kezunovic et al., 2011a,b; Urbano et al., 2014). Stimulation at this natural frequency range thus recruits the entire population to fire at its natural frequency, efficiently driving its projections.

Despite the need to establish low-threshold stimulation sites in order to conclusively identify this region, a multitude of investigators have made assertions about the location of the MLR without labeling stimulation sites for cholinergic neurons or without carrying out actual "controlled" locomotion on a treadmill studies (Noga et al., 2017; Takakusaki et al., 2003). Moreover, using intact (nontransected) animals likely will test the many ways in which arousal can induce "unspecific" behavioral approach and escape in intact animals, but will not identify the site (s) essential for "controlled" locomotion (Noga et al., 2017). Based on our suggestions over the years that stimulation of this region could be used therapeutically in PD (Garcia-Rill, 1986, 1991; Reese et al., 1995), there are now over 200 patients implanted with PPN deep brain stimulation (DBS) electrodes. These studies have shown salutary effects on gait, posture, sleep, and even cognitive measures (Garcia-Rill et al., 2014a). The most effective frequencies are not surprisingly in the 25–60 Hz (beta/gamma) range, but at 10–20 Hz as recently purported (Noga et al., 2017). Such low frequencies do not produce the full range of beneficial effects of PPN DBS. The most effective frequencies (beta/gamma) are in keeping with the natural frequencies of firing of PPN neurons when they are activated (Garcia-Rill et al., 2014b; Kezunovic et al., 2011a,b; Luster et al., 2015, 2016; Simon et al., 2010). Moreover, from a philosophical viewpoint, given the existence of spinal pattern generators for walking, why would there be a region high in the brain stem dedicated to locomotor control? We reiterate our position that the term "MLR" is outdated and should be retired.

Optogenetics

Optogenetic tools derived from microbial (type I) or animal (type II) opsins are widely used in a fast-growing area of neuroscience that combines several animal models in order to target subpopulations of neurons within particular tissues/brain areas (Fenno et al., 2011). While this is a powerful technology, there are a number of caveats that need to be remembered. (1) First, the combination of behavioral with optogenetic methods requires in many cases the implantation of an optical fiber to deliver blue light capable of reaching subcortical structures. However, recent studies show that blue light activates cells that do not express opsins (Cheng et al., 2016). (2) Then, there is the issue of latency. Although rapid resolution of opsins like channelrhodopsin or archaerhodopsin allows the study of fast interactions between cell groups, their association with behavioral events is not always rapid or direct. This makes it difficult to interpret multifactorial behavioral assays (i.e., sleep-wake rhythms, anxiety, fear, or depression tests) due to the long latency between stimulation and behavioral event. (3) Stimulation to "recruit" locomotion typically has a latency of ~1 s, but optogenetic stimulation normally takes 10–20 s to induce an effect (Garcia-Rill et al., 2018). This link would be particularly remote in assessing whether animals sleep or awaken for longer periods or not (Scammell et al., 2018). (4) Lastly, the methodology employs the genetic introduction of the opsin using a vector, usually viral or transgenic, attached to a promoter that overexpresses the protein. Therefore expression of opsins might permanently alter synaptic activity among brain nuclei by simply changing the lipid environment required for the normal membrane expression of channels and/or receptors (Brady et al., 2004).

Many of these issues are particularly critical when studying the PPN. For example, in some groups of neurons like the PPN, high-frequency subthreshold oscillations require the use of a ramp-like, sustained depolarization that cannot be achieved using trains of light stimulation. Therefore the kind of light-stimulated activity induced may be far from that induced by the normal synaptic activation of calcium channels gradually recruiting intrinsic oscillations. Moreover, these channels are expressed all along the dendrites of PPN neurons (Hyde et al., 2013a,b). The initial overexpression of the channelrhodopsin could interfere with the lipid rafts that organize, for example, high-threshold, voltage-dependent channels (Guéguinou et al., 2015). This would markedly change the gamma band responsiveness of PPN neurons. Finally, since expressing these proteins in a cell could (a) alter protein expression, (b) opsins could induce heat that changes the cell, and (c) channel and pump activity can be altered (Allen et al., 2015), the long-term insertion of these calcium leak channels could gradually change calcium intracellular concentrations, leading to abnormal responses; furthermore, it needs to be determined if such alterations can lead to cell death in the long run.

Until the appropriate controls are carried out, which include determining if high-threshold calcium currents (or for that matter other calcium currents) are changed by the genetic insertion of overexpressed calcium channels in the PPN, we can expect reports questioning well-established observations. For example, the role of PPN gamma activity during waking and REM sleep has been supported by many years of research in various species and preparations. However, optogenetic studies downplay the role of PPN neurons with these channels and differ in the established role they play in waking and REM sleep (Scammell et al., 2018). Another example is the reported lack of startle response prepulse inhibition by optogenetically modified cholinergic PPN neurons (Azzopardi et al., 2018), which counters well-established evidence. It is the responsibility of those carrying out optogenetic experiments that their preparation is not introducing uncontrolled conditions that create erroneous controversies.

Clinical implications

There are two clear options for known treatments that modulate gamma band activity. In terms of electric coupling, the mechanism of action of modafinil, an atypical stimulant used for treatment of narcolepsy and daytime sleepiness, was found to involve increased electric coupling across gap junctions (Urbano et al., 2007). The mechanism suggests that, since most electrically coupled neurons are GABAergic, increasing electric coupling will decrease input resistance and thus decrease activity. This leads to decreased GABA release, with resultant disinhibition of other transmitter systems. That is, its stimulatory effect is via overall disinhibition along with increased coherence due to increased coupling. Therefore modafinil will promote additional coherence at whatever frequencies the circuit is firing. Sleep/wake states will thus become more stable, and each state will be more easily maintained. Table 2.1 lists a number of disorders addressed in this book, the arousal symptoms manifested in each, and the potential effects of modafinil.

The other option is for modulating gamma band activity directly using stimulation. DBS has been used for a number of neurological and psychiatric disorders. The use of DBS in such regions as the subthalamic region involves high-frequency (>80/100 Hz) stimulation to alleviate motor symptoms. Some believe that such high frequencies basically depolarize block the region being stimulated, presumably ameliorating movements by inhibiting spurious signals. On the other hand, PPN DBS has come into more frequent use but using preferred frequencies at 25–40 Hz (Garcia-Rill et al., 2014a). Such frequencies of stimulation at the natural or preferred frequencies of the cells in the region promote circuit activity at physiologically functional levels, with considerable salutary effects by most groups (Garcia-Rill et al., 2018). Table 2.1 also lists neurological and psychiatric disorders and their gamma band pathologies that could be addressed by PPN DBS. Therefore both of these major discoveries regarding RAS function can be clinically targeted.

Future directions

A field of study that has not received sufficient attention is that of neuroepigenetics. Does the manifestation of intrinsic gamma band oscillations lead to gene transcription? Is gene transcription different during waking versus REM sleep? Is gene transcription different when the P/Q-type channel/CaMKII pathway is activated compared with the N-type channel/cAMP/PKA pathway? What purpose would such transcription possess? Is it a mechanism for rebuilding the channels activated during a particular state? Or is it a mechanism that allows the

TABLE 2.1 Potential effects of PPN stimulation or modafinil in various disorders

Disorder	Arousal symptom	Effect of PPN stimulation/modafinil
Parkinson's disease	Increased REM sleep	Decrease REM sleep
	Fragmented sleep	Normalize sleep
	Insomnia	Normalize sleep
	Cognition	Improve attention
	Gait and posture	Improve
Alzheimer's disease	Increased gamma	40–60 Hz negative effect
		>100 Hz positive effect
Epilepsy	Low-frequency EEG	Increase EEG frequency
Schizophrenia	Fragmented sleep	Normalize sleep
	Interrupted gamma	Increase gamma
	Increased REM sleep	Decrease REM sleep
Bipolar disorder	Fragmented sleep	Normalize sleep
	Interrupted gamma	Increase gamma
	Increased REM sleep	Decrease REM sleep
Neglect	Contralateral neglect	Increase gamma
	Decreased arousal	Increase arousal
Coma	Vegetative	Increase gamma
	Minimally conscious	Increase gamma
	Locked-in	Increase gamma

consolidation of the organism's identity? That is, after repeated bouts of SWS during the night, do we need a reminder of who we are? During SWS, what are cortical cells doing? The EEG tells us that large ensembles of neurons are firing in unison during SWS, but during waking, they are firing independently. What this means is that cortical columns, the constituent parts of the main mechanism regarded as responsible for perception and cortical function, are firing together during SWS. That is, there is no lateral inhibition since they are active synchronously. Thus inputs to individual columns go undetected; therefore no contrast exists between inputs to different columns that would allow perception. That is why there is no perception during SWS. On the other hand, independent firing is restored within a second or so upon waking, and the business of perception is once again active.

As mentioned earlier, increased blood flow in the brain stem and thalamus precedes that of the cortex upon waking. Yet we immediately and seamlessly know who we are as soon as we awaken. Does gene transcription during waking help the daily reformulation of the self? We will explore this suggestion in a later chapter, as we describe the third major advance we have made on the function of the RAS.

William James proposed that the "stream of consciousness" is "a river flowing forever through a man's conscious waking hours" (James, 1890). This pulsing stream persistently infiltrates our essence while waking, and yet we often fail to pay it heed, letting much sensory information go unnoticed within the innermost confines of our mind. Once beckoned into awareness, we can actively pay tribute to a particular piece of sensory information: we become fully "conscious" of the information (Civin and Lombardi, 1990). At the wellspring of this process is the RAS, a phylogenetically conserved area of the brain inundated by the continuous flow of internal (especially intrinsic oscillations) and external sensory information (impinging on dendrites). By traversing the RAS, this information modulates sleep-wake cycles, the startle response, and fight-or-flight responses, along with changes in muscle tone and locomotion. Accordingly, we speculate that activation of the RAS during waking induces coherent activity (through electrically coupled cells) and high-frequency oscillations (through P/Q-type calcium channel and subthreshold oscillations) to sustain gamma activity and support a persistent, reliable state for assessing our world.

Moreover, Penfield arrived at the conclusion that there was a mechanism for consciousness that was subcortical. He stated, "There is no place in the cerebral cortex where electrical stimulation will cause a patient to believe or to decide" (Penfield, 1970). In addition, he emphasized that, while cortical seizures localized to specific regions elicit sensory or motor effects but maintain consciousness, *petit mal* seizures in "mesothalamic" (midbrain and thalamus) regions always eliminate consciousness. Based on these results, Penfield proposed the presence of a "centrencephalic integrating system" that fulfills the role of sensorimotor integration necessary for consciousness. Perhaps gene transcription in the RAS allows the maintenance of the core identity of the being from day to day.

Acknowledgments

Supported by NIH award P30 GM110702 from the IDeA program at NIGMS. EGR would also like to express profound gratitude to all of the federal funding agencies, especially NIH and NSF, which have continuously funded his lab for the last 40 years. We also appreciate the assistance of S. Mahaffey with the figures in this and other chapters.

References

Allen, B., Singer, A.C., Boyden, E.S., 2015. Principles of designing interpretable optogenetic behavior experiments. Learn. Mem. 22, 232–238.

Azzopardi, E., Loutitt, A.G., DeOliveira, C., Laviolette, S.R., Scmid, S., 2018. The role of cholinergic midbrain neurons in startle and prepulse inhibition. J. Neurosci. 38, 8798–8808. epub.

Balkin, T.J., Braun, A.R., Wesensten, N.J., Jeffries, K., Varga, M., et al., 2002. The process of awakening: a PET study of regional brain activity patterns mediating the re-establishment of alertness and consciousness. Brain 125, 2308–2319.

Boer, D.L., Henriksen, S.J., Cravatt, B.F., 1998. Oleamide: an endogenous sleep-inducing lipid and prototypical member of a new class of biological signaling molecules. Curr. Pharm. Des. 4, 303–314.

Boucetta, S., Cisse, Y., Mainville, L., Morales, M., Jones, B.E., 2014. Discharge profiles across the sleep-waking cycle of identified cholinergic, gabaergic, and glutamatergic neurons in the pontomesencephalic tegmentum of the rat. J. Neurosci. 34, 4708–4727.

Brady, J., Rich, T., Le, X., Stafford, K., Fowler, C., et al., 2004. Functional role of lipid raft microdomains in cyclic nucleotide-gated channel activation. Mol. Pharmacol. 65, 503–511.

Castro, S., Falconi, A., Chase, M., Torterolo, P., 2013. Coherent neocortical 40-Hz oscillations are not present during REM sleep. Eur. J. Neurosci. 37, 1330–1339.

Cavelli, M., Castro, S., Schwartzkopf, N., Chase, M., Falconi, A., et al., 2015. Coherent cortical oscillations decrease during REM sleep in the rat. Behav. Brain Res. 281, 318–325.

Cheng, K.P., Kiernan, E.A., Eliceiri, K.W., Williams, J.C., Watters, J.J., 2016. Blue light modulates murine microglial gene expression in the absence of optogenetic protein expression. Sci. Rep. 6, 21172.

Civin, M., Lombardi, K.L., 1990. The preconscious and potential space. Psychoanal. Rev. 77, 573–585.

Colgin, L.L., Moser, E.I., 2010. Gamma oscillations in the hippocampus. Physiology (Bethesda) 25, 319–329.

Datta, S., 2002. Evidence that REM sleep is controlled by the activation of brain stem pedunculopontine tegmental kainate receptor. J. Neurophysiol. 87, 1790–1798.

Datta, S., Siwek, D.F., 2002. Single cell activity patterns of pedunculopontine tegmentum neurons across the sleep-wake cycle in the freely moving rats. J. Neurosci. Res. 70, 79–82.

Datta, S., Spoley, E.E., Patterson, E.H., 2001. Microinjection of glutamate into the pedunculopontine tegmentum induces REM sleep and wakefulness in the rat. Am. J. Physiol. Regul. Integr. Comp. Physiol. 280, R752–R759.

Datta, S., Patterson, E.H., Spoley, E.E., 2002. Excitation of the pedunculopontine tegmental nmda receptors induces wakefulness and cortical activation in the rat. J. Neurosci. Res. 66, 109–116.

Datta, S., Siwek, D.F., Stack, E.C., 2009. Identification of cholinergic and non-cholinergic neurons in the pons expressing phosphorylated cyclic adenosine monophosphate response element-binding protein as a function of rapid eye movement sleep. Neuroscience 163, 397–414.

Eckhorn, R., Bauer, R., Jordan, W., Brosch, M., Kruse, W., et al., 1988. Coherent oscillations: a mechanism of feature linking in the visual system? Biol. Cybern. 60, 121–130.

Evans, W.H., Boitano, S., 2001. Connexin mimetic peptides: specific inhibitors of gap junctional intercellular communication. Biochem. Soc. Trans. 29, 606–612.

Fenno, L., Yizhar, O., Deisseroth, K., 2011. The development and application of optogenetics. Ann. Rev. Neurosci. 34, 389–412.

Fraix, V., Bastin, F., David, O., Goetz, L., Ferraye, M., et al., 2008. Pedunculopontine nucleus area oscillations during stance, stepping and freezing in Parkinson's disease. PLoS One 8, e83919.

Garcia-Rill, E., 1986. The basal ganglia and the locomotor regions. Brain Res. Rev. 11, 47–63.

Garcia-Rill, E., 1991. The pedunculopontine nucleus. Prog. Neurobiol. 36, 363–389.

Garcia-Rill, E., 2015. Waking and the Reticular Activating System. Academic Press, New York, p. 330.

Garcia-Rill, E., 2017. Bottom-up gamma and stages of waking. Med. Hypotheses 104, 58–62.

Garcia-Rill, E., Skinner, R.D., 1988. Modulation of rhythmic function in the posterior midbrain. Neuroscience 17, 639–654.

Garcia-Rill, E., Skinner, R.D., Fitzgerald, J.A., 1985. Chemical activation of the mesencephalic locomotor region. Brain Res. 330, 43–54.

Garcia-Rill, E., Houser, C.R., Skinner, R.D., Smith, W., Woodward, D.J., 1987. Locomotion-inducing sites in the vicinity of the pedunculopontine nucleus. Brain Res. Bull. 18, 731–738.

Garcia-Rill, E., Reese, N.B., Skinner, R.D., 1996. Arousal and locomotion: from schizophrenia to narcolepsy. In: Holstege, G., Saper, C. (Eds.), The Emotional Motor System, Prog. Brain Res. 107. pp. 417–434.

Garcia-Rill, E., Heister, D.S., Ye, M., Charlesworth, A., Hayar, A., 2007. Electrical coupling: novel mechanism for sleep-wake control. Sleep 30, 1405–1414.

Garcia-Rill, E., Charlesworth, A., Heister, D.S., Ye, M., Hayar, A., 2008. The developmental decrease in REM sleep: the role of transmitters and electrical coupling. Sleep 31, 673–690.

Garcia-Rill, E., Kezunovic, N., Hyde, J., Beck, P., Urbano, F.J., 2013. Coherence and frequency in the reticular activating system (RAS). Sleep Med. Rev. 17, 227–238.

Garcia-Rill, E., Hyde, J., Kezunovic, N., Urbano, F.J., Petersen, E., 2014a. The physiology of the pedunculopontine nucleus- implications for deep brain stimulation. J. Neural Transm. 122, 225–235.

Garcia-Rill, E., Kezunovic, N., D'Onofrio, S., Luster, B., Hyde, J., et al., 2014b. Gamma band activity in the RAS- intracellular mechanisms. Exp. Brain Res. 232, 1509–1522.

Garcia-Rill, E., D'Onofrio, S., Mahaffey, S., 2016a. Bottom-up gamma: the pedunculopontine nucleus and reticular activating system. Transl. Brain Rhythm. 1, 49–53.

Garcia-Rill, E., Virmani, T., Hyde, J., D'Onofrio, S., Mahaffey, S., 2016b. Arousal and the control of perception and movement. Curr. Trends Neurol. 10, 53–64.

Garcia-Rill, E., Mahaffey, S., Hyde, J., Urbano, F.J., 2018. Bottom-up gamma in various disorders. Neurobiology of Disease: Pedunculopontine Nucleus Deep Brain Stimulation. In Press.

Gigout, S., Louvel, J., Kawasaki, K., D'Antuono, M., et al., 2006. Effects of gap junction blockers on human neocortical synchronization. Neurobiol. Dis. 22, 496–508.

Goetz, L., Piallat, B., Bhattacharjee, M., Mathieu, H., David, O., et al., 2016. The primate pedunculopontine nucleus region: towards a dual role in locomotion and waking state. J. Neural Transm. 123, 667–678.

Gray, C.M., Singer, W., 1989. Stimulus-specific neuronal oscillations in orientation columns of cat visual cortex. Proc. Natl. Acad. Sci. U. S. A. 86, 1698–1702.

Guéguinou, M., Gambade, A., Félix, R., Chantôme, A., Fourbon, Y., et al., 2015. Lipid rafts, KCa/ClCa/Ca^{2+} channel complexes and EGFR signaling: novel targets to reduce tumor development by lipids? Biochim. Biophys. Acta. 1848 (10 Pt B), 2603–2620.

He, D.S., Burt, J.M., 2000. Mechanism and selectivity of the effects of halothane on gap junction channel function. Circ. Res. 86, 1–10.

Heister, D.S., Hayar, A., Charlesworth, A., Yates, C., Zhou, C., et al., 2007. Evidence for electrical coupling in the subcoeruleus (SubC) nucleus. J. Neurophysiol. 97, 3142–3314.

Hyde, J., Kezunovic, N., Urbano, F.J., Garcia-Rill, E., 2013a. Spatiotemporal properties of high speed calcium oscillations in the pedunculopontine nucleus. J. Appl. Physiol. 115, 1402–1414.

Hyde, J., Kezunovic, N., Urbano, F.J., Garcia-Rill, E., 2013b. Visualization of fast calcium oscillations in the parafascicular nucleus. Pflugers Arch. 465, 1327–1340.

James, W., 1890. In: Holt, H. (Ed.), The Principles of Psychology. vol. 1. Henry Holt and Co., New York, pp. 225.

Jenkinson, N., Kuhn, A.A., Brown, P., 2013. Gamma oscillations in the human basal ganglia. Exp. Neurol. 245, 72–76.

Jiang, X., Lautermilch, N.J., Watari, H., Westenbroek, R.E., Scheuer, T., et al., 2008. Modulation of $Ca_v2.1$ channels by Ca^+/calmodulin-dependent kinase II bound to the C-terminal domain. Proc. Natl. Acad. Sci. U. S. A. 105, 341–346.

Kayama, Y., Ohta, M., Jodo, K., 1992. Firing of 'possibly' cholinergic neurons in the rat laterodorsal tegmental nucleus during sleep and wakefulness. Brain Res. 569, 210–220.
Kezunovic, N., Hyde, J., Simon, C., Urbano, F.J., Williams, D., et al., 2011a. Gamma band activity in the developing parafascicular nucleus (Pf). J. Neurophysiol. 107, 772–784.
Kezunovic, N., Urbano, F.J., Simon, C., Hyde, J., Smith, K., et al., 2011b. Mechanism behind gamma band activity in the pedunculopontine nucleus (PPN). Eur. J. Neurosci. 34, 404–415.
Leznik, E., Llinas, R., 2005. Role of gap junctions in synchronized oscillations in the inferior olive. J. Neurophysiol. 94, 2447–2456.
Llinas, R., Yarom, Y., 1986. Oscillatory properties of guinea-pig inferior olivary neurons and their pharmacological modulation: an in vitro study. J. Physiol. 376, 163–182.
Llinas, R., Leznik, E., Urbano, F.J., 2002. Temporal binding via cortical coincidence detection of specific and nonspecific thalamocortical inputs: a voltage-dependent dye-imaging study in mouse brain slices. Proc. Natl. Acad. Sci. U. S. A. 99, 449–454.
Luster, B., D'Onofrio, S., Urbano, F.J., Garcia-Rill, E., 2015. High-threshold Ca^{2+} channels behind gamma band activity in the pedunculopontine nucleus (PPN). Physiol. Rep. 3, e12431.
Luster, B., Urbano, F.J., Garcia-Rill, E., 2016. Intracellular mechanisms modulating gamma band activity in the pedunculopontine nucleus (PPN). Physiol. Rep. 4, e12787.
Middleton, S.J., Racca, C., Cunningham, M.O., Traub, R.D., Monyer, H., et al., 2008. High-frequency network oscillations in cerebellar cortex. Neuron 58, 763–774.
Moruzzi, G., Magoun, H.W., 1949. Brain stem reticular formation and activation of the EEG. Electroencephalogr. Clin. Neurophysiol. 1, 455–473.
Murillo-Rodriguez, E., Blanco-Centurion, C., Sanchez, C., Piomelli, D., et al., 2003. Anandamide enhances extracellular levels of adenosine and induces sleep: an in vivo microdialysis study. Sleep 26, 943–947.
Noga, B., Sanchez, M., Villamil, L., O'Toole, C., Kasicki, L., et al., 2017. LFP oscillations in the mesencephalic locomotor region during voluntary locomotion. Front. Neural Circuits 11, 34.
Penfield, W., 1970. The Mystery of the Mind. Princeton University Press, Princeton, New Jersey, p. 157.
Philips, S., Takeda, Y., 2009. Greater frontal-parietal synchrony at low gamma-band frequencies for inefficient then efficient visual search in human EEG. Int. J. Psychophysiol. 73, 350–354.
Reese, E.N.B., Garcia-Rill, E., Skinner, R.D., 1995. The pedunculopontine nucleus-auditory input, arousal and pathophysiology. Prog. Neurobiol. 47, 105–133.
Rozental, R., Srinivas, M., Spray, D.C., 2001. How to close a gap junction channel. Methods Mol. Biol. 154, 447–477.
Sakai, K., El Mansari, M., Jouvet, M., 1990. Inhibition by carbachol microinjections of presumptive cholinergic PGO-on neurons in freely moving cats. Brain Res. 527, 213–223.
Scammell, T., Arrigoni, E., Lipton, L., 2018. Neural circuitry of wakefulness and sleep. Neuron 93, 747–765.
Shik, M.L., Severin, S.V., Orlovskii, G.N., 1966. Control of walking and running by means of electric stimulation of the midbrain. Biofizika 11, 659–666.
Simon, C., Kezunovic, N., Ye, M., Hyde, J., Hayar, A., et al., 2010. Gamma band unit activity and population responses in the pedunculopontine nucleus. J. Neurophysiol. 104, 463–474.
Skinner, R.D., Garcia-Rill, E., 1990. Brainstem modulation of rhythmic functions and behaviors. In: Klemm, W.R., Vertes, R.P. (Eds.), Brainstem Mechanisms of Behavior. John Wiley & Sons, New York, pp. 419–445.
Sperry, R.W., 1969. A modified concept of consciousness. Psychol. Rev. 76, 532–536.
Stea, A., Soomg, T.W., Snutch, T.P., 1995. Determinants of PKC-dependent modulation of a family of neuronal Ca^{2+} channels. Neuron 15, 929–940.

Steriade, M., 1999. Cellular substrates of oscillations in corticothalamic systems during states of vigilance. In: Lydic, R., Baghdoyan, H.A. (Eds.), Handbook of Behavioral State Control. Cellular and Molecular Mechanisms. CRC Press, New York, pp. 327–347.

Steriade, M., Llinás, R., 1988. The functional states of the thalamus and the associated neuronal interplay. Physiol. Rev. 68, 649–742.

Steriade, M., Datta, S., Pare, M., Oakson, G., Curro Dossi, R.C., 1990a. Neuronal activities in brain-stem cholinergic nuclei related to tonic activation processes in thalamocortical systems. J. Neurosci. 10, 2541–2559.

Steriade, M., Paré, D., Datta, S., Oakson, D., Curro Dossi, R., 1990b. Different cellular types in mesopontine cholinergic nuclei related to ponto-geniculo-occipital waves. J. Neurosci. 10, 2560–2579.

Takakusaki, K., Habaguchi, T., Ohtinata-Sugimoto, J., Saitoh, K., Sakamoto, T., 2003. Basal ganglia efferents to the brainstem centers controlling postural muscle tone and locomotion: a new concept for understanding motor disorders in basal ganglia dysfunction. Neuroscience 119, 293–308.

Thiagarajan, T.C., Lebedev, M.A., Nicolelis, M.A., Plenz, D., 2010. Coherence potentials: loss-less, all-or-none network events in the cortex. PLoS Biol. 8, e1000278.

Torterolo, P., Castro-Zaballa, S., Cavelli, M., Chase, M., Falconi, A., 2015. Neocortical 40 Hz oscillations during carbachol-induced rapid eye movement sleep and cataplexy. Eur. J. Neurosci. 281, 318–325.

Urbano, F.J., Leznik, E., Llinas, R., 2007. Modafinil enhances thalamocortical activity by increasing neuronal electrotonic coupling. Proc. Natl. Acad. Sci. U. S. A. 104, 12554–12559.

Urbano, F.J., Kezunovic, N., Hyde, J., Simon, C., Beck, P., et al., 2012. Gamma band activity in the reticular activating system (RAS). Front. Neurol.; Sleep Chronobiol. 3, 6. 1–16.

Urbano, F.J., D'Onofrio, S., Luster, B., Hyde, J., Bisagno, V., et al., 2014. Pedunculopontine nucleus gamma band activity- preconscious awareness, waking, and REM sleep. Front. Sleep Chronobiol. 5, 210.

Vanderwolf, C.H., 2000a. Are neocortical gamma waves related to consciousness? Brain Res. 855, 217–224.

Vanderwolf, C.H., 2000b. What is the significance of gamma wave activity in the pyriform cortex? Brain Res. 877, 125–133.

Watson, R.T., Heilman, K.M., Miller, B.D., 1974. Neglect after mesencephalic reticular formation lesions. Neurology 24, 294–298.

Further reading

Mena-Segovia, J., Bolam, P., 2017. Rethinking the pedunculopontine nucleus: from cellular organization to function. Neuron 94, 7–18.

Palva, S., Monto, S., Palva, J.M., 2009. Graph properties of synchronized cortical networks during visual working memory maintenance. Neuroimage 49, 3257–3268.

Sherman, D., Fuller, P., Marcus, J., Yu, J., Zhang, P., et al., 2015. Anatomical location of the mesencephalic locomotor region and its possible role in locomotion, posture, cataplexy, and Parkinsonism. Front. Neurol. 6, 140.

Simon, C., Kezunovic, N., Williams, D., Urbano, F.J., Garcia-Rill, E., 2011. Cholinergic and glutamatergic agonists induce gamma frequency activity in dorsal subcoeruleus nucleus neurons. Am. J. Physiol. Cell Physiol. 301, C327–C335.

Chapter 3

Schizophrenia and arousal

Erick Messias*, Edgar Garcia-Rill†
*Department of Psychiatry, University of Arkansas for Medical Sciences, Little Rock, AR, United States, †Center for Translational Neuroscience, Department of Neurobiology and Developmental Sciences, University of Arkansas for Medical Sciences, Little Rock, AR, United States

Introduction

The term schizophrenia, derived from the Greek roots "schizo" for fractured and "phrenia" for mind, was coined by the Swiss psychiatrist Eugen Bleuler in 1911 to rename the category of "Dementia Praecox" proposed by Emil Kraepelin (Shorter, 2005). The key features of psychotic disorders are delusions, hallucinations, disorganized thinking, disorganized behavior, and negative symptoms (American Psychiatric Association, 2013).

Symptoms

Schizophrenia is a heterogeneous disorder marked by psychotic symptoms such as delusions and hallucinations, as well as attentional impairment, emotional withdrawal, apathy, and cognitive impairment (Andreasen and Flaum, 1991). More specifically, the positive symptoms (not present normally but patients have) include hallucinations, delusions, thought disorder, and agitation, while negative symptoms (present normally but patients lack) include the lack of affect, anhedonia, and withdrawal. Cognitive symptoms include poor executive function, the lack of attention, and disturbed working memory. In addition, abnormal movements have been described. Equally heterogeneous mechanisms have been advanced to explain the disease, including cortical atrophy, catecholaminergic abnormalities, and early brain injury.

Etiology

The factors involved in schizophrenia include genetic, environmental, and developmental. Some stimulants, specifically amphetamine and cocaine, can induce some of the symptoms of schizophrenia. Cannabis use has also been shown to be part of the component cause of schizophrenia and psychotic symptoms (Henquet et al., 2005). Since stimulant agents increase dopaminergic drive and

Arousal in Neurological and Psychiatric Diseases. https://doi.org/10.1016/B978-0-12-817992-5.00003-9

the use of dopamine receptor blockers has been found to alleviate some of the symptoms of schizophrenia, the "dopamine theory" of schizophrenia has some validity. In about 80% of patients the disorder develops after puberty, between the ages of 15 and 25. It is not clear that some aspects of the disorder were not present well before puberty in severe cases, but due to the reluctance of placing the diagnostic label of schizophrenia on children, the diagnosis is not offered even with fairly clear positive and negative symptoms. On the other hand, some cases of schizophrenia are postpubertal without prior history of related symptomatology. The incidence of schizophrenia is 0.5%–1.0% worldwide, with a male to female ratio of 1.4:1. In addition, schizophrenia is accompanied by comorbid depression (in 50% of patients), anxiety and posttraumatic stress disorder (PTSD) (in 30%), and obsessive-compulsive disorder (OCD) (in 25%) (Buckley et al., 2009).

As far as genetic factors are concerned, there is increased risk ~40% in monozygotic twins of also developing the disorder and an even greater risk if both parents are schizophrenic (Craddock and Owen, 2010; Picchioni and Murray, 2007). A number of different genes have been implicated in schizophrenia, many of which are also implicated in bipolar disorder and autism, yet there are enough differences to warrant separate diagnostic categories (Cradock et al., 2005). Because of our discovery of gamma band activity in the reticular activating system (RAS) being mediated by high-threshold, voltage-dependent calcium channels, we will concentrate on what is known regarding the genetics of calcium channels in schizophrenia.

As far as environmental factors are concerned, schizophrenia is more common in cities, stressful environments, and exposure to drug abuse (Picchioni and Murray, 2007). As far as developmental factors are concerned, abnormal fetal development or perinatal injury can increase risk and developmental changes at puberty (Feinberg, 1969; Weinberger, 1982). Epidemiological studies in six countries have established that, in about 20% of schizophrenics, the mother had an influenza attack in the second trimester (Mednick et al., 1988). Such a viral insult would induce increased levels of gamma interferon, a potent mitogen that can influence such agents as tumor necrosis factor and fibroblast growth factor (FGF). We identified FGF as the survival molecule for cholinergic pedunculopontine nucleus (PPN) neurons (Garcia-Rill et al., 1991). In cell culture studies, nerve growth factor did not influence the survival of PPN cells but maintained basal forebrain cholinergic cells, whereas PPN neurons were maintained by FGF but not by nerve growth factor. This is particularly important because those with schizophrenia have simplified whorls in their fingerprints (Bracha et al., 1991), and they show disturbed fibroblast metabolism (Hashimoto et al., 2003). This indicates a developmental dysregulation of FGF metabolism in at least some patients with schizophrenia. Interestingly, FGF is the survival molecule for subsets of cells in the same regions that have been implicated in schizophrenia, including the cortex (Akbarian et al., 1993), cerebellum (Nasrallah et al., 1991), hippocampus (Kovelman and Scheibel, 1984), and RAS (Garcia-Rill et al., 1991). That is, dysregulation of FGF metabolism during development may be responsible for the subsequent symptoms in at least some patients with schizophrenia. Moreover, FGF receptors can act as portals to viral infection (Kaner et al., 1990), and exposure to viruses has been proposed in the etiology of schizophrenia (Yolken, 2004). Whether or not the involvement of FGF in schizophrenia is through developmental (FGF and calcium control axon growth in the brain) or genetic dysregulation and/or via viral infection remains to be determined.

EEG, reflexes, and P50 potential

Schizophrenia is characterized by abnormalities in wake-sleep control, including hypervigilance, decreased slow-wave sleep (SWS) especially deep sleep stages, increased rapid eye movement (REM) sleep drive, and fragmented sleep (Caldwell and Domino, 1967; Feinberg et al., 1969; Itil et al., 1972; Jus et al., 1973; Zarcone et al., 1975). These wake-sleep abnormalities reflect increased vigilance and REM sleep drive, that is, overactive RAS output. The increased REM sleep drive has been proposed to account for REM sleep intrusion during waking, that is, in eliciting hallucinations (Dement, 1967; Mamelak and Hobson, 1989). This suggests that, normally, gamma band activity during waking is separate from gamma band activity during REM sleep. In fact, gamma band activity during waking may be modulated by the CaMKII pathway, and gamma band activity during REM sleep may be mediated by the cAMP/protein kinase (PKA) pathway (Garcia-Rill et al., 2014). We hypothesize that these two pathways are activated individually and may, in fact, inhibit each other. However, in schizophrenia, the two pathways are activated simultaneously, thus leading to REM sleep intrusion during waking, that is, one way to conceptualize hallucinations.

The sleep disturbances are correlated better with negative rather than positive symptoms (Ganguli et al., 1987). Other studies emphasized the relationship between negative symptoms in schizophrenia and increased cholinergic output (Janowsky et al., 1979; Tandon et al., 1993). There are also postural and motor abnormalities (King, 1974; Manschrek, 1986), as well as eye movement dysregulation (Holzman et al., 1973; Karson et al., 1990). These findings are in keeping with the modulation of motor control by the RAS described elsewhere in this book. In addition, these patients suffer from sensory gating deficits determined using habituation of the blink reflex (Geyer and Braff, 1982). One of the first studies using the P50 potential in clinical conditions showed that habituation of the P50 potential using a paired stimulus paradigm was decreased in schizophrenia (Freedman et al., 1983). These sensory gating deficits demonstrate a lack of inhibition of responses to repetitive stimuli, which can be manifested in exaggerated fight-or-flight responses upon sudden arousal. The P50 potential, as established in human and animal studies, is generated by the PPN and is manifested at the vertex.

In addition, aberrant gamma band activity and coherence during cognitive tasks or attentional load have been reported in schizophrenia (Uhlhaas and Singer, 2010). Several human studies demonstrated frequency-specific deficits in the coherence and maintenance of gamma oscillations in patients with schizophrenia (Spencer et al., 2003). These results suggest that the generation and maintenance of gamma band activity may be abnormal in schizophrenia. As described in Chapter 2, the RAS generates gamma band activity related to awareness. We discovered a potential mechanism for the decreases in gamma band activity and maintenance in schizophrenia that is described in the succeeding text. In addition, schizophrenic patients suffer from hypofrontality, which contributes not only to the sensory gating deficits but also to the lack of critical judgment. This sign may also exacerbate the intensity of hallucinations, such that the patient is deluded into obeying the voices, with dangerous and unfortunate repercussions.

The results of EEG, reflex, and P50 potential testing all point to increased arousal and increased REM sleep drive in schizophrenia. That is, the PPN is overactive in schizophrenia, but it is overactive in a specific manner. Responses to repetitive stimuli are increased, and reflexes are exaggerated suggesting that *phasic* responses to brief stimuli are dysregulated. However, the decreased and interrupted gamma band activity also suggests that gamma oscillations are not properly maintained on a *tonic* basis. This combination of short-term hyperexcitability and long-term diminution of RAS activity is functionally devastating. This is in agreement with findings described earlier showing that anticholinergic agents appear to alleviate some of the negative symptoms of schizophrenia (Janowsky et al., 1979; Tandon et al., 1993).

Treatment

We should point out that the treatments for schizophrenia are only marginally effective, with as many as one-half of patients not responsive to even the latest generation agents. Few patients, even those responding, ever regain a normal living and social life. Therefore, while we have done well to do away with padded rooms and lobotomy, modern therapies have a long way to go. Early treatment was directed at the positive symptoms, specifically delusions and hallucinations. The first generation antipsychotics included chlorpromazine and haloperidol, which showed DA D_2 receptor blockade (Snyder, 2006). A next-generation antipsychotic named clozapine was found to act on the same DA receptors in addition to muscarinic cholinergic and serotonergic receptors. In fact, clozapine was initially developed as an antimuscarinic cholinergic agent intended to balance the decrease in dopamine present in the Parkinson's disease, that is, by decreasing cholinergic tone, thus rebalancing the striatum. This proved untenable due to its rare but potentially lethal side effect of agranulocytosis (decreasing white blood cells) but became a uniquely effective antipsychotic that reduced negative symptoms while also addressing positive ones.

Fig. 3.1 is a diagram of the sites of action at which clozapine can act in the RAS. While increased DA drive to the striatum is a major component of the disease, the substantia nigra (SN) is activated by the PPN, especially by muscarinic cholinergic input. Clozapine appears to partially (~40% penetrance) block muscarinic input to the SN and DA input to the striatum. In addition, clozapine acts as a partial serotonin reuptake receptor blocker, thereby increasing inhibition of the PPN, further downregulating the RN-PPN-SN-striatum pathway. Drug companies have attempted to eliminate the side effects of clozapine, but only olanzapine—and to an extent quetiapine—has retained antimuscarinic properties, and those are among most widely used second-generation antipsychotics.

A large number of other transmitter modulators are being tested without clear success. However, it is likely that more than one drug, along with intensive therapy, will probably become the treatment of choice in the future. Hopefully, the intensive research being performed on this disease will render better options sooner rather than later. One approach that seems to be lacking is the more extended use of physiological measures such as EEG to determine changes in wake-sleep patterns, startle response, P50 potential habituation, and frontal lobe blood flow. These quantitative assessments are rarely used in clinical trials, which themselves are hampered by differences in dosage and testing instruments.

Some investigators believe that nicotinic agonists may be a potential treatment for the symptoms of schizophrenia since these patients tend to chain-smoke. The idea was advanced that this is tantamount to self-medication and

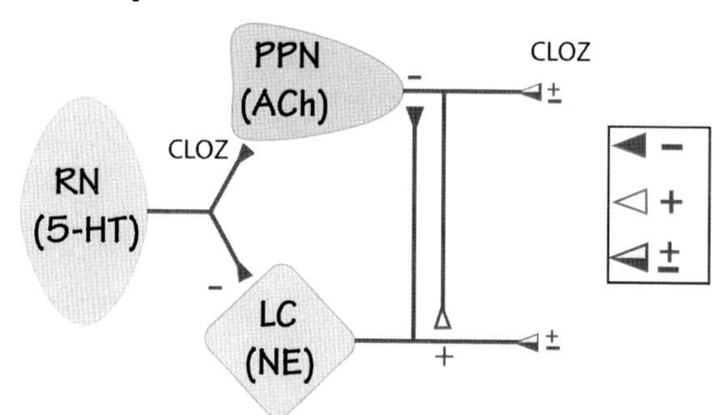

FIG. 3.1 Mesopontine projections in schizophrenia. Serotonergic (5-HT) raphe nucleus (RN) neurons send inhibitory projections to the cholinergic (ACh) PPN and noradrenergic (NE) locus coeruleus (LC). Clozapine (CLOZ) acts as a serotonin reuptake inhibitor to increase inhibition of the PPN and LC. PPN projects to the SN exciting through muscarinic inputs that are inhibited by clozapine. In addition, clozapine inhibits dopaminergic input to the striatum (not shown). These effects suggest that, in schizophrenia, there is decreased serotonergic inhibition of PPN and increased cholinergic excitation of nigra by PPN, as well as the well-known excess dopaminergic input to the striatum from the nigra. By partially increasing serotonergic PPN inhibition, partially decreasing cholinergic activation of nigra and decreasing dopaminergic drive of the striatum, clozapine, and similar agents may downregulate the RN-PPN-SN-striatum pathway to normalize arousal.

nicotine was found to normalize a rodent model of sensory gating. However, the fact is that the P50 potential and the sleep dysregulation are not sensitive to nicotinic agents, but respond to muscarinic agents. Basically, smoking induces a powerful, short-latency, and short-duration anxiolytic response. It is this brief respite from the overwhelming intrusion wrought by decreased sensory gating that the patients are seeking. Unfortunately, tobacco then induces a longer latency anxiogenic response, so that, by the end of the cigarette, the smoker is prompted to light up again. This is a very addictive mechanism in smokers, which is even more powerful in those with schizophrenia. In fact, most of the disorders described that include hypofrontality will be less likely to kick the habit than normal individuals.

Neuronal calcium sensor protein 1 in schizophrenia

Studies on postmortem human brains found increased expression of neuronal calcium sensor protein 1 (NCS-1) protein in a portion of bipolar disorder and schizophrenia patients compared with the brains from normal controls and major depression patients (Bergson et al., 2003; Koh et al., 2003). The levels of NCS-1 in different subjects within each group suggested that some patients had a 50% increase in expression, while others manifested normal levels. Thus it appears that gamma band activity is reduced or disrupted in the same disorders that exhibit NCS-1 overexpression in the brain (Leicht et al., 2015; Senkowsky and Gallinat, 2015; Wynn et al., 2015). However, other workers found that persons with schizophrenia show increased gamma band activity (Hirano et al., 2015), pointing to a well-known heterogeneity in patients with the disease. We tested the hypothesis that PPN neuron calcium channels that manifest gamma band oscillations are modulated by NCS-1 and that increased levels of NCS-1, such as those expected with overexpression, would reduce or block gamma band oscillations in these cells (D'Onofrio et al., 2015).

Recordings in PPN neurons were found to increase the amplitude and frequency of ramp-induced oscillations within ~25 min of diffusion of 1 μM NCS-1 into the cell. We used recordings of ramp-induced membrane potential oscillations in PPN neurons in the presence of synaptic blockers and tetrodotoxin. Soon after patching, low-amplitude oscillations in the beta/gamma range were induced by the ramps. After 25 min of recording, NCS-1 at 1 μM significantly increased the amplitude of the oscillations. Cells recorded without NCS-1 in the pipette showed no significant changes throughout the recording period of ~30 min. These values were similar to those at the 0 min recordings using pipettes with NCS-1 or before NCS-1 diffused into the cell; therefore the 0 min recordings were an accurate representation of control levels.

We then carried out a study to determine the effects of NCS-1 concentration on PPN cell ramp-induced oscillations (D'Onofrio et al., 2016). Fig. 3.2 shows graphs of the peak power of the ramp-induced oscillations in PPN neurons at various frequencies, low alpha (10–15 Hz), high alpha (15–20 Hz), beta

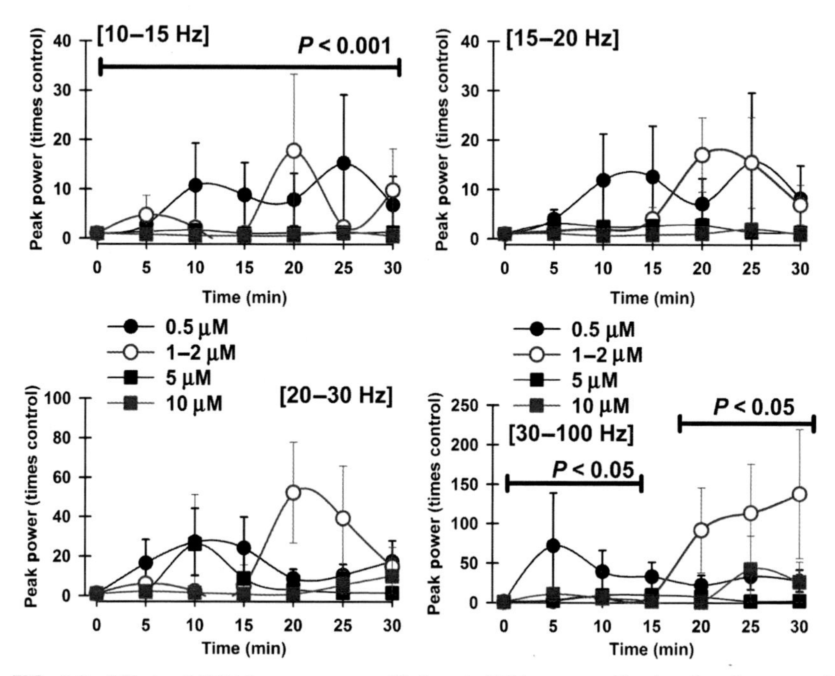

FIG. 3.2 Effects of NCS-1 on gamma oscillations in PPN neurons. Graphs of peak power of ramp-induced intrinsic membrane oscillations in PPN neurons at low alpha (10–15 Hz, top left), high alpha (15–20 Hz, top right), beta (20–30 Hz, bottom left), and gamma (30–100 Hz, bottom right). Various concentrations of NCS-1 (0.5 μM *filled black circles*, 1–2 μM *open circles*, 5 μM *filled black squares*, and 10 μM *filled red squares*) were used and the effects recorded for 30 min after patching the neuron. Significant changes were observed at low alpha (top left) but only after 20–25 min and only at low concentrations (0.5 and 1–2 μM). Significant changes were observed at gamma (bottom right) but only at low concentrations (0.5 μM at 5 min and at 20–30 min at 1–2 μM). In all cases, high concentrations of NCS-1 (5 and 10 μM) suppressed oscillation peak power at all frequencies (see D'Onofrio et al., 2015).

(20–30 Hz), and gamma (30–100 Hz) using various concentrations of NCS-1. There was an increase in peak power in the low alpha range (top left) but only after 20–25 min (i.e., a delayed effect) and only at low concentrations (0.5 and 1–2 μM). There were no significant changes at high alpha and beta, but at gamma frequencies, low concentrations (0.5 and 1–2 μM) increased peak power. High concentrations of NCS-1 (5 and 10 μM) showed suppressive effects on peak power at all frequencies, suggesting that overexpression of NCS-1 would have inhibitory effects on the manifestation of intrinsic membrane oscillations in PPN neurons, that is, would decrease the maintenance of gamma oscillations.

The studies on the human brains suggest that only some patients with schizophrenia show significant overexpression of NCS-1 (Koh et al., 2003), which may be manifested as decreased gamma band activity only in a subpopulation of patients. Studies on humans have never measured gamma band activity and correlated it with NCS-1 levels. This is not feasible because serum sampling does not reflect brain levels of NCS-1. Actually, NCS-1 levels in leukocytes are instead decreased in schizophrenic patients (Torres et al., 2009). In the future, clinical trials in untreated patients with schizophrenia may be able to determine a significant decrease in gamma band activity prior to pharmacotherapy, which may also help address the heterogeneity of schizophrenia and facilitate the process of identifying more homogeneous groups within the syndrome (Picardi et al., 2012). It is precisely those patients with decreased gamma band activity that pharmacological therapy to increase gamma band activity may be of benefit. Our preliminary findings suggest that the stimulant modafinil may partly compensate for excessive amounts of NCS-1. After exposure to modafinil, we found a partial return of gamma oscillations that had been suppressed by high levels of NCS-1 (Garcia-Rill et al., 2014).

The postmortem findings described earlier revealed that, while schizophrenia and bipolar disorder are similar in their NCS-1 overexpression, major depression patients do not manifest NCS-1 overexpression. Therefore it may be the mania and hyperactivation symptoms that schizophrenia and bipolar disorder share that are affected by excessive NCS-1. Interestingly, we will see in Chapter 4 that lithium acts by decreasing the effects of NCS-1, shifting the balance to normalize gamma oscillations, and is a frontline treatment for bipolar disorder. However, lithium is not effective in schizophrenia, which suggests that other sites are dysregulated in schizophrenia, all of which must be controlled to reduce symptoms. Much research remains to be done in order to develop new treatments for schizophrenia, especially those that can help patients carry out normal lives.

Clinical implications

PPN projections modulate cortical arousal in addition to posture and locomotion. In responding to varying sensory inflow, the PPN also maintains tonic gamma band activity during waking. These membrane oscillations are mediated by voltage-dependent high-threshold N- and P/Q-type calcium channels (see Chapter 2). It appears that these two types of channels with separate intracellular pathways are involved in selectively controlling high-frequency activity. P/Q-type channels are modulated by CaMKII during waking, while N-type channels are modulated by cyclic adenosine monophosphate (cAMP) during REM sleep (Garcia-Rill et al., 2014). Fig. 3.3 is a diagram of the intracellular pathways involved in the action of NCS-1 on these calcium channels. Briefly, P/Q-type calcium channels are modulated by CaMKII, while N-type calcium channels are modulated by cAMP/PKA. Transmitter activation, for example, via acetylcholine, activates muscarinic receptors that are G protein coupled to act through phospholipase C and phospholipid phosphatidylinositol bisphosphate to generate inositol triphosphate (IP3). This is released into the cytoplasm to activate IP3 receptors in the endoplasmic reticulum to release calcium intracellularly that is bound, among other proteins, by NCS-1. Normally, NCS-1 positively

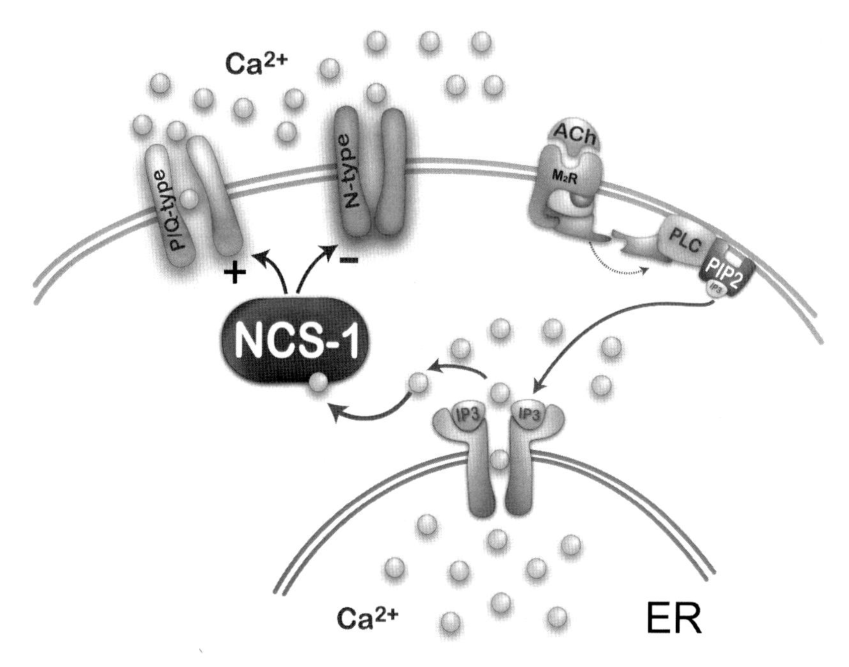

FIG. 3.3 Intracellular pathways mediating NCS-1 modulation of intracellular calcium and P/Q-type calcium channels. Representation of effects of acetylcholine (ACh) activation of a muscarinic 2 cholinergic receptor (M_2R) acting through G protein coupling to phospholipase C (PLC) that in turn cleaves phospholipid phosphatidylinositol biphosphate (PIP_2) into inositol triphosphate (IP3). IP3 is released intracellularly and binds to IP3 receptors in the endoplasmic reticulum (ER) to release calcium (Ca^{2+}). One of the intracellular pathways activated involves NCS-1, which stimulates (+) P/Q-type calcium channels that are related to waking and somewhat inhibits (−) N-type calcium channels that are related to REM sleep. NCS-1 at low concentrations increases gamma oscillations, while NCS-1 at high concentrations, such as would occur with overexpression, blocks them.

modulates P/Q-type channels and may negatively modulate N-type channels but may block oscillations by inactivating P/Q-type channels and decreasing intrinsic gamma oscillations. In addition to intrinsic membrane oscillations, the maintenance of gamma band activity requires synaptic connectivity within the nucleus and between regions of the brain. PPN circuitry includes cholinergic, glutamatergic, and GABAergic neurons. Some GABAergic cells are electrically coupled to provide coherence (Garcia-Rill et al., 2013), and the nucleus may include functional cell clusters (Garcia-Rill et al., 2014).

As soon as we awaken, the PPN provides the necessary background of activity in order to preconsciously evaluate the world around us (Garcia-Rill et al., 2013, 2014; Urbano et al., 2014). That is, this mechanism is integral to the formulation of our perceptions and actions and modulates higher-level gamma activity via projections to the intralaminar thalamus, basal ganglia, hypothalamus, and basal forebrain. Therefore disruption of PPN function affects different functions, from waking and REM sleep to mood and perception and even homeostatic regulation. As a consequence, dysregulation of the PPN will be manifested in motor dysregulation, psychotic symptoms, and hypervigilance, in addition to sleep disturbances.

In summary, the implications of PPN dysregulation for schizophrenia suggest that the sleep-wake dysfunction in the disease arises from the nucleus. In some patients, overexpression of NCS-1 may downregulate high-frequency oscillations. This would lead to interrupted or decreased gamma band activity during waking, disturbing the maintenance of the state of waking and preconscious awareness. The recent discoveries described earlier provide novel therapeutic targets for alleviating some of the arousal and sleep-wake disturbances in this devastating disease.

References

Akbarian, S., Bunney, W.E., Potkin, S.G., Wigal, S.B., Hagman, J.D., et al., 1993. Altered distribution of nicotinamide-adenine dinucleotide phosphate-diaphorase cells in frontal lobe of schizophrenics implies disturbances in cortical development. Arch. Gen. Psychiatry 50, 169–177.

American Psychiatric Association (Eds.), 2013. Diagnostic and Statistical Manual of Mental Disorders: DSM-5, fifth ed. American Psychiatric Association, Washington, DC.

Andreasen, N.C., Flaum, M., 1991. Schizophrenia: the characteristic symptoms. Schizophr. Bull. 17, 27–49.

Bergson, C., Levenson, R., Goldman-Rakic, P., Lidow, M.S., 2003. Dopamine receptor-interacting proteins: the Ca^{2+} connection in dopamine signaling. Trends Pharmacol. Sci. 24, 486–492.

Bracha, H.S., Torrey, E.F., Bigelow, L.B., Lohr, J.B., Linington, B.B., 1991. Subtle signs of prenatal maldevelopment of the hand ectoderm in schizophrenia: a preliminary monozygotic twin study. Biol. Psychiatry 30, 719–725.

Buckley, P.F., Miller, B.J., Lehrer, D.S., Castle, D.J., 2009. Psychiatric comorbidities and schizophrenia. Schizophr. Bull. 35, 383–402.

Caldwell, D.F., Domino, E.F., 1967. Electroencephalographic and eye movement patterns during sleep in chronic schizophrenic patients. Electroencephalogr. Clin. Neurophysiol. 22, 414–420.

Craddock, N., Owen, M.J., 2010. The Kraepelinian dichotomy—going, going … but still not gone. Br. J. Psychiatry 196, 92–95.

Cradock, N., O'Donovan, M.C., Owen, M.J., 2005. The genetics of schizophrenia and bipolar disorder: dissecting psychosis. J. Med. Genet. 42, 193–204.

D'Onofrio, S., Kezunovic, N., Hyde, J.R., Luster, B., Messias, E., et al., 2015. Modulation of gamma oscillations in the pedunculopontine nucleus (PPN) by neuronal calcium sensor protein-1 (NCS-1): relevance to schizophrenia and bipolar disorder. J. Neurophysiol. 113, 709–719.

D'Onofrio, S., Urbano, F.J., Messias, E., Garcia-Rill, E., 2016. Lithium decreases the effects of neuronal calcium sensor protein 1 in pedunculopontine neurons. Physiol. Rep. 4, e12740.

Dement, W.C., 1967. Studies on the effects of REM deprovation in humans and animals. Res. Publ. Assoc. Res. Nerv. Ment. Dis. 43, 456–467.

Feinberg, I., 1969. Schizophrenia: caused by a fault in programmed synaptic elimination during adolescence? J. Psychiatr. Res. 17, 319–334.

Feinberg, I., Braun, M., Koresko, R.L., Gottleib, F., 1969. Stage 4 sleep in schizophrenia. Arch. Gen. Psychiatry 21, 262–266.

Freedman, R., Adler, L.E., Waldo, M.C., Pachtman, E., Franks, R.D., 1983. Neurophysiological evidence for a defect in inhibitory pathways in schizophrenia: comparison of medicated and drug-free patients. Biol. Psychiatry 18, 537–551.

Ganguli, R., Reynolds, D.F., Kupfer, D.F., 1987. Electroencephalographic sleep in young, never-medicated schizophrenics. Arch. Gen. Psychiatry 44, 36–44.

Garcia-Rill, E., Davies, D., Skinner, R.D., Biedermann, J.A., McHalffey, C., 1991. Fibroblast growth factor-induced increased survival of cholinergic mesopontine neurons in culture. Dev. Brain Res. 60, 267–270.

Garcia-Rill, E., Kezunovic, N., Hyde, J., Beck, P., Urbano, F.J., 2013. Coherence and frequency in the reticular activating system (RAS). Sleep Med. Rev. 17, 227–238.

Garcia-Rill, E., Kezunovic, N., D'Onofrio, S., Luster, B., Hyde, J., et al., 2014. Gamma band activity in the RAS-intracellular mechanisms. Exp. Brain Res. 232, 1509–1522.

Geyer, M.A., Braff, D.U., 1982. Habituation of the blink reflex in normals and schizophrenic patients. Psychophysiology 19, 1–6.

Hashimoto, K., Shimizu, E., Komatsu, N., Nakazato, M., Okamura, N., et al., 2003. Increased levels of serum basic fibroblast growth factor in schizophrenia. Psychiatry Res. 120, 211–218.

Henquet, C., Murray, R., Linszen, D., van Os, J., 2005. The environment and schizophrenia: the role of cannabis use. Schizophr. Bull. 31, 608–612.

Hirano, Y., Oribe, N., Kanba, S., Onitsuka, T., Nestor, P.G., et al., 2015. Spontaneous gamma activity in schizophrenia. JAMA Psychiat. 72, 1067–1081.

Holzman, P.S., Proctor, L.R., Hughes, D.W., 1973. Eye tracking patterns in schizophrenia. Science 181, 179–181.

Itil, T.M., Hsu, W., Klingenberg, W., Saletu, B., Gannon, P., 1972. Digital computer-analyzed all-night sleep EEG patterns (sleep prints) in schizophrenics. Biol. Psychiatry 4, 3–16.

Janowsky, D.S., Davis, J.M., Huey, L., Judd, L.L., 1979. Adrenergic and cholinergic drugs as episode and vulnerability markers of affective disorders and schizophrenia. Psychoparmacol. Bull. 15, 33–34.

Jus, K., Bouchard, M., Jus, A., Villeneuve, A., Lachance, R., 1973. Sleep EEG studies in untreated long-term schizophrenic patients. Arch. Gen. Psychiatry 29, 286–290.

Kaner, R.J., Baird, A., Mansukhani, A., Basilico, C., Summers, B.D., et al., 1990. Fibroblast growth factor receptor is a portal of cellular entry for herpes simplex virus type I. Science 248, 1410–1413.

Karson, C.N., Dykman, R.A., Paige, S.R., 1990. Blink rates in schizophrenia. Schizophr. Bull. 16, 345–354.

King, L.J., 1974. A sensory-integrative approach to schizophrenia. Am. J. Occup. Ther. 28, 529–536.

Koh, P.O., Undie, A.S., Kabbani, N., Levenson, R., Goldman-Rakic, P., Lidow, M.S., 2003. Up-regulation of neuronal calcium sensor-1 (NCS-1) in the prefrontal cortex of schizophrenic and bipolar patients. Proc. Natl. Acad. Sci. 100, 313–317.

Kovelman, J.A., Scheibel, A.B., 1984. A neurohistochemical correlate of schizophrenia. Biol. Psychiatry 19, 1601–1621.

Leicht, G., Andreou, C., Polomac, N., lanig, C., Schottle, D., et al., 2015. Reduced auditory evoked gamma band response and cognitive processing deficits in first episode schizophrenia. World J. Biol. Psychiatry 16, 1–11.

Mamelak, A.N., Hobson, J.A., 1989. Dream bizarreness as the cognitive correlate of altered neuronal brain in REM sleep. J. Cogn. Neurosci. 1, 201–222.

Manschrek, T.C., 1986. Motor abnormalities in schizophrenia. In: Nasrallah, H.A., Weinberger, D.R. (Eds.), Handbook of Schizophrenia. Elsevier, Amsterdam, pp. 65–96.

Mednick, S.A., Machon, R.A., Huttunen, M.O., Bonett, D., 1988. Adult schizophrenia following prenatal exposure to an influenza epidemic. Arch. Gen. Psychiatry 45, 189–192.

Nasrallah, H.A., Schwarzkopf, S.B., Olson, S.C., Coffman, J.A., 1991. Perinatal brain injury and cerebellar vermal lobules I-X in schizophrenia. Biol. Psychiatry 29, 567–574.

Picardi, A., Viroli, C., Tarsitani, L., Miglio, R., de Girolamo, G., et al., 2012. Heterogeneity and symptom structure of schizophrenia. Psychiatry Res. 198, 386–394.

Picchioni, M.M., Murray, R.M., 2007. Schizophrenia. Br. Med. J. 335, 91–96.

Senkowsky, D., Gallinat, J., 2015. Dysfunctional prefrontal gamma-band oscillations reflect working memory and other cognitive deficits in schizophrenia. Biol. Psychiatry 77, 1010–1019.

Shorter, E. (Ed.), 2005. A Historical Dictionary of Psychiatry. Oxford University Press, Oxford, New York.

Snyder, S., 2006. Dopamnie receptor excess and mouse madness. Neuron 49, 484–485.

Spencer, K.M., Nestor, P.G., Niznikiewicz, M.A., Salisbury, D.F., Shenton, M.E., et al., 2003. Abnormal neural synchrony in schizophrenia. J. Neurosci. 23, 7407–7411.

Tandon, R., Greden, J.F., Haskett, R.F., 1993. Cholinergic hyperactivity and negative symptoms: behavioral effects of physostigmine in normal controls. Schizophr. Res. 9, 19–23.

Torres, K.C., Souza, B.R., Miranda, D.M., Sampiao, A.M., Nicolato, R., et al., 2009. Expression of neuronal calcium sensor-1 (NCS-1) is decreased in leukocytes of schizophrenia and bipolar disorder patients. Prog. Neuro-Psychopharmacol. Biol. Psychiatry 33, 229–234.

Uhlhaas, P.J., Singer, W., 2010. Abnormal neural oscillations and synchrony in schizophrenia. Nat. Rev. Neurosci. 11, 100–113.

Urbano, F.J., D'Onofrio, S.M., Luster, B.R., Hyde, J.R., Bisagno, V., Garcia-Rill, E., 2014. Pedunculopontine nucleus gamma band activity-preconscious awareness, waking, and REM sleep. Front. Neurol. 5, 210.

Weinberger, D.R., 1982. Implications of normal brain development for the pathogenesis of schizophrenia. Arch. Gen. Psychiatry 44, 660–669.

Wynn, J.K., Roach, B.J., Lee, J., Horan, W.P., Ford, J.M., et al., 2015. EEG findings of reduced neural synchronization during visual integration in schizophrenia. PLoS One 10, e0119849.

Yolken, R., 2004. Viruses and schizophrenia: a focus on herpes simplex virus. Herpesviridae 11 (S2), 83A–88A.

Zarcone, V., Azumi, K., Dement, W., Gulevich, G., Kraimer, H., Pivik, R., 1975. REM phase deprivation and schizophrenia II. Arch. Gen. Psychiatry 32, 1431–1436.

Chapter 4

Bipolar disorder, depression, and arousal

Stasia D'Onofrio, Edgar Garcia-Rill
Center for Translational Neuroscience, Department of Neurobiology and Developmental Sciences, University of Arkansas for Medical Sciences, Little Rock, AR, United States

Bipolar disorder

Symptoms

Bipolar disorder is characterized by periods of elevated mood or mania and periods of depression that significantly impair function at work and socially. Mania can be milder and is termed hypomania, although it can progress to full mania. In many cases with hypomania, patients live for many years before a diagnosis is made, with recurrent adverse consequences. The depressive episodes are longer lasting, and the cycling between elevated and depressed moods can repeat regularly, with more rapid cycling with age being indicative of increased disease severity. Stress and anxiety can trigger symptoms, which are similar to those in attention deficit hyperactivity disorder (ADHD), but the label of bipolar disorder is generally avoided in children, unless they manifest intense episodes of mania or depression. Bipolar disorder patients also show comorbid anxiety disorder and may even manifest psychotic symptoms. Bipolar disorder patients are more likely to commit suicide than major depression patients, emphasizing the need for effective treatment (Nordentoft et al., 2011). The symptoms of bipolar disorder are similar to those in schizophrenia, with the major difference being a history of mania or hypomania (Anderson et al., 2012). The presence of psychotic symptoms in bipolar disorder can be mistaken for schizophrenia. The incidence of bipolar disorder is ~1% and also has a postpubertal age of onset, with most patients being diagnosed in their early 20s, with symptoms present as teenagers (Merikangas et al., 2011).

Etiology

Physical or sexual abuse and environmental factors all contribute to increase incidence (Etain et al., 2008). However, genetic factors contribute the most to the

Arousal in Neurological and Psychiatric Diseases. https://doi.org/10.1016/B978-0-12-817992-5.00004-0

incidence of bipolar disorder, which also has genetic components overlapping with schizophrenia (Sullivan et al., 2012). Interestingly, fibroblast growth factor (FGF) metabolism is also disturbed in bipolar disorder and major depression, as it is in schizophrenia (Liu et al., 2014; Persson et al., 2009). As we saw in Chapter 3, there is a developmental dysregulation of FGF metabolism, which may account for some of the neuronal abnormalities present in these disorders.

EEG, reflexes, and P50 potential

As expected, the wake-sleep patterns manifested in the EEG of bipolar disorder patients including fragmented sleep, decreased SWS, increased vigilance, and increased rapid eye movement (REM) sleep drive (Kadrmas and Winokur, 1979; Kupfer et al., 1978). As in schizophrenia, bipolar disorder patients show decreased habituation of the P50 potential in a paired stimulus paradigm (Olincy and Martin, 2005; Schulze et al., 2007), along with exaggerated startle response (Perry et al., 2001), and dysregulation of blink reflexes (Depue et al., 1990).

Reduced gamma band activity has been reported in bipolar disorder patients (Özerdem et al., 2011), similarly to that reported in schizophrenia (Uhlhaas and Singer, 2010). Decreased gamma band activity can account for many of the symptoms in these disorders, including wake-sleep, arousal, and cognitive symptoms. The mechanism behind the decrease in gamma band activity and maintenance was unknown until recently. Human postmortem studies reported increased expression of neuronal calcium sensor-1 (NCS-1) protein in the brains of bipolar disorder and schizophrenic patients compared with normal controls and major depression patients (Bergson et al., 2003; Koh et al., 2003). That is, gamma band activity is reduced or disrupted in precisely the same disorders that show brain NCS-1 overexpression.

Neuronal calcium sensor protein 1

We tested the hypothesis that NCS-1 modulates calcium channels in pedunculopontine nucleus (PPN) neurons that generate gamma band oscillations (see Chapter 2) and that excessive levels of NCS-1, as would be expected with overexpression, reduce or block gamma band oscillations in these cells.

We carried out recordings in PPN neurons and found that 1-μM NCS-1 increased the amplitude and frequency of ramp-induced oscillations within 20–25 min of diffusion into the cell. Fig. 4.1A is a representative example of ramp-induced membrane potential oscillations in a PPN neuron in the presence of synaptic blockers and tetrodotoxin (Garcia-Rill et al., 2014; D'Onofrio et al., 2015). Shortly after patching, the ramp typically induced low-amplitude oscillations in the beta/gamma range (light gray record in Fig. 4.1A and light gray line in Fig. 4.1B). Fig. 4.1A dark gray record shows that, after 10 min of recording, some increase in the oscillation amplitude and frequency was present (also evident in Fig. 4.1B as a dark gray line in the power spectrum). After 25 min of

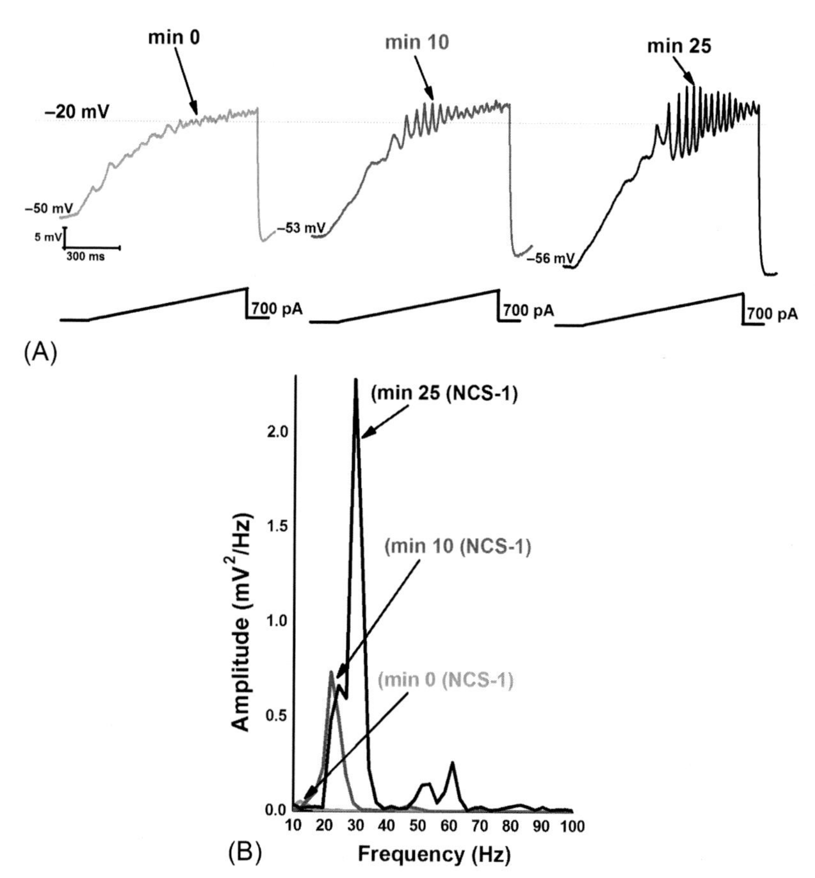

FIG. 4.1 Effects of NCS-1 at 1 μM on intrinsic gamma oscillations in PPN. (A) Ramp-induced gamma oscillations were recorded from patch-clamped PPN neurons in the presence of fast synaptic blockers and tetrodotoxin to record intrinsic properties (*light gray* record). After 10 min of exposure to NCS-1, oscillation amplitude and frequency increased (*dark gray* record), reaching a maximal increase in frequency and amplitude after 25 min (*black* record). (B) Power spectrum of the recordings shown in A, with low-amplitude and low-frequency (delta) oscillations initially *(light gray line)*, which increased in amplitude and frequency (beta) after 10 min *(dark gray line)*, until reaching a maximal increase in amplitude and frequency (beta/gamma) after 25 min of NCS-1 at 1 μM *(black line)*.

recording, NCS-1 at 1 μM significantly increased the amplitude and frequency of oscillations (back record in Fig. 4.1A and as the black line in the power spectrum). There were no significant changes in amplitude or frequency throughout the 30-min recording period in control cells recorded without NCS-1 in the pipette. These values were similar to the 0-min recordings with NCS-1 in the pipette, that is, before NCS-1 diffused into the cell to induce significant effects. Therefore the 0-min recordings were an accurate representation of control oscillations.

We wanted to determine the effects of varying NCS-1 concentration on PPN cell ramp-induced oscillations. With 0.5-μM NCS-1 in the pipette, we observed no significant changes in amplitude, suggesting that this concentration does not significantly affect oscillation amplitude. However, with 1-μM NCS-1 in the pipette, the oscillation amplitude increased significantly by 20 min and subsequently suggesting a gradual increase in amplitude as more NCS-1 diffused into the cell. With 5-μM NCS-1 in the pipette, there was a significant increase in amplitude at 5 min but not afterward. There were no subsequent changes observed, suggesting that the effect at 5 min had saturated. When using 10-μM NCS-1, the oscillation amplitude immediately increased beyond control levels and then gradually decreased until it was significantly reduced after 30 min. These findings suggest a rapid effect on amplitude by very high levels of NCS-1 that gradually led to increasing blockade. In conclusion, 1-μM NCS-1 appeared to be the most critical concentration for promoting gamma oscillations (Garcia-Rill et al., 2014; D'Onofrio et al., 2015).

In order to determine the effects of NCS-1 on high-threshold voltage-dependent calcium currents (I_{Ca}) manifested in PPN neurons, square voltage steps were used in combination with high cesium/QX314 intracellular pipette solution and synaptic receptor blockers. Results showed the time course of mean reduction of I_{Ca} by either 0.5-μM or 1-mM NCS-1. Both curves were well fitted to a single exponential ($R^2 > 0.99$), yielding *tau* (t) values of 9.4 or 7.8 min for 0.5-μM or 1-μM NCS-1 curves, respectively. These results suggest that NCS-1 decreased calcium currents and increased series resistance suggesting long-term effects on intracellular calcium metabolism that require further study (D'Onofrio et al., 2015).

The human brain postmortem findings originally described (Koh et al., 2003) showed that only some patients with bipolar disorder may suffer from significant overexpression of NCS-1, which may be manifested as decreased gamma band activity only in a subpopulation of patients. Human studies have not measured gamma band activity and correlated it with NCS-1 levels. The reason is that serum sampling does not reflect brain levels and with NCS-1 levels in leukocytes actually decreased in bipolar disorder patients (Torres et al., 2009). In the future, clinical trials in patients with bipolar disorder could first determine if there is a significant decrease in gamma band activity, in order to identify potential subgroups of patients. It is in those patients that pharmacological targeting to increase gamma band activity may be of benefit. Preliminary results from our labs suggest that the stimulant modafinil could somewhat compensate for excessive levels of NCS-1. We observed a partial return of gamma oscillations that had been suppressed by high levels of NCS-1, after exposure to modafinil (Garcia-Rill et al., 2014).

Treatment

Serendipitously, the mood disturbances in bipolar disorder were treated effectively using lithium, one of the best treatment options, although it does have side effects (Brown and Tracy, 2012). Lithium has also been used with some success for regulating mood in schizophrenia and also proposed as a neuroprotective agent. Lithium has been proposed to act by inhibiting the interaction between NCS-1 and inositol 1, 4,5-trisphosphate receptor protein (InsP) (Schlecker et al., 2006), which acts in concert with NCS-1, which is overexpressed in bipolar disorder (Bergson et al., 2003; Koh et al., 2003). A number of other agents are also used successfully, including anticonvulsants and antipsychotics. Fortunately, for many patients, such therapy does allow them to function more normally and lead productive lives. We emphasize that NCS-1 enhances the activity of InsP (Kasri et al., 2004), which is manifested in the PPN (Rodrigo et al., 1993). Lithium, as mentioned earlier, is thought to inhibit the effects of NCS-1 on InsP (Schlecker et al., 2006). A diagram of these intracellular pathways appears in Fig. 4.2. That is, lithium may reduce the effects of overexpressed NCS-1 in bipolar disorder, thereby normalizing gamma band oscillations mediated by high-threshold calcium channels modulated by NCS-1. That is, the effects of overexpression of NCS-1 in bipolar disorder may be decreased by lithium. We found that excessive NCS-1 decreased gamma oscillations; therefore lithium may prevent the downregulation of gamma band activity and restore normal levels of gamma band oscillations. These findings taken together resolve the 60-year mystery of how lithium works in bipolar disorder and schizoaffective disease. An interesting observation is that NCS-1 downregulates N-type calcium channels, at least in some cell lines (Gambino et al., 2007). This may mean that under some circumstances, NCS-1 may inhibit N-type channel function, while promoting P/Q-type channel function.

Fig. 4.2 reiterates the concept discussed in Chapter 2, in which gamma band activity in the cortical EEG during waking, which is modulated by P/Q-type calcium channels under the control of CaMKII, retains coherence across distant sites. However, gamma band activity in the cortical EEG during REM sleep, which is generated by N-type calcium channels that are under the control of cAMP/PKA, does not retain coherence across distant sites. The role of lithium appears to be to maintain appropriate levels of NCS-1, since too low and too high levels downregulate gamma band activity. This is perhaps why the blood levels of lithium must be maintained at effective concentrations.

Major depression

Symptoms

The most striking symptom in major depression is the persistent low mood, which is accompanied by anhedonia and low self-esteem. In severe cases, patients experience psychosis, which can include delusions and hallucinations. Changes in wake-sleep patterns and appetite are common, suggesting that other homeostatic systems besides wake-sleep control are impaired. About half of patients show comorbid anxiety (Kessler et al., 1996), which can increase the risk of suicide (Hirschfeld, 2001), and patients with major depression can have a

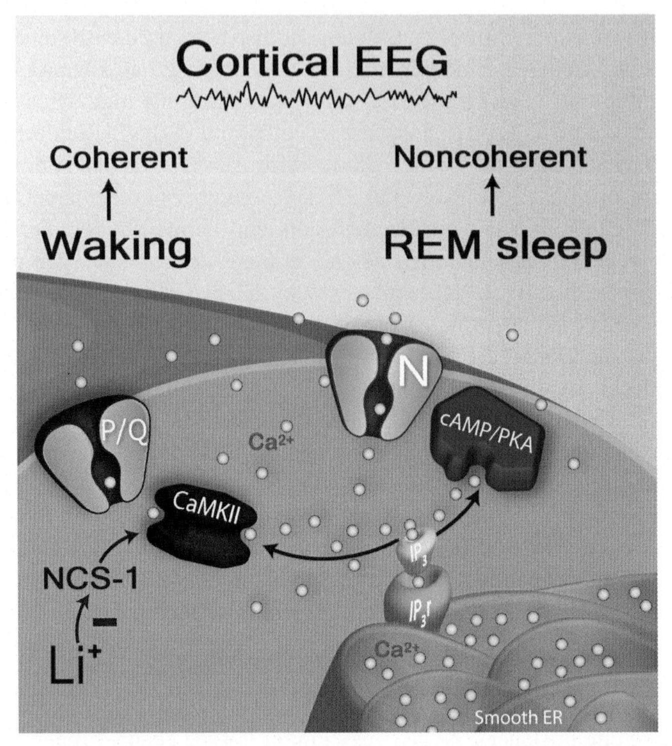

FIG. 4.2 Intracellular pathways involved in the action of NCS-1 and lithium. P/Q-type calcium channels *(red channel)* are modulated by CaMKII *(red structure)*, while N-type calcium channels *(blue channel)* are modulated by cAMP/PKA *(blue structure)*. Smooth endoplasmic reticulum (ER, *purple*) releases calcium intracellularly from IP3. NCS-1 modulates CaMKII so that when it is overexpressed in bipolar disorder and schizophrenia (but not major depression), gamma activity generated by P/Q-type calcium channels is decreased. However, appropriate concentrations of lithium (Li^+) will reduce the levels of NCS-1, allowing the manifestation of gamma band activity. The gamma activity generated by the RAS, termed "bottom-up gamma," leads to a cortical EEG marked by coherence between distant sites during waking (P/Q-type channels), while gamma activity generated during REM sleep (N-type channels) does not lead to coherence across distant sites in the cortical EEG.

number of systemic disorders like cardiovascular disease, chronic pain, and diabetes. The incidence of depression ranges from 8% to 12% worldwide. Suicidal ideation is perhaps the most astounding delusion. The person believes that the world is better off without them. That means that the survival instinct has been completely abrogated. If someone tries to throw you off the top of a building, you fight for your life. Yet, a deeply depressed person is willing to jump off the building voluntarily. That this disorder can eliminate one of our strongest instincts, the survival instinct, makes major depression among the most serious medical conditions. Such delusions have to be taken very seriously, because the sufferer is probably deeply incapacitated. Under such circumstances, decision-making must be suspect. Because terminally ill patients can develop depression and seek assisted suicide, regardless of the level of acceptance for the procedure, it is advisable to ensure that the person is not clinically depressed and hypofrontal when deciding to end their existence.

Etiology

Genetic factors are perhaps the most common (Kendler et al., 2006), with indications of dysregulation in serotonin transporter metabolism (Caspi et al., 2003). The monoamine hypothesis for major depression has been proposed (Nutt, 2008). In addition, poverty, social isolation, and early abuse all are linked with developing major depression later in life (Heim et al., 2008). A number of metabolic disturbances produce depressive symptoms, including thyroid problems, calcium imbalance, and chronic disease. A significant difference between major depression and bipolar disorder is that major depression patients do not manifest overexpression of NCS-1 (Koh et al., 2003). In fact, NCS-1 levels are close to those in normal brains, suggesting that NCS-1 may not be involved in the etiology of the disease and accounting for the lack of beneficial effect of lithium in major depression.

EEG, reflexes and P50 potential

Insomnia is a hallmark of major depression, in addition to wake-sleep cycle dysregulation, that includes decreased SWS, increased REM sleep drive, and frequent awakenings (Arfken et al., 2014; Kudlow et al., 2013; Seifritz, 2001). The startle response (Carroll et al., 2007; Kohl et al., 2013) and blink reflexes (Kohl et al., 2013) are exaggerated in major depression. We found that the P50 potential exhibited decreased habituation in depression (Garcia-Rill et al., 2002). Briefly, the habituation at the 250- and the 500-ms interstimulus intervals (ISIs) was decreased compared with normal, age- and gender-matched controls (Fig. 4.3). That is, in the normal condition, the response to the second stimulus of a pair is reduced compared with the response to the first stimulus. This is considered to be due to descending modulation inhibiting rapidly occurring inputs so that the early stimuli can be detected, a sort of filtering function. In the case of depression, the response to the second stimulus of a pair is more than twice the amplitude as the response to the first stimulus. This allows the second stimulus to have excessive weight, generating a condition promoting distraction and lowered ability to properly assess the first sensory event. This is known as a sensory gating deficit, a sign of hyper responsiveness. The results on depression suggest that the changes in wake-sleep, startle/reflex responses, and P50 potential in major depression are all in the same direction as in schizophrenia and bipolar disorder. Interestingly, major depression patients do not show a decrement in gamma band activity like bipolar disorder and schizophrenic patients (Liu et al., 2012).

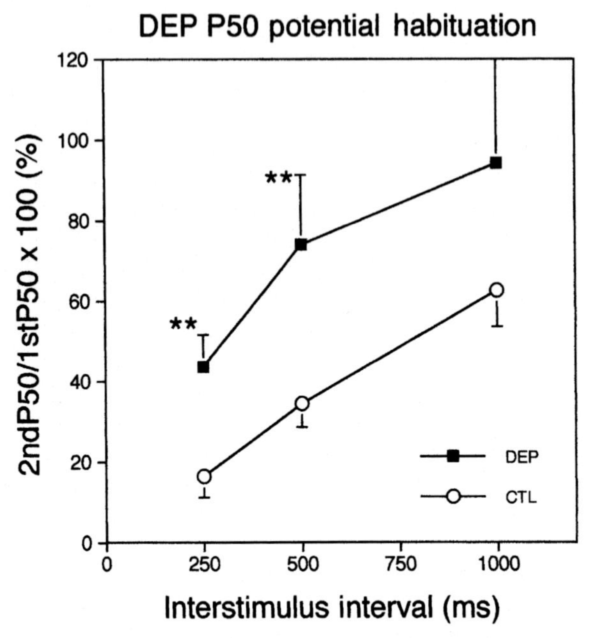

FIG. 4.3 Midlatency auditory P50 potential in major depression. Graph of the average percent (±SE) of P50 potential habituation (sensory gating) using a paired stimulus paradigm in depressed (DEP) *(filled squares)* compared with control (CTL) *(open circles)* groups. Three ISIs were tested, 250, 500, and 1000 ms. For each ISI, the amplitude of the P50 potential following the second stimulus was calculated as a percent of the amplitude of the P50 potential following the first stimulus. The percent habituation was significantly ($*P<0.01$) higher in the DEP compared with the CTL group at the 250- and 500-ms ISI. There was no statistically significant difference in the percent habituation of the groups at the 1000-ms ISI.*(Data from Garcia-Rill, E., Skinner, R.D., Clothier, J., Dornhoffer, J., Uc, E., et al. 2002. The sleep state-dependent midlatency auditory evoked P50 potential in various disorders. Thalamus Relat. Syst. 2, 9–19.)*

Treatment

The most common form of treatment aside from psychotherapy is antidepressant intervention, usually in the form of specific serotonin reuptake inhibitors (SSRIs). The site of action of SSRIs, as far as the RAS is concerned, is an increase in serotonergic inhibition of the PPN and locus coeruleus (LC). That is, as evident on the left side of Fig. 4.4 (depression), the inhibitory inputs to both the PPN and LC would increase, thereby decreasing the outputs of these nuclei. That is, normally, these nuclei are inhibited by serotonin. In depression, the overall output of serotonin decreases, thereby releasing the activity of the PPN and LC. This may account for the hyperexcitability and disturbances in sleep-wake cycles in the disorder. In the presence of SSRIs, by inhibiting the reuptake of serotonin, the levels of serotonin at the synapses on the PPN and LC will increase, restoring normal inhibition. This would have the effect of decreasing some of the hypervigilance and increased REM sleep drive present in major

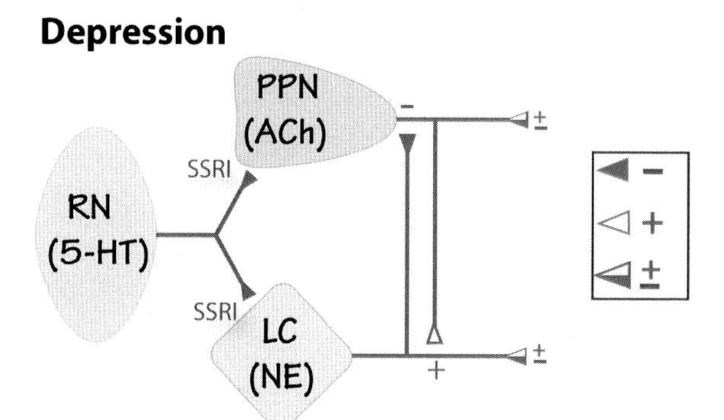

FIG. 4.4 Synaptic interactions in the RAS in major depression. The raphe nuclei (RN) that contain mostly serotonergic neurons (5-HT) project to both PPN and LC, inhibiting both targets *(filled black synapses)*. In turn, the PPN sends excitatory projections to LN *(open synapse)*, while the LC sends inhibitory input to the PPN *(half-filled black synapse)*. The PPN and LC send both excitatory and inhibitory projections to various ascending targets. Figure legend (filled black synapse = inhibitory −; open synapse = excitatory +; half-filled black synapse = both inhibitory and excitatory ±).

depression. Electroconvulsive therapy and, more recently, transcranial magnetic stimulation (TMS) have been approved for the treatment of depression. TMS appears to be quite effective in alleviating the symptoms in most, but not all, patients. Unfortunately, most TMS treatments produce only a partial alleviation of symptoms in many patients, with ~10% of patients unresponsive to any therapy. These numbers underscore the lack of truly effective treatment that can allow all patients to regain productive lives.

Hypofrontality

Both bipolar disorder and major depression share a major symptom that of hypofrontality. The decrease in frontal lobe blood flow may account for the lack of critical judgment in the decisions made by each sufferer. In the case of bipolar disorder, the risky and erratic behavior and the exaggerated fight-or-flight responses are periodic. The patient may behave in seductive and risky behavior without regard for the consequences and, when stressed or pressed, escape in spectacular fashion. In the case of major depression, escape responses are just as salient, placing the patient in equal peril. A major concern with these disorders is suicidality. A problem arises in the treatment of bipolar disorder misdiagnosed as major depression, in that therapy with SSRIs may in some cases induce suicidal ideation. Therefore proper diagnosis by a physician is essential. The fact that these disorders can abrogate the survival instinct speaks to the startling significance of suicidality and the need to seek rapid treatment. Whether hypofrontality is a cause of RAS dysregulation or RAS dysregulation leads to hypofrontality remains to be determined. However, successful treatment of any of these disorders should be marked by normalization of wake-sleep rhythms and reflexes and normalization of frontal lobe blood flow.

Acknowledgments

Supported by NIH award P30 GM110702 from the IDeA program at NIGMS.

References

Anderson, I.M., Haddad, P.M., Scott, J., 2012. Bipolar disorder. Brit. Med. Assoc. 345, e8508.

Arfken, C.L., Joseph, A., Sandhu, G.R., Roehrs, T., Douglass, A.B., Boutros, N.N., 2014. The status of sleep abnormalities as a diagnostic test for major depressive disorder. J. Affect. Disord. 156, 36–45.

Bergson, C., Levenson, R., Goldman-Rakic, P.S., Lidow, M.S., 2003. Dopamine receptor-interacting proteins: the Ca^{2+} connection in dopamine signaling. Trends Pharmacol. Sci. 24, 486–492.

Brown, K.M., Tracy, D.K., 2012. Lithium: the pharmacodynamics actions of the amazing ion. Ther. Adv. Psychopharmacol. 3 (32012), 163–176.

Carroll, C.A., Vohs, J.L., O'Donnell, B.F., Shekhar, A., Hetrick, W.P., 2007. Sensorimotor gating in manic and mixed episode bipolar disorder. Bipolar Disord. 9, 221–229.

Caspi, A., Sugden, K., Moffitt, T.E., Taylor, A., Craig, I.W., et al., 2003. Influence of life stress on depression: moderation by a polymorphism in the 5-HTT gene. Science 301, 386–389.

D'Onofrio, S., Kezunovic, N., Hyde, J.R., Luster, B., Messias, E., et al., 2015. Modulation of gamma oscillations in the pedunculopontine nucleus (PPN) by neuronal calcium sensor protein-1 (NCS-1): relevance to schizophrenia and bipolar disorder. J. Neurophysiol. 113, 709–719.

Depue, R.A., Arbisi, P., Krauss, S., Iacono, W.G., Leon, A., et al., 1990. Seasonal independence of low prolactin concentration and high spontaneous blink rates in unipolar and bipolar II seasonal affective disorder. Arch. Gen. Psychiatry 47, 356–364.

Etain, B., Henry, C., Bellivier, F., Mathieu, F., Leboyer, M., 2008. Beyond genetics: childhood affective trauma in bipolar disorder. Bipolar Disord. 10, 867–876.

Gambino, F., Pavlowsky, A., Béglé, A., Dupont, J.L., Bahi, N., et al., 2007. IL1-receptor accessory protein-like 1 (IL1RAPL1), a protein involved in cognitive functions, regulates N-type Ca^{2+}-channel and neurite elongation. Proc. Natl. Acad. Sci. 104, 9063–9068.

Garcia-Rill, E., Skinner, R.D., Clothier, J., Dornhoffer, J., Uc, E., et al., 2002. The sleep state-dependent midlatency auditory evoked P50 potential in various disorders. Thalamus Relat. Syst. 2, 9–19.

Garcia-Rill, E., Kezunovic, N., D'Onofrio, S., Luster, B., Hyde, J., et al., 2014. Gamma band activity in the RAS-intracellular mechanisms. Exp. Brain Res. 232, 1509–1522.

Heim, C., Newport, D.J., Mletzko, T., Miller, A.H., Nemeroff, C.B., 2008. The link between childhood trauma and depression: insights from HPA axis studies in humans. Psychoneuroendocrinology 33, 693–710.

Hirschfeld, R.M., 2001. The comorbidity of major depression and anxiety disorder: recognition and management in primary care. Prim. Care Companion J. Clin. Psychiatry 3, 244–254.

Kadrmas, A., Winokur, G., 1979. Manic depressive illness and EEG abnormalities. J. Clin. Psychiatry 40, 306–307.

Kasri, N.N., Holmes, A.M., Bultynck, G., Parys, J.B., Bootman, M.D., et al., 2004. Regulation of $InsP_3$ receptor activity by neuronal Ca^{2+}-binding proteins. EMBO J. 23, 312–321.

Kendler, K.S., Gatz, M., Gardner, C.O., Pedersen, N.L., 2006. A Swedish national twin study of lifetime major depression. Am. J. Psychiatry 163, 109–114.

Kessler, R.C., Nelson, C.B., McGonagle, K.A., Lui, J., Swartz, M., Blazer, D.G., 1996. Comorbidity of DSM-III-R major depressive disorder in the general population: results from the US National Comorbidity Survey. Br. J. Psychiatry Suppl. *30* (*1996*), 17–30.

Koh, P.O., Undie, A.S., Kabbani, N., Levenson, R., Goldman-Rakic, P.S., Lidow, M.S., 2003. Upregulation of neuronal calcium sensor-1 (NCS-1) in the prefrontal cortex of schizophrenic and bipolar patients. Proc. Natl. Acad. Sci. 100, 313–317.

Kohl, S., Heekeren, K., Klosterkötter, J., Kuhn, J., 2013. Prepulse inhibition in psychiatric disorders—apart from schizophrenia. J. Psychiatr. Res. 47, 445–452.

Kudlow, P.A., Cha, D.S., Lam, R.W., McIntyre, R.S., 2013. Sleep architecture variation: a mediator of metabolic disturbance in individuals with major depressive disorder. Sleep Med. 14, 943–949.

Kupfer, D.J., Foster, F.G., Coble, P., McPartland, R.J., Ulrich, R.F., 1978. The application of EEG sleep for the differential diagnosis of affective disorders. Am. J. Psychiatry 135, 69–74.

Liu, T.Y., Hsieh, J.C., Chen, Y.S., Tu, P.C., Su, T.P., Chen, L.F., 2012. Different patterns of abnormal gamma oscillatory activity in unipolar and bipolar disorder patients during an implicit emotion task. Neuropsychologia 50, 1514–1520.

Liu, X., Zhang, T., He, S., Hong, B., Chen, Z., et al., 2014. Elevated serum levels of FGF-2, NGF and IGF-1 in patients with manic episode bipolar disorder. Psychiatry Res. 218, 54–60.

Merikangas, K.R., Jin, R., He, J.P., Kessler, R.C., Lee, S., et al., 2011. Prevalence and correlates of bipolar spectrum disorder in the world mental health survey initiative. Arch. Gen. Psychiatry 68, 241–251.

Nordentoft, M., Mortensen, P.B., Pedersen, C.B., 2011. Absolute risk of suicide after first hospital contact in mental disorder. Arch. Gen. Psychiatry 68, 1058–1064.

Nutt, D.J., 2008. Relationship of neurotransmitters to the symptoms of major depressive disorder. J. Clin. Psychiatry 69, 4–7.

Olincy, A., Martin, L., 2005. Diminished suppression of the P50 auditory evoked potential in bipolar disorder subjects with a history of psychosis. Am. J. Psychiatry 162, 43–49.

Özerdem, A., Güntenkin, B., Atagün, I., Turp, B., Başar, E., 2011. Reduced long distance gamma (28-48 Hz) coherence in euthymic patients with bipolar disorder. J. Affect. Disord. 132, 325–332.

Perry, W., Minassian, A., Feifel, D., Braff, D.L., 2001. Sensorimotor gating deficits in bipolar disorder patients with acute psychotic mania. Biol. Psychiatry 50, 418–424.

Persson, M.L., Johansson, J., Vumma, R., Raita, J., Bjerkenstedt, L., et al., 2009. Aberrant amino acid transport in fibroblasts from patients with bipolar disorder. Neurosci. Lett. 457, 49–52.

Rodrigo, J., Suburu, A.M., Bentura, M.L., Fernandez, T., Nakade, S., et al., 1993. Distribution of the inositol 1,4,5-triphosphate receptor, P400, in adult rat brain. J. Comp. Neurol. 337, 493–517.

Schlecker, C., Boehmerle, W., Jeromin, A., DeGray, B., Varshney, A., et al., 2006. Neuronal calcium sensor-1 enhancement of $InsP_3$ receptor activity is inhibited by therapeutic levels of lithium. J. Clin. Invest. 116, 1668–1674.

Schulze, K.K., Hall, M.H., McDonald, C., Marshall, N., Walshe, M., et al., 2007. P50 auditory evoked potential suppression in bipolar disorder patients with psychotic features and their unaffected relatives. Biol. Psychiatry 62, 121–128.

Seifritz, E., 2001. Contribution of sleep physiology to depressive pathophysiology. Neuropsychopharmacology 25, S85–S88.

Sullivan, P.F., Daly, M.J., O'Donovan, M., 2012. Genetic architectures of psychiatric disorders: the emerging picture and its implications. Nat. Rev. Genet. 13, 537–551.

Torres, K.C., Souza, B.R., Miranda, D.M., Sampiao, A.M., Nicolato, R., et al., 2009. Expression of neuronal calcium sensor-1 (NCS-1) is decreased in leukocytes of schizophrenia and bipolar disorder patients. Prog. Neuropsychopharmacol. Biol. Psychiatry 33, 229–234.

Uhlhaas, P.J., Singer, W., 2010. Abnormal neural oscillations and synchrony in schizophrenia. Nat. Rev. Neurosci. 11, 100–113.

Chapter 5

Posttraumatic stress and anxiety, the role of arousal

Edgar Garcia-Rill
Center for Translational Neuroscience, Department of Neurobiology and Developmental Sciences, University of Arkansas for Medical Sciences, Little Rock, AR, United States

Introduction

Symptoms

Posttraumatic stress disorder (PTSD) is an anxiety disorder caused by an event in which there is fear of death. Symptoms are classified around three clusters: (1) hypervigilance that includes insomnia, exaggerated startle responses, and attentional problems; (2) reexperiencing that includes recurring vivid memories and dreams of the event; and (3) avoidance that involves avoiding places, people, and discussion of the event. These will be considered in detail later. In conjunction with the stress response and how its mechanisms are at the root of the pathology of this disorder.

Etiology

About one in three combat veterans will develop the disorder, with epilepsy, concussion, or other brain insult as potential predisposing factors. The incidence of PTSD in Vietnam era combat veterans was higher (~30%) (O'Toole et al., 1996; Roy-Byrne et al., 2004) than in Gulf War era combat veterans (~20%) (Goodman et al., 2006), but this may be an underestimate. PTSD can also be caused by sexual or physical abuse during childhood or adulthood, but it is not known what percent of abused children and adults (including battered women) will develop the condition. On the other hand, we do know that over one-half of rape victims will develop PTSD (Kilpatrick et al., 1987). Hostage or crime situations, torture, and even witnessing man-made or natural disasters that involve the loss of life, death, or serious injury to family members or close associates all can lead to this disorder. Estimates of the lifetime prevalence of PTSD from these causes range from 1% to 14% in the general population (Breslau and Davis, 1992). The most important predictor of PTSD is the number of stressors (Bramsen et al., 2000),

Arousal in Neurological and Psychiatric Diseases. https://doi.org/10.1016/B978-0-12-817992-5.00005-2

with childhood trauma and preexisting depression as likely factors (Breslau and Davis, 1992; Breslau et al., 1997) in facilitating the onset of PTSD in adulthood following a second stressor. Women are twice as likely to develop PTSD as a result of equivalent trauma as men (Breslau et al., 1991), with lifetime prevalence of 5% and 10% for men and women, respectively.

The stress-response

All of the anxiety disorders are associated with a stressful event or events, either real or imagined. The event or the memory of the event induces a specific kind of physiological arousal called the stress response (Sapolsky, 1992). Normally, the stress response represents a survival mechanism. However, when the stress response becomes repetitive or chronic, it becomes deleterious to health. The stress response is related to the fight-or-flight response. It is important to understand what actually happens to the body when the stress response is activated. This is a kind of survival response, conserved in evolution because animals with a good fight-or-flight response tend to survive moments of danger or emergency.

Try to imagine a humanoid a million years ago, foraging for food, and he suddenly comes face to face with a lion. What does the man's body need to do to survive? Run away, quickly. His muscles need all the energy they can muster, energy produced by burning glucose. Therefore the stress response consists of squeezing glucose out of fat cells, liver cells, etc. for use by the muscles. This is done by cortisol, more on that later. To deliver all the necessary glucose to the muscles, the heart rate and blood pressure need to rise, and the need for oxygen increases so that respiration also increases. Along with the responses in the man's body, there are changes in his brain. The senses become sharper, and memory improves during this response. This is due to an increase in awareness, because the reticular activating system (RAS) is alerted, and blood flow to the brain is boosted, especially to the frontal lobes. This cognitive component is needed in order to remember that man cannot outrun a lion; therefore the man needs to devise an escape, perhaps by climbing a tree to escape the lion.

There are also a number of energy-consuming processes that the man will want to shut down in order to optimize energy use to meet the demands of this emergency. He will not need to use up energy digesting the food he just ate, so his body shuts down digestion. The man will not want to spend energy repairing tissue, reproducing, having sexual drive, or fighting infections. He can do all that later, if he is still alive. Therefore, during the stress response, his body stops tissue repair and curtails the production of hormones related to sperm (or egg) production, and the immune system is suppressed. Pain is blunted, so that if he is clawed by the lion, he may still be capable of escaping. The pain pathway is thus suppressed. The manifestation of the stress response is well and good in an emergency, but one would not want to keep doing this all the time. The stress response is normally short-lasting and occasional. It is not intended for repeated, long-lasting responsiveness to the environment (Fig. 5.1).

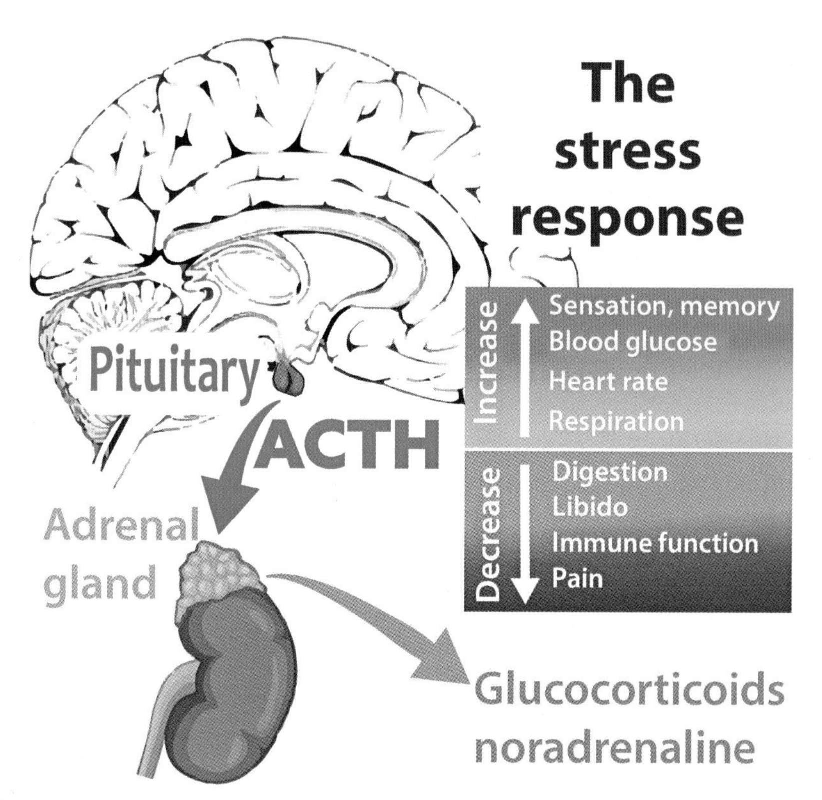

FIG. 5.1 The stress response. A stressor activates the RAS that in turn triggers the hypothalamus to begin the cascade of the stress response. Secretion into the bloodstream of ACTH activates the adrenal glands to secrete a short-acting activator, noradrenaline, and a long-acting activator, cortisol, a glucocorticoid. Processes that are required for fight or flight include increased blood glucose to increase energy expenditure by muscles, respiration to oxygenate blood to feed muscles, heart rate to carry oxygenated blood to muscles, and sensation and memory to detect threats and deal effectively with them. Processes that are not needed and are decreased include digestion, libido, immune function, and pain. Long-lasting or repetitive stress can result in degradation of health akin to premature aging through the modulation of all of these processes.

On a chronic basis, the stress response can kill just as surely as a lion but more slowly. If the stress response is elicited repetitively for prolonged periods, the increases in blood pressure can lead to heart disease, including heart attacks and strokes from the runaway blood pressure. During chronic stress, one does not repair the wear and tear of everyday aging. A simple cut can take days to heal, and it will scar easily. Glucose utilization problems can lead to a form of diabetes; egg or sperm count drops; libido decreases; and the body being less able to fight infections, including cancer. Now, suppose that you are a modern individual, rarely having to face a lion but living in a crowded environment, toiling daily at a stressful job, facing large debts, shrinking income, increasing time demands, and family problems, none of which can be resolved quickly.

Living daily with chronic stress produces a candidate for hypertension, diabetes, diminished sex drive (including impotence), unsuccessful reproduction, and increased chance of suffering from cancer, and to top it all off, the changes in the brain are making you dumber. This is akin to premature aging. When you are on edge and stressed, the RAS is constantly on edge, ready to be provoked. It seems as if the next little thing, like the faucet sprouting a leak, will be the last straw. That last stimulus produces a full-blown fight-or-flight response in which you will strike out (verbally or physically) or escape (into the mind-numbing television set), reflexively, without "thinking." Even so, this still only represents an excessive fight-or-flight response. That is, the severity of these reactions is within "normal" limits and does not represent an anxiety disorder yet.

The anxiety disorders

There are three major categories of anxiety disorders, namely, panic disorder, obsessive-compulsive disorder (OCD), and PTSD. Panic disorder is an anxiety disorder in which there are panic attacks, short periods of intense fear associated with impending doom. During these panic attacks, the heart starts to pound; there is sweating, trembling, shortness of breath, chest pain, dizziness, numbness, chills, or hot flashes; essentially, this is an unchecked episode of increased physiological arousal and stress. Panic attacks represent an arousal response to a thought, to a fear- inducing idea, and not necessarily to the appearance of a physical predator. Panic disorder usually first manifests itself during adolescence. A smaller population develops the disease in their mid-30s. There is a genetic contribution to the disorder, sometimes running in families and being prevalent in twins at a higher rate than in the population at large.

OCD also usually begins at or around puberty, especially in males, with gradually increasing symptoms that can also become incapacitating. Obsessions are persistent ideas, thoughts, impulses, or images that are considered intrusive and produce anxiety and distress. Compulsions are repetitive behaviors like arranging things and handwashing or mental acts like praying or silently repeating words, designed to reduce anxiety or distress. Just as avoidance is a survival strategy used by panic disorder patients, OCD patients use compulsive actions and thoughts to ward off impending danger. For example, OCD patients will wash their hands until the skin is macerated, hating every minute of it. These unfortunate individuals are slaves to their compulsions and obsessions knowing that these are inappropriate behaviors. However, if they do not perform the clearly embarrassing ritual and if they do not succumb to the urges, their anxiety becomes unbearable. The fear and impending doom are intolerable unless these tasks are performed, over and over again. These individuals are enslaved by their brain's dysfunction.

All anxiety disorders manifest a sensory gating deficit, which is an abnormality in the process we use to gate, or filter out, sensory input. Normally, we habituate, that is, we have a reduced response, to repetitive stimuli. For example, if we hear two loud sounds spaced, say, one-half second apart, we will startle less to the second stimulus than to the first. The response to the second sound will be about 15%–20% of the response to the first. We will not jump anywhere near as high to the second sound as we did to the first. On the other hand, people with an anxiety disorder will show a response to the second stimulus of about 60%–80%, jumping almost as high to the second sound as to the first. These individuals have an inability to gate or filter, to habituate, to repeated sensory inputs, thus the term sensory gating deficit. A sensory gating deficit implies excessive distractibility. This means that the reactivity to stimuli in their world never wanes. All light, sound, and touch are intrusive, continuous, and punishing, driving you well ...crazy.

We are very good at detecting change. That is, we perceive change in the world around us and habituate, respond less and less, to stimuli that are repetitive. This is not just some quirk of nature. This capacity has great survival value and was thus preserved in evolution. After all, if you are continuously responding to the sound of the wind in the trees, you are less likely to detect the sound of an approaching predator. In anxiety disorder, the exaggerated responsiveness to all sensory inputs not only serves to overwhelm the afflicted individual but also induces a great deal of fear and stress, a feeling of inability to cope and to detect that approaching predator. It induces a constant state of impending doom.

What causes this over reactivity to sensations? Obviously, the brain system controlling arousal, the startle response, and fight-or-flight responses is disturbed. This means that the RAS is not working properly in anxiety disorders. Studies have confirmed the well-established observation that these people also have disturbed sleep patterns. These individuals are not only hyperreactive to sensory input but also hyperaroused in general. These people have difficulty falling asleep and staying asleep. When they do sleep, they awaken tired and unrefreshed. Their dreams are vivid; their nightmares are intense, indicative of increased rapid eye movement (REM) sleep drive. These are all symptoms of increased arousal and hypervigilance, of a problem in the RAS. One of the major problems in this disease is that the RAS, allowed to throb unchecked, propels the brain into exaggerated responses, delusions, and hallucination.

PTSD is perhaps the most common anxiety disorder in this trilogy. Experiencing or witnessing an event involving death, injury, or threat, coupled with the intense fear that the event generated, along with a feeling of helplessness, induces a characteristic set of symptoms. The first symptom cluster is *avoidance* of thoughts, feelings, places, situations, or activities that remind the person of the traumatic event. This symptom is characteristic not only of PTSD but also of all of the anxiety disorders. Avoidance of places, people, and things, of situations or events from which you cannot escape or flee, would seem to be a natural tendency. After all, it is a strategy designed for survival, to save oneself from the unseen danger. Fear of something causes avoidance. However, fear of everything can be terribly incapacitating. For those unafflicted, this is difficult to understand. A sudden, unexplainable, intense fear will induce an escape response, a flight, and an avoidance. Forcing a victim of a panic attack to remain in that situation, to face the fear, and to fight the unseen predator is not appropriate. These individuals consider their lives to be in danger and will fight to survive and even strike out in order to flee. A calm voice and a quiet environment is one way to try to ameliorate the situation. By keeping the sensory input down, at least the world outside will not add to the roar inside the brain. By itself, avoidance can be terribly disabling in one's life, but unfortunately, it is only the beginning.

A second major symptom cluster is hyperarousal and hypervigilance, which are often present along with disturbed sleep. This symptom obviously indicates dysregulation of the RAS. There is a third and particularly damaging symptom cluster to this disorder, a persistent reexperiencing of the traumatic event. Every time the traumatic event is recollected, it triggers a physiological arousal and stress response, intense fear, and anxiety of the event. Our memories are recollections from which we cannot escape. A color, a light, a sound, a word, and almost anything can trigger a reminder of that feared incident, causing the same stress response every time, reliving the entire event in our memories. The fear, pain, helplessness, and all of the physical reactions that accompanied the initial event, including the stress response, are all reexperienced, along with the inevitable consequences.

How does this mechanism produce brain damage? The stress response is a chain reaction of hormonal activity. A stressor, any stressor, (1) induces physiological arousal in the RAS that then (2) triggers activity in the hypothalamus and then (3) signals the pituitary gland to then (4) secrete ACTH, a hormone, into the circulation, which then (5) reaches the adrenal gland, causing (6) production of adrenaline, the substance that leads to the stress response (i.e., glucose mobilization and accelerated heartbeat). Adrenaline has a rapid effect on the various organs on which it acts and then dissipates or is broken down quickly. Adrenaline is the ideal substance for a getaway. The adrenal glands also produce glucocorticoids that have longer actions on these organs, prolonging the stress response beyond the few seconds during which adrenaline is active. Cortisol is the best known of these "stress hormones."

The brain has an efficient mechanism to terminate the stress response. Being able to end this response allows the body to go back to its housekeeping duties, like digesting, reproducing, and fighting infection. Circulating glucocorticoids in the blood reach the hypothalamus and there shut off the further release of ACTH. That is, the same substance that is released as a result of the stress response feeds back to the regions that started the process. Too much glucocorticoid in the blood leads to feedback inhibition of the hypothalamus, leading to the shutdown of ACTH release from the pituitary, which no longer induces the adrenals to release glucocorticoids. The termination of the stress response is important also because excess glucocorticoids can have a deleterious effect on the brain. The occasional stress response is not going to do any harm, but if you are under continuous stress, high circulating levels of glucocorticoids may lead to the shrinking of dendrites, probably wherever there are glucocorticoid receptors in the brain.

The parts of the brain with significant numbers of these receptors are the hypothalamus, the hippocampus, and the RAS, specifically the LC (see later). These are the three brain regions primarily involved in anxiety disorders. There is no clear scientific evidence to suggest that the difference between a normal individual living in a stressful society and an individual with anxiety disorder is the extent of damage or dysfunction in these regions of the brain. Obviously, something else happens to these brain centers in a person with anxiety disorder, especially after puberty. Likewise, something else leads some combat-exposed soldiers to develop PTSD, while their fellow combatants remain normal. Why do only one-half or so of rape victims manifest this disorder but the rest do not? There must be other factors that, when combined with repeated or excessive stress responses, may lead to sufficient damage by high circulating levels of glucocorticoids to kill or render inactive many of these neurons (hippocampal, hypothalamic, and RAS).

The exaggerated effect of glucocorticoids is probably a major element contributing in the disorder. One of the duties of glucocorticoids is to block glucose uptake into tissues and release it into the bloodstream. If glucocorticoids block glucose uptake into neurons, will these cells not die or be weakened by a lack of energy production? One possible predisposing factor placing individuals at risk for developing an anxiety disorder is head injury. If the brain has already been racked by a concussion or two or if there is another disease like epilepsy in that tissue, are these not "debilitated" cells? The chances that the loss of glucose is due to glucocorticoid exposure, sending these cells "over the edge" of recovery, are probably fairly high. Attempting to confirm or deny these effects is a very active area of research at the present time. It is also a very important area of research, one with implications for our overall health in today's stressful world.

During the stress response, the music of the brain is driven by the percussion section and the drumbeat rising in volume with every stimulus. In anxiety disorder, the sensory gating abnormality ensures that the volume will keep increasing. If, as some researchers believe, there is damage to the noradrenergic neurons of the LC in anxiety disorder, their activity will decrease. This means that the neighboring cholinergic pedunculopontine nucleus (PPN) neurons will be less inhibited and less regulated by noradrenergic input (more on this later). The result is that arousal will be heightened during waking, and REM sleep will be intensified, producing more vivid dreams and nightmares. That is, it will exaggerate the output of the part of the RAS that controls waking and REM sleep (see Chapter 2). In some people, this overactivity could induce hallucinations, or dreaming while awake. The pounding of the bass drums of the cerebral orchestra would be wrenching.

On an everyday basis, these individuals would be less likely to withstand the psychological pressures of a stressful occupation, choosing to avoid noisy or disturbing environments and activities. These individuals would be the ones who scream and jump the highest at the action-suspense movies, if they view them at all. Those exposed to abuse or trauma would flinch at (and avoid) the violent scenes of a movie or news program. Fear and impending doom cast a dark shadow over their lives in which imagined catastrophes will kill them or their loved ones. The music of the brain would be constantly playing a requiem, albeit very loudly. The overwhelming pounding would lead to exaggerated, hair-trigger fight-or-flight responses. Chronic overactivity in the RAS can also produce "hypofrontality," decreased blood flow to the frontal lobes. That means that the part of the brain most responsible for critical judgment is less active, so the making of critical decisions is impaired. It should be noted that baseline cortical metabolism is decreased in PTSD patients, but these individuals will show increased blood flow compared with normal controls when exposed to trauma-related stimuli or abuse-related imagery (Liberzon et al., 1999; Shin et al., 1999; Zubieta et al., 1999).

An individual previously abused physically and mentally is likely to respond in the same manner when stressed, to respond violently and to lash out indiscriminately, reflexively. If beaten as a child for crying, that person as an adult could be more likely to abuse his own children when stressed. Those children could, in turn, grow up to beat their children, and the behavior becomes a "sick" family tradition. This could be one reason why anxiety disorders run in families. At some point, however, one of these individuals may seek help, stop responding excessively to stressful stimuli, and attempt to break the cycle of violence. Otherwise, the family legacy of abuse will be passed down from generation to generation.

If there is damage to the cells of the hippocampus, the music of the brain will be less expressive and less brilliant. Memory and learning will be somewhat impaired, but some memories will be repeated, reinforced, and amplified. Those vivid memories of the trauma, of the abuse, will be written in stone. Every time they are recalled, every time any little thing reminds them of that event, the memory will induce another full-blown stress response. The reexperiencing of the traumatic event in PTSD is extremely damaging because it will generate a stress response and lead to high circulating levels of glucocorticoids, putting even more brain cells at risk. If a few cells die, the next stress response will then be a little stronger, producing more damage. A vicious cycle of brain damage is caused by a bad memory from which they cannot escape. This may be one reason why PTSD can last a lifetime. While about 50% of patients with PTSD can be said to be in remission 3 years after the trauma (can live normal lives, even with occasional symptoms), over 40% are still considered to meet criteria for PTSD 10 years after the traumatic event (Kessler et al., 1995). That is, for many victims, this is a lifelong disorder (Zlotnick et al., 1999). PTSD can also have a delayed onset, sometimes of many years, such as in cases of combat veterans who developed PTSD after retirement (Van Dyke et al., 1985).

EEG, reflexes and P50 potential

Sleep patterns in PTSD are similar to those in schizophrenia and bipolar disorder, basically showing frequent awakenings, decreased SWS, increased REM sleep drive, and hypervigilance (Ross et al., 1989, 1999; Sandor and Shapiro, 1994; Woodward et al., 1996). In addition, PTSD patients have exaggerated startle responses and reflexes (Butler et al., 1990; Ornitz and Pynoos, 1989; Shalev et al., 2000). We described the manifestation of the P50 potential in PTSD. Briefly, we studied a group of combat-exposed veterans with PTSD and compared their P50 potential responses using a paired stimulus paradigm with those of normal age and gender matched controls and combat-exposed veterans without PTSD and alcoholics in remission. PTSD patients tend to drink heavily, which may also represent an attempt at self-medication for the sensory gating abnormality produced by PTSD, so that alcoholics in remission control subjects are essential. We found that the P50 potential was of significantly higher amplitude in PTSD after the second stimulus of a pair than in each of the other control groups (Gillette et al., 1997). Fig. 5.2 shows the results of that study. Normally, the amplitude of the response to the second stimulus of a pair is ~10%–25%, which was evident in normal controls, combat-exposed veterans without PTSD, and alcoholic patients. However, in combat-exposed veterans with PTSD, habituation was reduced, and the second response was ~60%. These results were best correlated with the reexperiencing symptom cluster in psychiatric testing instruments, suggesting that it is the recollection of the traumatic experience on a recurrent basis that leads to the sensory gating deficit.

We then replicated these findings in male combat veterans and also found that female rape victims suffered from the same decreased habituation (Skinner et al., 1999). The P50 midlatency auditory evoked potential was studied in female rape victims with PTSD and compared with an age-matched female control group and in male combat veterans with PTSD and compared with three groups of age-matched male control subjects. Sensory gating of the P50 potential was determined using a paired click stimulus paradigm in which the stimuli were presented at 250-, 500-, and 1000-ms interstimulus intervals (ISI). Results showed that sensory gating of the P50 potential was significantly decreased at the 250-ms ISI and that there was a numerical, but not a statistically significant, decrease in sensory gating at the other intervals tested in both male and female PTSD subjects compared with all control groups (Skinner et al., 1999). Since the P50 potential may be generated, at least in part, by the RAS, dysregulation of sensory processing by elements of this system may be present in PTSD.

We later found in a small sample of postmortem brain tissue from combat-exposed veterans with PTSD that there was a significant reduction in the number of neurons in the LC (Bracha et al., 2005). Because enhanced central nervous system (CNS) noradrenergic postsynaptic responsiveness has been previously shown to contribute to PTSD pathophysiology, we investigated whether combat-related PTSD is associated with a postmortem change

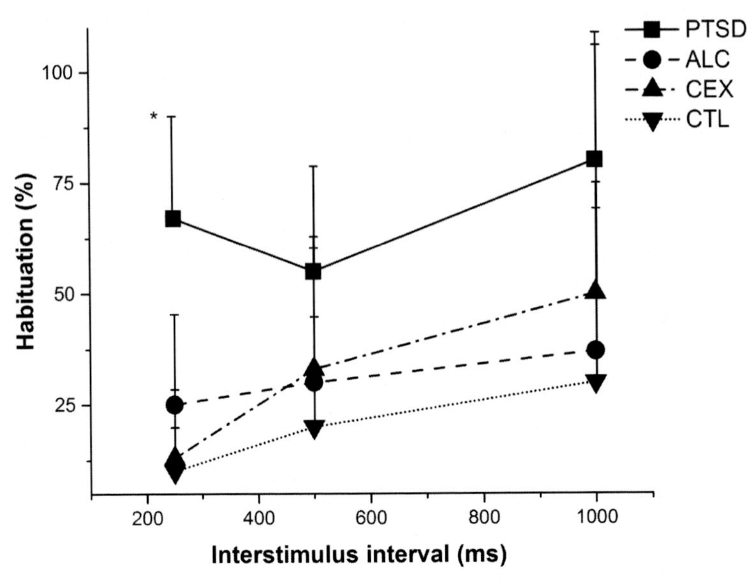

FIG. 5.2 Effects of PTSD on P50 potential habituation. Graph of the mean percent (±SE) P50 potential habituation (sensory gating) using a paired stimulus paradigm in PTSD (combat-exposed veterans with PTSD) *(filled black squares)*, compared with alcoholic patients in remission *(filled black circles*—ALC), with combat-exposed veterans without PTSD *(filled black up triangles*—CEX), and with control (CTL) *(filled black down triangle)* groups. Three ISIs were tested, 250, 500, and 1000 ms. For each ISI, the amplitude of the P50 potential following the second stimulus was calculated as a percent of the amplitude of the P50 potential following the first stimulus. The percent habituation was significantly (*$P < 0.05$) higher in the PTSD compared with all other groups at the 250-ms ISI. There was no statistically significant difference in the percent habituation of the groups at the 500-ms or 1000-ms ISI. *(Data from Gillette, G., Skinner, R.D., Rasco, L., Fielstein, E., Davis, D., Pawelak, J., Freeman, T., Karson, C.N., Boop, F.A., Garcia-Rill, E., 1997. Combat veterans with posttraumatic stress disorder exhibit decreased habituation of the P1 midlatency auditory evoked potential. Life Sci. 61, 1421–1434.)*

in neuronal counts in the LC. Using postmortem neuromorphometry, we counted the number of neurons in the right LC in seven deceased elderly male veterans. We classified three veterans as cases of probable or possible combat-related PTSD. All three veterans with probable or possible combat-related PTSD were found to have substantially lower LC neuronal counts compared with four controls (three nonpsychiatric veterans and one veteran with alcohol dependence and delirium tremens). To our knowledge, this case series is the first report of LC neuronal counts in patients with PTSD or any other DSM anxiety disorder. The finding reported here is consistent with the upregulation of norepinephrine (NE) biosynthetic capacity of surviving LC neurons in veterans who developed combat-related PTSD. If replicated, the finding we reported in combat-related PTSD may provide further explanation of the demonstrated effectiveness of PTSD treatment with propranolol and prazosin. Larger neuromorphometric studies of the LC in veterans with combat-related

PTSD and in other stress-induced and fear-circuitry disorders are warranted (Bracha et al., 2005).

We proposed that the mechanism involved in PTSD is mainly a result of high-amplitude pulses of cortisol induced by the repeated recollection of the trauma that travel across the blood-brain barrier to bind in regions that have glucocorticoid receptors such as the hippocampus, cortex, and LC. All of these regions have been reported to be damaged in PTSD. The repeated exposure to cortisol would ultimately lead to neuronal cell death. Since the LC normally inhibits the PPN (see Fig. 5.3), the loss of LC neurons would disinhibit the PPN and produce increased PPN output that would result in increased vigilance and REM sleep drive, in keeping with PPN functions. We also investigated the possibility that acute, high circulating levels of cortisol, possibly induced by the stress of reexperiencing the trauma, may lead to the LC cell death observed. A similar mechanism has been proposed for the effects of stress on hippocampal neurons (Sapolsky et al., 1984), which, like LC neurons, possess glucocorticoid receptors (Aronsson et al., 1988). In the rat, we found that neurotoxic lesions of LC led to decreased habituation of the P13 auditory response, the rodent equivalent of the human P50 potential, with a similar outcome resulting from daily administration of corticosterone (Miyazato et al., 2000).

These findings suggest that it is decreased, not increased, LC output to PPN that may be responsible for the symptoms of anxiety. The fact that yohimbine, an α-2 adrenergic receptor antagonist, induced anxiety in PTSD patients

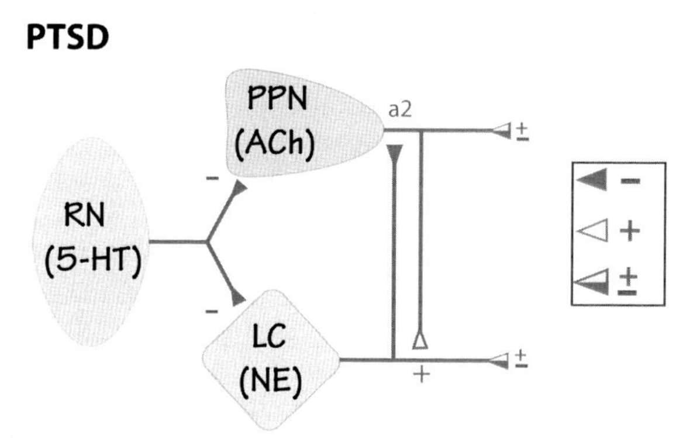

FIG. 5.3 RAS diagram showing sites of dysregulation in PTSD. The raphe nuclei (RN) that contain mostly serotonergic neurons (5-HT) project to both PPN and LC, inhibiting both targets *(filled black synapses)*. In turn, the PPN sends excitatory projections to LC *(open synapse)*, while the LC sends inhibitory input to the PPN *(half-filled black synapse)*. The PPN and LC send both excitatory and inhibitory projections to various ascending targets. Figure legend (filled black synapse = inhibitory −; open synapse = excitatory +; half-filled black synapse = both inhibitory and excitatory ±). In PTSD, decreased LC inhibition of the PPN appears to be a determinant of the hypervigilance and increased REM sleep drive, especially following LC cell loss, leading to disinhibition of arousal through the PPN. Administration of α-2 adrenergic receptor agonists such as clonidine will inhibit PPN, downregulating these symptoms, while α-2 adrenergic receptor antagonists such as yohimbine will exacerbate arousal and REM sleep drive dysregulation.

(Charney et al., 1987), while the adrenergic agonist clonidine reversed the effect (Charney and Heninger, 1986), suggests that the drug may be blocking, not LC autoreceptors, but synaptic input from LC to PPN (as in Fig. 5.3). This would disinhibit cholinergic neurons and thus account for the wake-sleep symptoms observed and the changes in the P50 potential we reported. Along these lines, we determined that injections of yohimbine directly into the PPN increase the amplitude and decrease the habituation of the P13 potential in the rat, an effect reversed by injections of clonidine (Miyazato et al., 1999). These results provided further support for the mechanism of PTSD being due to decreased LC inhibition of PPN, leading to hypervigilance and exaggerated reflexes.

Treatment

After a stressful day at the office, we reach the relative safety of home to the news that the daughter has a new tattoo, the son has been suspended from school, the dog threw up on our favorite chair, and the spouse was fired. It is time to open the bar. One of the easiest ways to calm down the RAS is with alcohol (Brust, 2004). Alcohol turns down the volume of the music. The conduction of impulses, of action potentials, along the multiple synapses of the RAS is slowed by alcohol. Just as one might use alcohol to relax and to recover from the stress response, so does the anxiety disorder patient self-medicate with alcohol. In virtually all disorders in which the output of the RAS is amplified, schizophrenia, anxiety disorder, and depression and bipolar disorder patients will self-medicate excessively with alcohol (Khantzian, 1985; McFarlane, 1989; Weiss et al., 1992).

This is not necessarily an addiction to alcohol, for these individuals will tell you that they hate the hangovers, the loss of control, but alcohol turns down the blaring music, the intrusive world. Sometimes this creates a problem in diagnosis. For example, if a teenager has a silly idea that she does not want to leave the house and she drinks a lot, her parents may conclude she has a drinking problem. Or does she? Drinking may be masking a panic disorder. Only a competent clinician may be able to arrive at an accurate diagnosis. Before embarking on an expensive treatment regimen for alcoholism that will clearly fail in this case, alternative diagnoses need to be considered. Alcohol is a very good anxiolytic drug. Unfortunately, it also has serious side effects. Alcohol preferentially affects granule cells located in the cortex, cerebellum, and hippocampus. What happens then? The music is missing the high notes, and the control is gone. The cortex is playing simpler music, so one is not a great intellect when drunk. The hippocampus is having a hard time remembering anything. The cerebellum is where the real danger lies. The regulation of rapid, highly integrated movements by the cerebellum is impaired, along with balance and equilibrium. Performing accurate movements after consuming alcohol is difficult at best. Driving is impaired and should not be attempted. The policeman who stops the drunk driver basically gives him a neurological exam of cerebellar function, including walking a straight line heel-to-toe, placing the feet together and arms apart, then touching the nose, closing the eyes (so that visual input cannot be used to orient), tilting the head back (to test balance control), and standing on one leg (all tests used by the neurologist when assessing cerebellar control).

There are other drugs that can be used to turn down the volume of the RAS. One type of agent is a benzodiazepine, like clonazepam, that amplifies the action of inhibitory neurons all over the brain, including those in the RAS. The good news is that it calms down the drumbeat of the brain. The bad news is that the use of benzodiazepines calms down activity in many other regions, interfering with cognitive processes, producing drowsiness, sleepiness, and lethargy, leading patients to resist therapy. This is a typical problem in the treatment of anxiety, when treatment includes the use of tranquilizers, sedatives, and barbiturates to calm the agitated patient. These drugs basically turn down all of the RAS, making it difficult to stay awake, much less carry out everyday activities. More recently, compounds with more specific actions have been used in the treatment of anxiety disorder. Xanax is a benzodiazepine-like compound that also has antidepressant effects. The good news is that Xanax relieves anxiety without making people sleepy, but the bad news is that it is very short-lasting, requiring repeated dosing, and it can be very addictive, producing repeated withdrawal symptoms as each dose wears off. The search is on for a longer-lasting, less addictive substitute. There are some agents currently being tested that may act more specifically in the RAS, with promising anxiolytic actions.

Thus treatment of PTSD is limited to anxiolytic and therapy, with mixed success. We hypothesize that it may be possible to successfully treat and perhaps even prevent this disorder. For example, it would appear that the ideal anxiolytic compound would be one able to be an agonist at the α-2 adrenergic receptor on PPN neurons (without the peripheral effects of agents such as clonidine). In addition, it may be possible to block the potentially toxic effects of cortisol on brain cells by using a glucocorticoid receptor blocker, such as RU-486, soon after the trauma. This would be a sort of preventive treatment in cases of recurrent increased levels of cortisol, presumably rescuing some neurons. Recent successes using propranolol, the β-adrenergic blocker, as an amnestic agent soon after the trauma suggest that this is another option for the reexperiencing consequences of PTSD (Pittman et al., 2002; Vaiva et al., 2003).

References

Aronsson, M., Fuxe, K., Dong, Y., Agnati, L.F., Okret, S., Gustafsson, J.A., 1988. Localization of glucocorticoid receptor mRNA in the male rat brain by in situ hybridization. Proc. Natl. Acad. Sci. 85, 9331–9335.

Bracha, H.S., Garcia-Rill, E., Mrak, R.E., Clothier, J., Karson, C.N., Skinner, R.D., 2005. Post-mortem locus coeruleus neuron count: a case report of a veteran with probable combat-related posttraumatic stress disorder. J. Neuropsychiatry Clin. Neurosci. 17, 503–509.

Bramsen, I., Dirkzwager, A.J., van der Ploeg, H.M., 2000. Predeployment personality traits and exposure to trauma as predictors of posttraumatic stress symptoms: a prospective study of former peacekeepers. Am. J. Psychiatry 157 (2000), 1115–1119.

Breslau, N., Davis, G.C., 1992. Posttraumatic stress disorder in an urban population of young adults: risk factors for chronicity. Am. J. Psychiatry 149, 671–675.

Breslau, N., Davis, G.C., Andreski, P., Peterson, E., 1991. Traumatic events and posttraumatic stress disorder in an urban population of young adults. Arch. Gen. Psychiatry 48, 216–222.

Breslau, N., Davis, G.C., Peterson, E.L., Schultz, L., 1997. Psychiatric sequelae of posttraumatic stress disorder in women. Arch. Gen. Psychiatry 54, 81–87.

Brust, C.M., 2004. Ethanol. In: Neurological Aspects of Substance Abuse. Butterworth-Heinemann, pp. 317–426.

Butler, R.W., Braff, D.L., Pausch, J.L., Jenkins, M.A., Sprock, J., Geyer, M.A., 1990. Physiological evidence of exaggerated startle response in a subgroup of Vietnam veterans with combat-related PTSD. Am. J. Psychiatry 147, 1308–1312.

Charney, D., Heninger, G., 1986. Abnormal regulation of noradrenergic function panic disorders. Arch. Gen. Psychiatry 43, 1042–1054.

Charney, D.S., Woods, S.W., Goodman, W.K., Heninger, G.R., 1987. Neurobiological mechanisms of panic anxiety: biochemical and behavioral correlates of yohimbine-induced panic attacks. Am. J. Psychiatry 144, 1030–1036.

Gillette, G., Skinner, R.D., Rasco, L., Fielstein, E., Davis, D., Pawelak, J., Freeman, T., Karson, C.N., Boop, F.A., Garcia-Rill, E., 1997. Combat veterans with posttraumatic stress disorder exhibit decreased habituation of the P1 midlatency auditory evoked potential. Life Sci. 61, 1421–1434.

Goodman, L.R., et al., Committee on Gulf War Health, 2006. Gulf War and Health: Vol. 4, Health Effects of Serving in the Gulf War. National Academies Press, p. 292.

Kessler, R.C., Sonnega, A., Bromet, E., Hughes, M., Nelson, C.B., 1995. Posttraumatic stress disorder in the national comorbidity survey. Arch. Gen. Psychiatry 52, 1048–1060.

Khantzian, E.J., 1985. The self-medication hypothesis of addictive disorders: focus on heroin and cocaine dependence. Am. J. Psychiatry 142, 1259–1264.

Kilpatrick, D.G., Saunders, B.E., Veronen, L.J., Best, C.L., Von, J.M., 1987. Criminal victimization: lifetime prevalence, reporting to police and psychological impact. Crime Delinq. 33, 479–489.

Liberzon, I., Taylor, S.F., Amdur, R., Jung, T.D., Chamberlain, K.R., Minoshima, S., koeppe, R.A., Fig, l.M., 1999. Brain activation in PTSD in response to trauma-related stimuli. Soc. Biol. Psychiatry 45, 817–826.

McFarlane, A.C., 1989. Epidemiological evidence about the relationship between PTSD and alcohol abuse: the nature of the association. Addict. Behav. 23, 813–825.

Miyazato, H., Skinner, R.D., Garcia-Rill, E., 1999. Neurochemical modulation of the P13 midlatency auditory evoked potential in the rat. Neuroscience 92, 911–920.

Miyazato, H., Skinner, R.D., Garcia-Rill, E., 2000. Locus coeruleus involvement in the effects of immobilization stress on the P13 midlatency auditory evoked potential in the rat. Prog. Neuropsychopharmacol. Biol. Psychiatry 24, 1177–1201.

Ornitz, E.M., Pynoos, R.S., 1989. Startle modulation in children with psottraumatic stress disorder. Am. J. Psychiatry 146, 866–870.

O'Toole, B.I., Marshall, R.P., Grayson, D.A., Schureck, R.J., Dobson, M., Ffrench, M., Pulvertaft, B., Meldrum, L., Bolton, J., Vennard, J., 1996. The Australian Vietnam Veterans Health Study: III. Psychological health of Australian Vietnam veterans and its relationship to combat. Int. J. Epidemiol. 25, 331–340.

Pittman, R.K., Sanders, K.M., Zusman, R.M., Healy, A.R., Cheema, F., Lasko, N.B., Cahill, L., Orr, S.P., 2002. Pilot study of secondary prevention of posttraumatic stress disorder with propranolol. Biol. Psychiatry 51, 189–192.

Ross, R.J., Ball, W.A., Sullivan, K.A., Caroff, S.N., 1989. Sleep disturbance as the hallmark of posttraumatic stress disorder. Am. J. Psychiatry 146, 697–707.

Ross, R.J., Ball, W.A., Dinges, D.F., Kribbs, N.B., Morrison, A.R., Silver, S.M., Mulvaney, F.D., 1999. Rapid eye movement sleep changes during the adaptation night in combat veterans with posttraumatic stress disorder. Biol. Psychiatry 45, 938–941.

Roy-Byrne, P., Arguelles, L., Vitek, M.E., Goldberg, J., Keane, T.M., True, W.R., Pitman, R.K., 2004. Persistence and change of PTSD symptomatology—a longitudinal co-twin control analysis of the Vietnam Era Twin Registry. Social Psychiatry Psychiatr. Epidemiol. 39, 681–685.

Sandor, P., Shapiro, C.M., 1994. Sleep patterns in depression and anxiety: theory and pharmacological effects. J. Psychosom. Res. 38, 125–139.

Sapolsky, R.M., 1992. The stress-response and the emergence of stress-related disease. In: Stress, the Aging Brain, and the Mechanisms of Neuron Death. MIT Press, Cambridge, MA, pp. 429.

Sapolsky, R.M., Krey, L.C., McEwen, B.S., 1984. Prolonged glucocorticoid exposure reduces hippocampal neuron number. Implications for aging. J. Neurosci. 5, 1222–1227.

Shalev, A.Y., Freedman, S., Peri, T., Brandes, D., Sahar, T., Orr, S.P., Pitman, R.K., 2000. Auditory startle response in trauma survivors with posttraumatic stress disorder: a prospective study. Am. J. Psychiatry 157, 255–261.

Shin, L.M., McNally, R.j., Kosslyn, S.M., Thompson, W.L., Rauch, S.L., Alpert, N.M., Metzger, L.J., Lasko, N.B., Orr, S.P., Pittman, R.K., 1999. Regional cerebral blood flow during script-driven imagery in childhood sexual abuse-related PTSD: a PET investigation. Am. J. Psychiatry 156, 575–584.

Skinner, R.D., Rasco, L., Fitzgerald, J., Karson, C.N., Matthew, M., Williams, D.K., Garcia-Rill, E., 1999. Reduced sensory gating of the P1 potential in rape victims and combat veterans with post traumatic stress disorder. Depress. Anxiety 9, 122–130.

Vaiva, G., Ducrocq, F., Jezequel, K., Averland, B., Lestavel, P., et al., 2003. Immediate treatment with propranolol decreases posttraumatic stress disorder two months after trauma. Biol. Psychiatry 54, 947–999.

Van Dyke, C., Zilberg, N.J., McKinnon, J.A., 1985. Posttraumatic stress disorder: a thirty year delay in a World War II veteran. Am. J. Psychiatry 142, 1070–1073.

Weiss, R.D., Griffin, M.L., Mirin, S.M., 1992. Drug abuse as self-medication for depression: an empirical study. Am. J. Drug Alcohol Abuse 18, 121–129.

Woodward, S.H., Friedman, M.J., Bliwise, D.L., 1996. Sleep and depression in combat-related PTSD inpatients. Biol. Psychiatry 39, 182–192.

Zlotnick, C., Warshaw, M., Shea, M.T., Allsworth, J., Pearlstein, T., Keller, M.B., 1999. Chronicity in posttraumatic stress disorder (PTSD) and predictors of course of comorbid PTSD in patients with anxiety disorders. J. Trauma. Stress 12, 89–100.

Zubieta, J.K., Chinitz, J.A., Lombardi, U., Fig, L.M., Cameroc, O.G., Liberzon, I., 1999. Medial frontal cortex involvement in PTSD symptoms: a SPECT study. J. Psychiatry Res. 33, 259–264.

Chapter 6

Autism and arousal

James Hyde*, Edgar Garcia-Rill†
*Southern Arkansas University, Magnolia, AR, United States, †Center for Translational Neuroscience, Department of Neurobiology and Developmental Sciences, University of Arkansas for Medical Sciences, Little Rock, AR, United States

Introduction

Autism spectrum disorder (ASD) is a neurological condition with no known cause or cure and generates lifelong effects. Autism is a highly heterogeneous condition with a spectrum of phenotypes including poor social skills, disrupted communication skills, and aberrant behavior. These individuals show an equivalent variety in neuroanatomical and physiological differences across the prefrontal cortex (PFC), limbic system, temporal lobe, and basal ganglia. Underconnectivity between and within these regions and abnormalities within the areas themselves lead to a variety of interpersonal deficits such as the inability to perceive social cues, misunderstandings with nonverbal communication, and recognizing people. Generally, autism is considered to be a multifaceted disorder with both genetic and environmental influences. Many of these genes have been involved in both axon motility and synaptogenesis (Chang et al., 2015; Geschwind, 2011; Voineagu et al., 2011). While ASD has a wide spectrum of physiological and developmental variabilities, it is acknowledged that some individuals are able to lead full and independent lives, while the impact on others on the spectrum may be severe with significant impacts to their quality of life (Farley et al., 2009). This spectrum can include a wide variety of comorbid afflictions including anxiety, depression, seizures, and intellectual disabilities (Gillott et al., 2001; Newschaffer et al., 2007; Tuchman and Rapin, 2002). Among these issues are a variety of arousal regulation problems, attention reorientation deficits, and sensory modulation difficulties.

Current Center for Disease Control (CDC) data suggest that 1 out of 59 children receive an autism diagnosis in the United States (Baio, 2018). ASD prevalence has increased from approximately 1 in 150 children from 2000 to 2002 to 1 in 68 from 2010 to 2012 in 8-year-old children across US communities (CDC, 2007a,b, 2009, 2014; Christensen et al., 2016). This represents a near doubling of the ASD prevalence over that time period. While ASD prevalence

Arousal in Neurological and Psychiatric Diseases. https://doi.org/10.1016/B978-0-12-817992-5.00006-4

has increased over this time, this may be partially due to diagnosing ASD at younger ages and changes in DSM diagnostic criteria. However, increased risk factors are still likely present (Elsabbagh et al., 2012; Fombonne, 2009). On average, males have shown to be affected by ASD at rates two to three times greater than females that may be due to an underlying diagnostic bias leading to underdiagnosis of females with ASD (Baron-Cohen et al., 2011; Kim et al., 2011; Mattila et al., 2011; Saemundsen et al., 2007). The lifetime financial costs of ASD can reach nearly $2.4 million per family with psychiatric therapy and special education (Buescher et al., 2014). Due to the increasing prevalence of ASD, greater clinical and research efforts have been placed on identifying and treating patients while attempting to understand the underlying physiology.

Background

The term "autism" was first coined by Swiss psychiatrist, Paul Eugen Bleuler, to describe schizophrenia symptoms in 1912 (Bleuler and Bleuler, 1912). This was later revised closer to its modern definition by Hans Asperger when he used it to describe child psychology in 1938. Asperger later reported this pattern of behaviors as "autistic psychopathy." These behaviors included stereotypical movement and habits, the lack of association with their peer group, and little regard to adult authority in what ultimately became known as the Asperger syndrome (Asperger, 1944). The modern sense of autism was described in 1943 by Leo Kanner who described children with an inability to form "biologically provided affected contact with people" (Kanner, 1943).

ASD is a class of neurodevelopmental disorders chiefly characterized by impairments in communication and social skills and repetitive behaviors. Particularly, low-level abnormalities are present such as trouble-modulating sensory input (i.e., sensory gating deficits), irregular attention, and difficulty with regulating arousal levels. These individuals may range from having no direction whatsoever to being acutely overfocused (Allen and Courchesne, 2001; Hermelin and O'Connor, 1970). Overall, autism is a heterogeneous and pervasive developmental disorder that has extensive effects on sensory and arousal regulation systems. The DSM-5 characterizes ASD by early-onset deficits in social communication and interaction in the presence of repetitive, restricted behaviors, interests, or activities (American Psychiatric Association, 2013). An additional cooccurring condition is atypical language development. The DSM-5 broadened the category of "autism" into a wider spectrum with low functioning individuals on one end exhibiting general cognitive impairment to high functioning individuals at the other end of the spectrum exhibiting trouble with social interaction. This umbrella spectrum includes other previously separate disorders such as the Asperger syndrome. Many reports note the lowest functioning sufferers exhibit poor social communication while normally being nonverbal. These individuals exhibit a variety of self-injurious behavior, aggression, and tantrums toward their caretakers (Adler et al., 2015; Boser et al., 2002; Le and Lohr, 2012).

A variety of risk factors have been associated with autism including biological, genetic, and environmental aspects. The underlying etiology of ASD is multifaceted and results from an interaction of both genetic abnormalities and environmental factors. Premature birth and prenatal stress are correlated with increased autism risk. Additional risk factors include maternal consumption of thalidomide and valproic acid (Christensen et al., 2013; Strömland et al., 1994). Many of the genes involved in autism are generally related to forming, stabilizing, and maintaining working synapses. A combination of gene defects and mutations is been found in 10%–20% of ASD-afflicted patients (Herman et al., 2007; Miles, 2011).

These potentially susceptible genes are mainly found on chromosomes 2q, 7q, 15q, and 16p according to linkage studies (Wassink et al., 2004). Pathological conditions involving altered brain growth, synaptic and neurite morphology, and neuronal connectivity have been implicated by these chromosomal disturbances (Casas et al., 2004; Dykens et al., 2004; Weiss et al., 2008). Genes that are considered potential factors for ASD include gamma-aminobutyric acid (GABA)–associated genes, serotonin transporter genes, and the ubiquitin-protein ligase (UBE3A locus). Some work has suggested a correlation between ENGRAILED 2, a developmental patterning gene responsible for cerebellum foliation across the mediolateral axis, and increased ASD susceptibility in up to 40% of cases (Gharani et al., 2004). Monozygotic and dizygotic twin studies and family studies suggest heritability estimates of approximately 70%–80% (Bailey et al., 1995; Rosenberg et al., 2009). Concordance rates between monozygotic twins approach 82%–92%, while dizygotic twins are approximately 1%–10% (Hallmayer et al., 2011).

What is truly fascinating about ASD is that no specific gene is able to account for the bulk of ASD cases with the most common genetic mutations only accounting for 1%–2% of ASD cases (Abrahams and Geschwind, 2008). This results in an incomplete understanding of ASD's extreme heterogeneity with underlying molecular mechanisms that include changes to synaptic vesicle release and neurotransmission, RNA processing and splicing, synaptic structure, protein translation, neuronal migration, and cell adhesion. Many of the mutations that have been associated with ASD most frequently affect neural development from early in utero stages to early childhood. These mutations operate by interfering with synaptogenesis and axon motility that ultimately results in abnormalities across all levels, from microstructural to macrostructural with the accompanying functional deficits. Among the extensive variety of mutations, we are still uncertain whether there is some set of common molecular pathways that result in ASD or if these mutations represent a diverse array of underlying pathological etiologies. Regardless of the underlying molecular mechanism(s), these disruptions ultimately result in abnormal activity in brain regions responsible for language, social cognition, and behavioral flexibility. However, many of these ASD-associated genes do not show expression that is restricted to just these regions so that other factors must be in play (Geschwind and Levitt, 2007).

These regions typically require high speed and coherent integration across multiple other regions, thus making them highly susceptible to relatively minor but widespread aberrations. These slight fluctuations in neural transmission can result in a variety of conditions such as those described earlier and lead to gross developmental abnormalities (Geschwind and Levitt, 2007).

A variety of environmental elements have also been associated with ASD including prenatal, perinatal, and postnatal factors (London, 2000). Teratogen exposure, some viral infections, and maternal anticonvulsants are a few prenatal ASD-associated factors (Kern and Jones, 2006; Kolevzon et al., 2007). Perinatal factors such as low birth weight, birth asphyxia, and short gestation length may contribute to ASD (Kolevzon et al., 2007). Paternal age has also been offered as a risk factor for autism where advanced paternal age increases autism likelihood through genetic and epigenetic mechanisms (Durkin et al., 2008; Janecka et al., 2017). While environmental pollutant research has suggested increased autism prevalence, direct association has not been proved (Roberts et al., 2007). Autoimmune disease, hypoxia, viral infection, environmental toxins such as mercury, and stress may contribute to postnatal factors (Ashwood and Van de Water, 2004; Davidson et al., 2004; Kern and Jones, 2006). In addition, epigenetics may also play a role in ASD through DNA methylation and genomic imprinting. Environmentally mediated epigenetic modifications of ASD genes may also be a link between a genetic predisposition to ASD and environmental factors (Miyake et al., 2012).

ASD sufferers typically present with decreased executive functioning and emotional regulation with inflexibility and lacking social reciprocity (Stuss and Knight, 2012). These issues are generally associated with the frontal lobe and its connectivity with other regions such as the striatum, basal ganglia, and insula. Neuroimaging and neurophysiological work has been done to differentiate the various regions of PFC responsible for these different executive functioning aspects when they are affected by ASD. Some reports have demonstrated that excessive brain growth rates are found in infants with ASD. Autistic children have shown to have roughly 90% larger average brain volume by age 2–4 years when compared with control children through magnetic resonance imaging (Akshoomoff et al., 2002; Courchesne, 2004; Sparks et al., 2002). In particular, this excessive growth is focused in the frontal cortex (Courchesne et al., 2011b; Herbert et al., 2004), with enlargement being particularly prevalent in the dorsolateral PFC (Carper and Courchesne, 2005; Hua et al., 2013). Postmortem and MRI results have revealed that the PFC of young autistic children contains excess neurons with overall 12% greater cerebral gray matter volume than neurotypical children. In neurotypical children, cortical development typically continues through early childhood with the frontal cortex reaching its maximum volume between 10 and 12 years of age as it slowly matures. This maturation does not happen in ASD children. Their cortex will rapidly overexpand during early childhood, but further growth and maturation will be arrested as the child grows (Akshoomoff et al., 2002; Courchesne, 2004). These children have also shown increased cortical and cerebellar white matter when compared with controls (Courchesne et al., 2011a; Schumann et al., 2010). Normally, this proliferation of neurons and synapses is eliminated through a process of pruning as the brain normally develops. This process finely tunes neural circuits and supports learning and motor functions as a child grows. This fine-tuning is disrupted in autism cases. The overproliferation of neurons and their associated processes fails to be pruned and ultimately produce faulty connectivity during adolescence and adulthood (Courchesne et al., 2011a).

In addition to increased brain volume in children, ASD individuals (3–39) show accelerated cortical thinning based on cortical thickness measures. This suggests accelerated expansion during early childhood followed by accelerated thinning in later childhood and decelerated thinning in early adulthood (Zielinski et al., 2014). ASD individuals tend to have age-related and region-specific thinner cortices with correspondingly reduced surface area based on volumetric analysis. This suggests differences in developmental trajectories during maturation in ASD individuals (Ecker et al., 2014).

The frontal cortex

The frontal cortex is particularly involved in social cognition in conjunction with the anterior cingulate cortex, superior temporal sulcus, and temporal poles. The frontal cortex helps to control and monitor action, to monitor the results of punishment and rewards, and in social cognition (Amodio and Frith, 2006). Individuals with deficits in frontal lobe function exhibit language, social, and higher-order cognitive dysfunction (Stuss and Knight, 2012). The PFC has been reported to be related to ASD and includes Brodmann areas 8, 9, 10, 11, 44, 45, 46, and 47 (Preuss, 1995). This region receives multiple inputs from brain stem arousal systems and sends out a multitude of connections to other cortical, subcortical, and brain stem regions (Alvarez and Emory, 2006; Robbins and Arnsten, 2009). The PFC itself is subdivided into the medial and lateral PFC. The medial PFC has projections to emotion-associated regions such as the amygdala, memory-associated regions such as the hippocampus, and supplementary sensory regions such as the temporal cortex (Wood and Grafman, 2003). This extensive interconnected network between the medial PFC and other regions makes the medial PFC a major factor in understanding the underlying physiology of how social interaction and cognition are affected by ASD. In addition, the medial PFC is involved in fear learning through synaptic projections to the basolateral amygdala (Quirk and Mueller, 2008; Sotres-Bayon and Quirk, 2010). The medial PFC in conjunction with the amygdala may be implicated in ASD-associated fear and anxiety as well as deficits in memory processing (Bishop, 2007; Fuchs et al., 2007; Mashhoon et al., 2010; Quirk and Beer, 2006).

Diffusion tensor imaging (DTI) shows white matter differences present in ASD that persist into adult life. These differences are localized to major association and commissural tracts of the frontal lobe (Catani et al., 2016). DTI has also shown that the frontal fiber tracts display abnormal fractional anisotropy and volume in intrafrontal and interfrontal pathways involving speech, cognition, language, social, and behavior control (Solso et al., 2016). These abnormalities may then impede neuronal integration between the left and right frontal lobes. Other studies have identified abnormalities in cortical folding and enlarged gyrification of the frontal lobe in adolescents with ASD (Hardan et al., 2004; Levitt et al., 2003). Cytoarchitectural analysis has pointed to patches of both prefrontal and temporal cortices of boys and girls that exhibit abnormal cortical structure with some layers missing and altered or missing neuronal subtypes (Stoner et al., 2014).

The amygdala

The amygdala in particular has been affected in ASD cases through multiple neuroimaging and neuropathological studies. These studies have given strong implications that link the amygdala with abnormal aggressive and social behaviors in ASD individuals (Sah et al., 2003). The basolateral amygdala itself comprises reciprocal connections with the orbitofrontal cortex, anterior cingulate cortex, and medial PFC and responds to the actions of others as well as faces (Adolphs, 2010; Brothers et al., 1990). The centromedial amygdala in turn outputs to the hypothalamus, thalamus, ventral tegmental area, and reticular formation (Sah et al., 2003). Individuals with ASD demonstrate similarity to individuals with amygdala lesions in that emotional memory content, eye gaze, and fear processing are modulated. This is because the amygdala receives a variety of somatosensory, auditory, and visual inputs while sending efferents to the stria terminalis and ventral amygdalofugal pathways. Finally, the amygdala plays a major role in the limbic system as a loop in the cortico-striato-thalamo-cortical circuit (Alexander et al., 1986). The PFC is generally believed to regulate the amygdala, and disruption of PFC-amygdala pathways may result in autistic behavioral states and disrupted memory processing (Bishop, 2007; Quirk and Beer, 2006). Postmortem studies have compared the amygdala of ASD individuals with age- and sex-matched controls to show increased cell density and diminished neuron size in both medial, central, and cortical amygdala nuclei (Kemper and Bauman, 1998; Bauman and Kemper, 2006; Rapin and Katzman, 1998).

Direct activation of the basolateral amygdala in rats by enhancing glutamate or corticotropin-releasing factor or by blocking tonic GABAergic inhibition results in a reduction of social behaviors (Sajdyk and Shekhar, 2000). However, blocking amygdala excitability with glutamate antagonists or lesioning the region results in increased paired social interactions (Sajdyk and Shekhar, 1997). Similarly, humans with amygdala lesions demonstrate impaired social judgment analogous to that seen in ASD individuals (Adolphs et al., 1994; Young et al., 1996). Temporal lobe tumors have also suggested links between the amygdala and ASD as patients with temporal lobe damage express autistic symptoms (Hoon and Reiss, 1992; Taylor et al., 1999). MRI studies have demonstrated decreased amygdala, hippocampal, and parahippocampal gyrus volumes when comparing ASD individuals with gender-, age-, and verbal IQ–matched controls (Howard et al., 2000). Single-photon emission computed tomography has also showed reduced cerebral blood flow in the right amygdala and hippocampus as well as bilateral superior temporal gyri, insula, and left PFC (Ohnishi et al., 2000).

The nucleus accumbens

The nucleus accumbens (NAc) is a major component of the ventral striatum and is a key structure in mediating emotional and motivation processing, modulating reward and pleasure processing, and serving a key limbic-motor interface (Cohen et al., 2009; Salgado and Kaplitt, 2015). The NAc has a predominant role in fine-tuning reward and pleasure processing. The anticipation of reward stimulates the NAc in conjunction with other limbic regions, while pleasure activates the NAc, putamen, caudate, and ventromedial PFC (Ernst et al., 2004; Knutson et al., 2001; Wacker et al., 2009). Research investigating the neural correlates of reward learning in ASD children found decreased ventral striatal responses during social, but not monetary reward learning. This implied that activity in the ventral striatum predicted social reciprocity within the control group and not within the ASD group (Scott-Van Zeeland et al., 2010). In addition, autism mouse models with decreased serotonin drive from the raphe nuclei exhibit decreased social behavior. Modulating the serotonin release with optogenetic stimulation rescues these social deficits and demonstrates the requirement of serotonin in NAc for social behavior (Walsh et al., 2018). This also demonstrates the potential benefit of drugs that modulate serotonin levels in treating ASD. Moreover, these results point to the NAc as a potential target for deep brain stimulation (DBS) due to its central role in processing pleasure and reward (Cohen et al., 2009; Park et al., 2017).

The striatum itself has shown involvement in ASD due to motility disturbances demonstrated in neurological testing. ASD patients demonstrated signs of basal ganglia dysfunction including dystonia of the extremities, bradykinetic abnormalities, and choreoathetoid movements (Damasio and Maurer, 1978; Maurer and Damasio, 1982). MRI studies have demonstrated ASD-related changes in striatal volume and alterations in caudate size (Estes et al., 2011; Langen et al., 2007; Sears et al., 1999). Functional studies have pointed to increased connectivity between the caudate nucleus and cortical areas such as prefrontal, premotor, and parietal areas during task-specific and resting state paradigms (Di Martino et al., 2011; Turner et al., 2006). As striatal circuits are an intersection of sensory input representations, motor control, motivation, and prior experience as well as serving as a major regulator for midbrain dopamine output, simple disruptions may serve to initiate some of the behavioral changes seen in ASD (Kupchik et al., 2015; Leblois et al., 2010; Lerner et al., 2015; Schultz, 2007).

Arousal and attention

ASD is characterized by a constellation of abnormalities including sensory modulation difficulties, regulation problems, and difficulty with attention reorienting. In particular, ASD individuals may be overfocused on the one hand or have no direction on the other hand (Allen and Courchesne, 2001). These individuals may have abnormally low or elevated levels of autonomic and behavioral arousal (Hirstein et al., 2001; Kinsbourne, 1987). Generally, ASD individuals demonstrate difficulties with sensory modulation. These individuals exhibit hypo- or hyperresponsiveness to sensory stimuli across sensory modalities (Ben-Sasson et al., 2008). Arousal and attentional abnormalities are evident in the earliest stages of autism in infants with high ASD risk (Elison et al., 2013; Elsabbagh et al., 2013; Zwaigenbaum et al., 2005). The basic sensory processes appear to be intact in ASD individuals, but it is their abnormal responses to the detected sensory stimuli that point to irregularly exaggerated or reduced responses to environmental variation and stimuli.

Arousal can be divided into phasic and tonic states depending on reticular activating system (RAS) and thalamic input state (Sturm and Willmes, 2001). Phasic arousal typically defines rapid energetic reaction to specific stimuli, while tonic arousal is composed of relatively slow changes in arousal states during sleep and waking (Combs and Polich, 2006). While multiple arousal systems are involved in sleep and waking, they serve to regulate environmental awareness and activate and modulate corresponding behavior (Dringenberg and Vanderwolf, 1998; de Lecea et al., 2012; Robbins, 1997). With the complexity of these systems, it is unsurprising that they may be disturbed in ASD cases. ASD-affected individuals frequently demonstrated sleep disturbances (Tudor et al., 2012). Decreased or increased reactions to stimuli are frequently present in ASD patients along with abnormally elevated or diminished levels of vigilance during wakefulness (Hutt et al., 1964; Patel and Greydanus, 2012). The vigilance level might also fluctuate over time, possibly in conjunction with variable ascending activation from the RAS (Hermelin and O'Connor, 1970; Kinsbourne, 1987).

Arousal regulation abnormalities are likely present in ASD children well before traditional diagnosis and social irregularities show. Typically, these individuals will show higher incidences of sleep irregularities than neurotypical developing children or children exhibiting cognitive deficits without autism (Dahlgren and Gillberg, 1989). Toddlers who were later diagnosed with ASD exhibited elevated proneness and irritability in response to distress (Bryson et al., 2007). Infants who were in the neonatal intensive care unit and were later diagnosed with ASD exhibited an unusually high preference for more arousing visual stimulation regardless of any central nervous system injury (Cohen et al., 2013; Karmel et al., 2010).

ASD-affected individuals have been frequently described as lacking in attention to people and will either overfocus with difficulty redirecting or have no direction whatsoever (Allen and Courchesne, 2001; Ames and Fletcher-Watson, 2010; Grandin and Scariano, 1996). While difficulty reorienting attention is likely generalized due to issues with both social and nonsocial subjects (Baranek et al., 2013; Dawson et al., 1998; Harris et al., 1999; Townsend et al., 1996), some work has shown that ASD children have diminished ability to modulate their alertness levels and are slower to shift their attention (Keehn et al., 2010). Typically, issues with shifting attention and attention disengagement have been found in infants with high risk for ASD that ultimately predicted an ASD diagnosis (Elison et al., 2013; Elsabbagh et al., 2013; Zwaigenbaum et al., 2005). The attentional deficits present when attempting to orient to social or nonsocial stimuli may ultimately induce a variety of follow-on social problems with joint attention, language development, and communication (Baranek et al., 2013; Mundy and Jarrold, 2010; Mundy and Rebecca, 2000). The systems involved in disengagement and reorienting include the ventral frontal cortex and temporoparietal junction in combination with subcortical arousal systems. Through these systems, autonomic and motor responses are triggered through novel sensory stimuli to allow rapid and accurate responses to the environment (Corbetta et al., 2008; Corbetta and Shulman, 2011). Finally, while it appears that ASD deficits in attention disengagement is likely a widespread failure in attention networks since abnormalities are found in social and nonsocial stimuli, the widespread and nonspecific nature of these deficits suggests early modulation of neuronal development.

Sensory modulation

Individuals with ASD have repeatedly demonstrated variable sensory modulation ranging from hypersensitivity to hyposensitivity across multiple modalities (Crane et al., 2009; Harrison and Hare, 2004; O'Neill and Jones, 1997). Arousal itself is a general term to describe the degree of vigilance and alertness of an individual and can be described as a state of wakefulness with increased motor activation, emotional reactivity, enhanced cognitive processing, and responsiveness to sensory inputs (Pfaff and Kieffer, 2008). Perturbations in humans that trigger hyperarousal are considered symptoms of a variety of disorders such as schizophrenia, anxiety disorder, and ASD, whereas hypoarousal is a condition in aggressive behaviors and attention deficit hyperactivity disorder (Carlsson, 1995; Haller et al., 2005; Kirov et al., 2007). In contrast, ASD individuals may demonstrate both hyper- and hypoarousal. Ultimately, aberrant sensory processing can lead to hyperreactivity and subsequent negative responses such as hypervigilance, avoidance, and distress to common sensory stimuli (Mazurek et al., 2013; Reynolds and Lane, 2008).

Hypoactive responses, on the other hand, may result nonresponsiveness or unawareness to otherwise normal stimuli (Miller et al., 2007), and some of these individuals may seek more intense sensory experiences due to hypoactive sensory responses (Ornitz, 1974).

Arousal itself is regulated by multiple neural populations where increased activity directly correlates with increased arousal (Jones, 2003). These regions include hypocretin neurons in the lateral hypothalamus and histaminergic tuberomammillary nucleus neurons in the posterior hypothalamus. The brain stem contains noradrenergic locus coeruleus neurons, neuropeptide S neurons, and serotonergic dorsal raphe nucleus neurons. Cholinergic neurons in the pedunculopontine, laterodorsal tegmental nuclei, and basal forebrain are involved. Finally, glutamatergic and gamma-aminobutyric acidergic neurons located in both hypothalamic and brain stem regions play roles in arousal. All of these systems are fine-tuned to respond to diverse situation and sensory inputs. Ultimately, changes in the states and thresholds of these systems are at the core of psychiatric disorders with arousal involvement. The fine balance of inhibitory and excitatory networks allows these systems to balance out in both spatial and temporal domains.

The variability in sensory responsiveness found in ASD individuals can lean toward either ends of the spectrum of hyper- to hypoarousal and showing high variability within a single individual depending on stimulation, behavior, sensory modality, and context (Baranek et al., 2006; Hirstein et al., 2001). These individuals may be oversensitive to specific stimuli. Paradoxically, the same individual may underreact or not react at all to noxious stimuli such as painful injuries (Grandin and Sacks, 2006; Ornitz, 1988). Abnormal sensory gating has been reported in children as early as 6–12 months of age and may be one of the earliest signs of ASD (Freeman, 1993; O'Neill and Jones, 1997). Abnormal sensory gating persists and remains prevalent across the life span as these children age into adults (Crane et al., 2009; Leekam et al., 2007). The underlying neural basis for these paradoxical sensory reactions is poorly understood. Functional neuroimaging studies suggest increased activity in the sensory areas of the brain that are normally associated with stimulus-driven processing. Concurrently, decreased activity is found in areas normally associated with higher cognitive processing. These individuals show high activation in the ventral occipital areas with abnormally low activation in prefrontal and parietal regions (Ring et al., 1999).

MRI studies have shown that the brain stem and cerebellum in ASD children and adults develop rapidly and increase in size with age but are still significantly smaller compared with controls (Hashimoto et al., 1995). These data, coupled with data from auditory evoked potentials, implied developmental brain stem abnormalities as a core factor in aberrant ASD sensory gating (Klin, 1993; Ornitz, 1988). The brain stem itself is composed of many nuclei with widespread projections to both cortical and subcortical structures. The main subdivisions of the brain stem—the diencephalon, mesencephalon, pons, and medulla—are composed of cranial nerve nuclei, the reticular formation, and pontine nuclei. In addition, white matter fiber tracts convey sensory information to and motor information from the cerebrum. Finally, the brain stem itself is an important link between the cerebral hemispheres, the medulla, and the cerebellum to regulate many bodily functions. ASD-affected children display significantly longer pupillary light reflex latency in response to stimuli and modulated heart rate variability. These results suggested ASD abnormalities in the autonomic nervous system and, by extension, the brain stem (Daluwatte et al., 2013). Some models of behavior and emotion regulation have pointed to brain stem abnormalities as a key factor in both the core and associated symptoms of ASD. Behavioral regulation requires the effective integration of multiple system including the cerebral cortex and limbic system with the brain stem underlying everything (Panksepp, 2005; Tucker et al., 2005). Other conceptual models have integrated the cortex, brain stem, and limbic system by proposing three levels of processing and analysis: (1) brain stem–related physiological regulation of cyclic processes and sensory integration; (2) emotional and attention regulation capabilities that draw on brain stem integration with the limbic systems; and (3) higher-level outcomes such as socioemotional self-regulation, inhibitory control, and cognitive processing that draw on the intactness of the brain stem and limbic networks (Geva and Feldman, 2008). Given the central nature of the brain stem as a bridge for different neuronal structures and as both an early director of and processing region for sensory information, it is no surprise that brain stem insult could result in altered arousal and sensory perception in ASD patients.

Among altered sensory responses, altered sensory gating and arousal has also been described as a potential factor in stereotyped behavior seen in ASD-affected children. ASD children typically display stereotypical or repeated behaviors such as rocking, twirling, or spinning (Rogers and Ozonoff, 2005; Schaaf et al., 2011). Generally, the PFC, basal ganglia, and particularly the NAc are implicated as traditional targets for stereotyped behaviors found in many psychiatric disorders such as obsessive compulsive disorder, the Tourette syndrome, and potentially autism (Milad and Rauch, 2012; Rasmussen and Eisen, 1992; Saxena and Rauch, 2000). In ASD, repetitive movements may reflect attempts to lower or increase arousal. Some ASD children exhibiting stereotypical behaviors had greater sensory processing difficulties when compared with neurotypical children (Joosten and Bundy, 2010). In contrast, ASD children exhibiting rigid behaviors such as refusing to transition between activities and maintaining a rigid daily routine or a general preference for uniformity may be driven by hypo- or hyperarousal (Lane et al., 2010). These children may be bothered by noisy and complex environments or upset by tags on their clothing or simply unexpected touching (Liss et al., 2006). Additionally, the lack of response to sensory stimuli by hyperaroused ASD children may directly contribute to autism development by interrupting basic orienting responses that are normally foundational to joint attention skill development (Baranek et al., 2013). Research has also shown that low sensory responsiveness in ASD children was correlated with low mental age and lower levels of joint attention and language.

Finally, increased anxiety in ASD children can also be linked to hypervigilance to sensory stimuli. These children can easily become overaroused due to sensory hyperreactivity and thus have trouble regulating negative emotions and eventually result in anxiety (Bellini, 2006; Green and Ben-Sasson, 2010). Anxiety disorder rates for ASD children range from 18% to 87% (Gadow et al., 2004; Muris et al., 1998), while preadolescent rates are believed to be between 3% and 24% (Cartwright-Hatton et al., 2006).

Autism and the P50 potential

As previously described, ASD has widespread effects on both attention and arousal with over 96% of ASD children reporting hyper- or hyposensitivity across sensory modalities (Crane et al., 2009; Leekam et al., 2007; Minshew et al., 2002). Auditory sensation has been particularly noted for variable sensitivity in ASD individuals. This can be readily seen in the electroencephalogram (EEG) of affected patients. The P50 event-related potential (ERP) is a human midlatency auditory evoked response, recorded at the vertex and evoked by a click stimulus. The P50 generally occurs with a latency of 40–70 ms and is frequently referred to as the P1 potential as it is the first positive wave after early brain stem and primary auditory evoked responses (Skinner et al., 2004). The P50 response is directly related to the nonlemniscal ascending pathways and possesses three primary characteristics: (1) It is dependent on sleep/wake state. The P50 potential is only apparent during rapid eye movement (REM) sleep or waking when cortical EEG synchronization of fast 30–40 Hz oscillations is present. However, it is absent during slow-wave sleep or when cortical synchronization is <15 Hz (Erwin and Buchwald, 1986a). (2) The P50 is blocked by muscarinic cholinergic antagonists such as scopolamine, and so, it must be partially mediated by cholinergic neurons in the brain stem, specifically the pedunculopontine nucleus (PPN) (Buchwald et al., 1991; Garcia-Rill et al., 2011). (3) The P50 cannot be generated by stimulation rates >2 Hz and thus shows rapid habituation. This indicates that the P50 is likely generated by the nonspecific sensory system and particularly the RAS (Erwin and Buchwald, 1986a,b). The P50 potential is the only early latency auditory evoked potential that is diminished and disappears during deep sleep stages, which suggests that the P50 is functionally related to arousal states and likely generated by subcortical regions related to arousal (Kevanishvili and Von Specht, 1979). The RAS and the PPN in particular are among the key generators for this potential as the PPN represents the major concentration of cholinergic mesopontine cells and is preferentially active during waking and REM sleep while being inactive during slow-wave sleep (Steriade and McCarley, 1990).

P50 potential measures typically utilize a paired click stimulus with two clicks separated by a short interstimulus interval of 250–1000 ms, followed by a longer intertrial interval. The rapid habituation of the RAS causes a marked decrease in the amplitude of the response to the second click-evoked potential. The ratio of the amplitudes of the two click-evoked potentials provides a measure of sensory gating. Decreased amplitude correlates with increased inhibitory brain stem function and normal habituation (Adler et al., 1999; Nagamoto et al., 1989). Thus variation in the P50 potential is indicative of sensory and habituation abnormalities across various psychiatric disorders. For comparison, the P50 potential habituation is decreased in schizophrenia, indicating decreased habituation and sensory gating (Adler et al., 1982; Freedman et al., 1983). These individuals demonstrate hyperarousal, reduced slow-wave sleep, and decreased REM sleep latency while having an exaggerated startle response and hallucinations (Garcia-Rill, 1997; Reese et al., 1995). Similarly, decreased sensory gating is reported in posttraumatic stress disorder in both male combat veterans and female rape victims (Gillette et al., 1997; Skinner et al., 1999). Finally, the P50 potential can also be diminished in disorders such as the Alzheimer's disease and narcolepsy that are characterized by daytime somnolence, cataplexy, sleep paralysis, and hypnagogic hallucinations (Boop et al., 1994; Buchwald et al., 1989). In short, the P50 potential response is upregulated, as represented by increased amplitude and/or decreased sensory gating, in disorders with the upregulation of RAS output and downregulated in disorders presenting with decreased RAS output.

While research on P50 potential suppression is sparse, the P50 potential amplitude is typically suppressed in ASD children (3–8 years) and in low-IQ children (<72). However, this suppression improves with age (Buchwald et al., 1992; Orekhova et al., 2008). ASD children with normal IQs showed no P50 potential suppression. The decrease in P50 potential habituation amplitude may demonstrate impairments in the central inhibitory circuits that modulate sensory inputs. However, ASD individuals may have abnormally increased ratios of excitation to inhibition across sensory, mnemonic, social, and emotional systems (Rubenstein and Merzenich, 2003). While P50 potential studies in nonretarded children showed no P50 potential suppression, studies in ASD adults also did not show significant P50 potential changes (Kemner and van Engeland, 2002; Magnée et al., 2009). Later studies have shown that children with ASD had no significant differences in P50 potential suppression or amplitude. However, the same studies showed that the ASD subcategory, the Asperger disorder, demonstrated significantly decreased P50 potential habituation amplitude when compared with controls. The authors also showed a positive correlation between P50 potential habituation amplitude and anxiety (Madsen et al., 2015). Care must be taken to differentiate those with ASD and those with schizophrenia who typically exhibit marked P50 potential abnormalities (Chang et al., 2011). Considering the relatively few studies on diminished P50 potential habituation in ASD individuals, caution is needed when drawing conclusions.

Autism and sleep

ASD-affected individuals are characterized by a history of sleep pattern disruptions ranging from mild to extreme. These problems can vary from difficulty settling down for sleep to lengthy episodes of waking with confusion, screaming, and crying. In addition, these individuals can display bruxism, early morning waking, bed-wetting, and irregularities of the sleep/wake rhythm (Richdale and Prior, 1995; Wiggs and Stores, 2004). Generally, these disorders can be neurobiological, medical, or behavioral in origin. Neurobiologically, they may result from genetic or neurotransmitter variation such as modulation of the melatonin pathway. Medical basis for neural disruptions can include issues such as epilepsy and biological issues that disrupt sleeps such as gastrointestinal issues. Psychiatric conditions may include anxiety, attention deficit hyperactivity disorder, and depression that affect sleep patterns.

The most commonly reported sleep issues involve sleep onset and maintenances (Biederman et al., 2001; Elia et al., 2000; Godbout et al., 2000). General associations between sleep disturbances and cognitive and behavioral consequences are well established. Typically, disturbed sleep is associated with impaired vigilance, memory, learning, and creative thought across multiple studies (Gozal, 1998; Horne, 1988; Quine, 1991). Among the most commonly documented sleep problems are difficulty in falling asleep that is reported by nearly half of research respondents. This in turn fits with earlier estimates of 43%–83% of cases (Richdale, 1999). Moreover, studies on ASD children report 50%–80% with disturbed sleep, typically related to sleep onset or maintenance, in comparison with 11%–37% in neurotypical children (Couturier et al., 2005; Owens et al., 2000; Souders et al., 2009; Stein et al., 2001). Disorders of partial arousal such as sleep terrors and confusional arousal accompanied by screaming with subsequent resumption of sleep are among the parasomnias present in ASD children (Ming et al., 2009). These children typically had concurrent autonomic arousal such as tachycardia, elevated respiratory rate and amplitude, and limb movements. Other disorders of partial arousal such as sleep walking are not typically seen in ASD individuals.

Sleep state markers are noninvasively and objectively measured and referenced to typical developmental norms when analyzing sleep architecture. Polysomnography is usually used to study the basic components of sleep and employed across a variety of diseases and disorders including depression and narcolepsy (Kryger et al., 2011). While a complete understanding of REM sleep function and its overall relationship to cognitive function is still under study, REM sleep rebound after an intensive learning session implies that this stage is involved in cognitive processing and could be a potential indicator for brain plasticity (McGrath and Cohen, 1978; Mirmiran et al., 1982). Other work has implicated REM sleep in memory consolidation and suggests that REM sleep may be necessary to normal cognitive function and emotional processing in memory systems (Maquet, 2001; Stickgold, 2005). Converging evidence also suggests that REM sleep serves a developmental role as the total REM sleep time decreases over the human life span. Nearly 50% of an infant's sleep is composed of REM sleep as opposed to 20%–25% in adults (Hobson and Pace-Schott, 2002). This overabundance of REM sleep declines rapidly during the first few years of life and approaches adult levels by puberty (Hobson, 1995). REM deficiencies have also been found to have differential effects on age-dependent neuronal plasticity markers (Lopez et al., 2008). ASD-affected children generally show decreased REM sleep (Buckley et al., 2010). These individuals demonstrated immature sleep organization, decreased quantity, abnormal twitches, and undifferentiated sleep (Diomedi et al., 1999; Elia et al., 2000; Tanguay et al., 1976; Thirumalai et al., 2002).

Both children and adults with ASD have been reported to suffer from insomnia, which is defined as difficulty falling or staying asleep (Gail et al., 2004; Krakowiak et al., 2008; Limoges et al., 2013). Polysomnography reports indicated longer sleep latencies and lower sleep efficiency in ASD children who were described as "poor sleepers" by their parents (Malow et al., 2006). In addition, other results showed that ASD children woke up more during the night, fell asleep later, and woke very early or very late (Wiggs and Stores, 2004). These children also demonstrate disordered breathing by an increase in apnea-hypopnea index (AHI). This is compared with neurotypical children who may have less than one apnea/hypopnea per hour. Children with ASD show higher levels of snoring that point to mild obstructive sleep apnea (Ming et al., 2009). Older ASD individuals also show a variety of sleep disturbances based on questionnaires, sleep diaries, and narrative. In particular, these individuals have difficulty in initiating and maintaining continuous sleep (Tani et al., 2003). Work using actigraphy shows significantly lower sleep efficiency and longer sleep latencies, suggesting underreporting in adolescents, young adults, and possibly older populations (Øyane and Bjorvatn, 2005).

While a cause of delayed sleep is difficult to narrow down to simply insomnia or delayed sleep phase circadian rhythm disorder, ASD individuals typically do not fit traditional diagnostic criteria (American Academy of Sleep Medicine, 2014). ASD children are believed to be particularly vulnerable to circadian rhythm disorders as decreased sleep and difficulty waking in the morning are potential problems associated with circadian disturbances in the synchronization of neurobehavioral and endocrine functions. Data suggesting abnormal cortical and melatonin levels in children with ASD also support this idea of altered circadian rhythms (Corbett et al., 2006; Ćurin et al., 2003; Richdale and Prior, 1992; Ritvo et al., 1993).

When examining potential mechanisms for these sleep-wake disturbances, one must consider the role of brain stem circuitry in the regulation of waking and REM sleep. The control of waking and REM sleep arousal states is governed by a complex interplay of GABA, glutamate, serotonin from the dorsal raphe, norepinephrine from the locus coeruleus, and acetylcholine from the PPN. Acetylcholine is the primary regulator of REM sleep and waking and was proposed to work in conjunction with serotonin and norepinephrine to act on pontine and other midbrain structures via REM-on and REM-off cells (McCarley, 2007). However, the location of REM-off cells has not been established. In particular, the PPN possesses REM-on, wake-on, and wake/REM-on cells that are modulated by N-type, P/Q-type, or both types of voltage-gated calcium channels operating through the cAMP/protein kinase A (PKA) and Ca^{2+}/calmodulin-dependent protein kinase II (CaMKII) pathways (Luster et al., 2016). Imbalances in any step of these pathways could lead to destabilization of normal waking and REM sleep cycles. Thus the PPN and the cholinergic system are key areas of study when examining abnormal development in ASD-affected brains (Garcia-Rill, 2015; Hohmann and Berger-Sweeney, 1998). It should be noted that the PPN also contains glutamatergic and GABAergic cells (see Chapter 2).

In addition to sleep abnormalities, up to 30% or more ASD individuals present with some variety of EEG abnormality or manifest clinical seizures (Volkmar and Nelson, 1990). These individuals usually exhibit symptomatic epilepsy. Sleep deprivation is already known as a factor in lowering seizure thresholds. Considering the lack of sleep in ASD individuals, it is unsurprising that elevated seizure risk is present. Sleep deprivation typically causes a rebound in slow-wave sleep. Due to the low-frequency/high-amplitude, synchronized electric activity, slow-wave sleep can be closely associated with epileptogenesis with some types of epilepsy only generating seizures during this period (Steriade, 2000, 2006). The majority of the ASD individuals suffer clinically apparent seizures and developmental disorders during the day. However, electric status epilepticus (ESE) of sleep is typically associated with developmental regression (Volkmar et al., 2014). ESE is considered to be any increase in epileptic discharges during sleep. Slow-wave sleep will generally trigger frequent, bilateral 1.5–3-Hz spike-wave complexes (Değerliyurt et al., 2015). Of the various genetic mutations that can cause ASD, many carry an increased risk of epilepsy (Grayton et al., 2012). Examples include the single gene disorder, tuberous sclerosis complex, chromosomal duplication syndrome 15q, and de novo copy number variations, which all may manifest in ASD cases while also leading to seizures and epilepsy. For a more detailed description of epilepsy and arousal, see Chapter 12.

Treatments

Treatments for ASD are wide ranging and are just as heterogeneous as the disorder itself and, therefore, must be individualized. Educational and behavioral interventions are the primary treatment methods for the variety of ASD symptoms (Posey and McDougle, 2001). Simple treatments such as booklet-delivered behavioral interventions have demonstrated effectiveness (Gringras et al., 2012; Montgomery et al., 2004). Pharmacological treatments may include selective serotonin reuptake inhibitors or antidepressants, antipsychotics, atypical antipsychotics, α-2-adrenergic agonists, β-adrenergic antagonists, anticonvulsants, and mood stabilizers (Adler et al., 2015; Stigler et al., 2011). In spite of the long list of pharmacological agents used to treat ASD, no single agent has been effective in improving social communication. With the lifelong nature of ASD, the majority of patients are exposed to multiple combinations of interventions to address the variety of ASD symptoms. Most treatments are directed toward comorbid symptoms rather than the core symptoms associated with ASD (McPheeters et al., 2011). Among the major choices when planning pharmacological treatments for ASD is to balance the treated symptoms with potential other comorbid conditions such as obsessive compulsive disorder. Of these different treatments, antidepressants are the most commonly prescribed. Stimulants and antipsychotics are also prescribed as necessary. However, many treatments commonly used to treat ASD also affect sleep and induce hypersomnolence or insomnia (McPheeters et al., 2011). Among the commonly prescribed drugs are antipsychotics such as risperidone and selective serotonin reuptake inhibitors. Both of these can adversely affect sleep by inducing insomnia or hypersomnolence (Megens et al., 1994; Wichniak et al., 2017).

In spite of the variety of drugs tested with ASD patients, remarkably little work has been done to verify and study the benefit of various medications for ASD (McPheeters et al., 2011). With the variety of medications in use, the antipsychotic, risperidone, and atypical antipsychotic, aripiprazole, are the two best studied in ASD individuals. These drugs are typically prescribed to address problematic behaviors such as aggression, self-injury, and rapid mood changes (Hirsch and Pringsheim, 2016; Kirino, 2014). While not the primary treated behavior, they also decreased repetitive behaviors. However, as mentioned earlier, these medications have significant adverse effects such as weight gain, sedation, and extrapyramidal side effects such as tremors. Thus their use may be limited to ASD individuals with severe impairments or significant risk of injury (McPheeters et al., 2011).

Given the high comorbidity of ASD and sleep disorders, effective treatments are required to maintain the ASD individual's well-being. In addition, sleep disruption itself may provoke many deleterious ASD behaviors. Treatments can follow behavioral and/or pharmacological lines depending on symptoms, comorbid disorders, and other medications. Behavioral interventions including applied behavior analysis and parent-based education have proved beneficial with ASD sleep disorders (Cohen et al., 2014). Further behavioral treatments may include reinforcing correct sleep hygiene to promote sleep initiation and maintenance. Many of these practices focus on relaxing and calming the ASD individual (Dawson et al., 2012). ASD individuals may also suffer from sleep-disordered breathing problems such as obstructive sleep apnea, which will manifest as difficulty breathing, snoring, and repeated arousals from sleep (Gozal, 2001). First-line therapy may include adenotonsillectomy followed by continuous positive airway pressure (CPAP). However, ASD patients require desensitization to CPAP in order for treatment to be done successfully (Malow, 2004).

Pharmacological interventions for insomnia include melatonin treatment. Melatonin is a hormone that regulates sleep-wake cycles and is sold as an over-the-counter dietary supplement with a good safety record. It may be used as a sedative at low doses when given approximately 30 mins before bedtime (Braam et al., 2009; Malow et al., 2012a). Short-term side effects are typically mild, while long-term effects are unknown (Andersen et al., 2008). In addition, melatonin has also been used to treat circadian rhythm disorders when administered at specific time points to shift sleep onset. Other medications such as clonidine, mirtazapine, and gabapentin have been administered in smaller studies and are not approved for insomnia use in children. However, they have demonstrated variable efficacy (Malow et al., 2012a,b). Compounds such as gabapentin and pregabalin have antiepileptic and sedating effects that may help with treating ASD but can also amplify unidentified sleeping disorders (Robinson and Malow, 2013). Antidepressants are also sometimes used to increase total sleep time while decreasing sleep onset latency but are also powerful REM sleep suppressors (Malow et al., 2012a,b). Parasomnias such as partial arousal, night terrors, and REM sleep behavior disorder are typically treated with tricyclic antidepressants or benzodiazepines (Mason and Pack, 2007; Veatch et al., 2015). See Chapter 10 for a description of some of these parasomnias.

Intractable ASD symptoms and associated comorbidities may sufficiently degrade quality of life to the point alternative treatments such as DBS are called for. DBS is used to send controlled electric impulses to targeted brain regions. This technology was originally used to relieve the Parkinson's disease essential tremor but has branched out to treating a variety of symptoms such as obsessive compulsive disorder, the Tourette syndrome, and depression (Benabid, 2003; Hamani et al., 2014; Schrock et al., 2015; Taghva et al., 2013). DBS has been used to treat self-injurious behaviors. The first case targeted the paralaminar, basolateral, and central amygdala as well as the supra-amygdaloid projection system. The second case included the internal capsule and globus pallidus. The third case stimulated only the globus pallidus. All of the cases improved patient behavior. However, the patient receiving internal capsule stimulation regressed after a temporary improvement period (Stocco and Baizabal-Carvallo, 2014; Sturm et al., 2012). DBS in autism targeting the intralaminar thalamus has not apparently been used, despite moderate success in such disorders as the Tourette syndrome.

Ultimately, we see that autism is a highly variable and heterogeneous disorder with effects that vary across the entire central nervous system. Its effects on sleep and arousal are obvious considering the extreme changes to sensory processing systems and many reports of sleep disturbances. There is still an abundance of work remaining in the field of autism and the intersection of autism and arousal to identify and treat the underlying mechanisms and to alleviate the variety of sensory and sleep disturbances that affect the autism spectrum.

References

Abrahams, B.S., Geschwind, D.H., 2008. Advances in autism genetics: on the threshold of a new neurobiology. Nat. Rev. Genet. 9, 341–355.

Adler, L.E., Pachtman, E., Franks, R.D., Pecevich, M., Waldo, M.C., et al., 1982. Neurophysiological evidence for a defect in neuronal mechanisms involved in sensory gating in schizophrenia. Biol. Psychiatry 17, 639–654.

Adler, L.E., Freedman, R., Ross, R.G., Olincy, A., Waldo, M.C., 1999. Elementary phenotypes in the neurobiological and genetic study of schizophrenia. Biol. Psychiatry 46, 8–18.

Adler, B.A., Wink, L.K., Early, M., Shaffer, R., Minshawi, N., et al., 2015. Drug-refractory aggression, self-injurious behavior, and severe tantrums in autism spectrum disorders: a chart review study. Autism 19, 102–106.

Adolphs, R., 2010. What does the amygdala contribute to social cognition? Ann. N. Y. Acad. Sci. 1191, 42–61.

Adolphs, R., Tranel, D., Damasio, H., Damasio, A., 1994. Impaired recognition of emotion in facial expressions following bilateral damage to the human amygdala. Nature 372, 669–672.

Akshoomoff, N., Pierce, K., Courchesne, E., 2002. The neurobiological basis of autism from a developmental perspective. Dev. Psychopathol. 14, 613–634.

Alexander, G.E., DeLong, M.R., Strick, P.L., 1986. Parallel organization of functionally segregated circuits linking basal ganglia and cortex. Annu. Rev. Neurosci. 9, 357–381.

Allen, G., Courchesne, E., 2001. Attention function and dysfunction in autism. Front. Biosci. J. Virtual Libr. 6, D105–D119.

Alvarez, J.A., Emory, E., 2006. Executive function and the frontal lobes: a meta-analytic review. Neuropsychol. Rev. 16, 17–42.

American Academy of Sleep Medicine, 2014. International Classification of Sleep Disorders. American Academy of Sleep Medicine.

American Psychiatric Association (Eds.), 2013. Diagnostic and Statistical Manual of Mental Disorders. fifth ed. American Psychiatric Association.

Ames, C., Fletcher-Watson, S., 2010. A review of methods in the study of attention in autism. Dev. Rev. 30, 52–73.

Amodio, D.M., Frith, C.D., 2006. Meeting of minds: the medial frontal cortex and social cognition. Nat. Rev. Neurosci. 7, 268–277.

Andersen, I.M., Kaczmarska, J., McGrew, S.G., Malow, B.A., 2008. Melatonin for insomnia in children with autism spectrum disorders. J. Child Neurol. 23, 482–485.

Ashwood, P., Van de Water, J., 2004. Is autism an autoimmune disease? Autoimmun. Rev. 3, 557–562.

Asperger, H., 1944. Die "Autistischen Psychopathen" im Kindesalter. Arch. Psychiatr. Nervenkr. 117, 76–136.

Bailey, A., Le Couteur, A., Gottesman, I., Bolton, P., Simonoff, E., et al., 1995. Autism as a strongly genetic disorder: evidence from a British twin study. Psychol. Med. 25, 63–77.

Baio, J., 2018. Prevalence of autism spectrum disorder among children aged 8 years—autism and developmental disabilities monitoring network, 11 sites, United States, 2014. MMWR Surveill. Summ. 67, 1–23.

Baranek, G.T., David, F.J., Poe, M.D., Stone, W.L., Watson, L.R., 2006. Sensory Experiences Questionnaire: discriminating sensory features in young children with autism, developmental delays, and typical development. J. Child Psychol. Psychiatry 47, 591–601.

Baranek, G.T., Watson, L.R., Boyd, B.A., Poe, M.D., David, F.J., et al., 2013. Hyporesponsiveness to social and nonsocial sensory stimuli in children with autism, children with developmental delays, and typically developing children. Dev. Psychopathol. 25, 307–320.

Baron-Cohen, S., Lombardo, M.V., Auyeung, B., Ashwin, E., Chakrabarti, B., et al., 2011. Why are autism spectrum conditions more prevalent in males? PLoS Biol. 9, e1001081.

Bauman, M.L., Kemper, T.L., 2006. The Neurobiology of Autism, second ed. Johns Hopkins University Press, Baltimore.

Bellini, S., 2006. The development of social anxiety in adolescents with autism spectrum disorders. Focus Autism Other Dev. Disabil. 21, 138–145.

Benabid, A.L., 2003. Deep brain stimulation for Parkinson's disease. Curr. Opin. Neurobiol. 13, 696–706.

Ben-Sasson, A., Cermak, S.A., Orsmond, G.I., Tager-Flusberg, H., Kadlec, M.B., et al., 2008. Sensory clusters of toddlers with autism spectrum disorders: differences in affective symptoms. J. Child Psychol. Psychiatry 49, 817–825.

Biederman, J., Faraone, S.V., Hirshfeld-Becker, D.R., Friedman, D., Robin, J.A., et al., 2001. Patterns of psychopathology and dysfunction in high-risk children of parents with panic disorder and major depression. Am. J. Psychiatry 158, 49–57.

Bishop, S.J., 2007. Neurocognitive mechanisms of anxiety: an integrative account. Trends Cogn. Sci. 11, 307–316.

Bleuler, E., Bleuler, E., 1912. The Theory of Schizophrenic Negativism. The Journal of nervous and mental disease publishing company, New York.

Boop, B., Garcia-Rill, E., Dykman, R., Skinner, R.D., 1994. The P1: insights into attention and arousal. Pediatr. Neurosurg. 20, 57–62.

Boser, K., Higgins, S., Fetherston, A., Preissler, M.A., Gordon, B., 2002. Semantic fields in low-functioning autism. J. Autism Dev. Disord. 32, 563–582.

Braam, W., Smits, M.G., Didden, R., Korzilius, H., Van Geijlswijk, I.M., et al., 2009. Exogenous melatonin for sleep problems in individuals with intellectual disability: a meta-analysis. Dev. Med. Child Neurol. 51, 340–349.

Brothers, L., Ring, B., Kling, A., 1990. Response of neurons in the macaque amygdala to complex social stimuli. Behav. Brain Res. 41, 199–213.

Bryson, S.E., Zwaigenbaum, L., Brian, J., Roberts, W., Szatmari, P., et al., 2007. A prospective case series of high-risk infants who developed autism. J. Autism Dev. Disord. 37, 12–24.

Buchwald, J.S., Erwin, R.J., Read, S., Van Lancker, D., Cummings, J.L., 1989. Midlatency auditory evoked responses: differential abnormality of P1 in Alzheimer's disease. Electroencephalogr. Clin. Neurophysiol. 74, 378–384.

Buchwald, J.S., Rubinstein, E.H., Schwafel, J., Strandburg, R.J., 1991. Midlatency auditory evoked responses: differential effects of a cholinergic agonist and antagonist. Electroencephalogr. Clin. Neurophysiol. 80, 303–309.

Buchwald, J.S., Erwin, R., Van Lancker, D., Guthrie, D., Schwafel, J., et al., 1992. Midlatency auditory evoked responses: P1 abnormalities in adult autistic subjects. Electroencephalogr. Clin. Neurophysiol. 84, 164–171.

Buckley, A.W., Rodriguez, A.J., Jennison, K., Buckley, J., Thurm, A., et al., 2010. Rapid eye movement sleep percentage in children with autism compared with children with developmental delay and typical development. Arch. Pediatr. Adolesc. Med. 164, 1032–1037.

Buescher, A.V.S., Cidav, Z., Knapp, M., Mandell, D.S., 2014. Costs of autism spectrum disorders in the United Kingdom and the United States. JAMA Pediatr. 168, 721–728.

Carlsson, A., 1995. Neurocircuitries and neurotransmitter interactions in schizophrenia. Int. Clin. Psychopharmacol. 10 (Suppl. 3), 21–28.

Carper, R.A., Courchesne, E., 2005. Localized enlargement of the frontal cortex in early autism. Biol. Psychiatry 57, 126–133.

Cartwright-Hatton, S., McNicol, K., Doubleday, E., 2006. Anxiety in a neglected population: prevalence of anxiety disorders in pre-adolescent children. Clin. Psychol. Rev. 26, 817–833.

Casas, K.A., Mononen, T.K., Mikail, C.N., Hassed, S.J., Li, S., et al., 2004. Chromosome 2q terminal deletion: report of 6 new patients and review of phenotype-breakpoint correlations in 66 individuals. Am. J. Med. Genet. A 130A, 331–339.

Catani, M., Dell'Acqua, F., Budisavljevic, S., Howells, H., Thiebaut de Schotten, M., et al., 2016. Frontal networks in adults with autism spectrum disorder. Brain J. Neurol. 139, 616–630.

CDC, 2007a. Prevalence of autism spectrum disorders—autism and developmental disabilities monitoring network, 14 sites, United States, 2002. Morb. Mortal. Wkly. Rep. Surveill. Summ. Wash. DC 2002 56, 12–28.

CDC, 2007b. Prevalence of autism spectrum disorders—autism and developmental disabilities monitoring network, six sites, United States, 2000. Morb. Mortal. Wkly. Rep. Surveill. Summ. Wash. DC 2002 56, 1–11.

CDC, 2009. Prevalence of autism spectrum disorders—autism and developmental disabilities monitoring network, United States, 2006. Morb. Mortal. Wkly. Rep. Surveill. Summ. Wash. DC 2002 58, 1–20.

CDC, 2014. Prevalence of autism spectrum disorder among children aged 8 years—autism and developmental disabilities monitoring network, 11 sites, United States, 2010. Morb. Mortal. Wkly. Rep. Surveill. Summ. Wash. DC 2002 63, 1–21.

Chang, W.-P., Arfken, C.L., Sangal, M.P., Boutros, N.N., 2011. Probing the relative contribution of the first and second responses to sensory gating indices: a meta-analysis. Psychophysiology 48, 980–992.

Chang, J., Gilman, S.R., Chiang, A.H., Sanders, S.J., Vitkup, D., 2015. Genotype to phenotype relationships in autism spectrum disorders. Nat. Neurosci. 18, 191–198.

Christensen, J., Grønborg, T.K., Sørensen, M.J., Schendel, D., Parner, E.T., et al., 2013. Prenatal valproate exposure and risk of autism spectrum disorders and childhood autism. JAMA 309, 1696–1703.

Christensen, D.L., Baio, J., Van Naarden Braun, K., Bilder, D., Charles, J., et al., 2016. Prevalence and characteristics of autism spectrum disorder among children aged 8 years—autism and developmental disabilities monitoring network, 11 sites, United States, 2012. Morb. Mortal. Wkly. Rep. Surveill. Summ. Wash. DC 2002 65, 1–23.

Cohen, M.X., Axmacher, N., Lenartz, D., Elger, C.E., Sturm, V., et al., 2009. Neuroelectric signatures of reward learning and decision-making in the human nucleus accumbens. Neuropsychopharmacol. Off. Publ. Am. Coll. Neuropsychopharmacol. 34, 1649–1658.

Cohen, I.L., Gardner, J.M., Karmel, B.Z., Phan, H.T.T., Kittler, P., et al., 2013. Neonatal brainstem function and 4-month arousal-modulated attention are jointly associated with autism. Autism Res. Off. J. Int. Soc. Autism Res. 6, 11–22.

Cohen, S., Conduit, R., Lockley, S.W., Rajaratnam, S.M., Cornish, K.M., 2014. The relationship between sleep and behavior in autism spectrum disorder (ASD): a review. J. Neurodev. Disord. 6, 44.

Combs, L.A., Polich, J., 2006. P3a from auditory white noise stimuli. Clin Neurophysiol. 117, 1106–1112.

Corbett, B.A., Mendoza, S., Abdullah, M., Wegelin, J.A., Levine, S., 2006. Cortisol circadian rhythms and response to stress in children with autism. Psychoneuroendocrinology 31, 59–68.

Corbetta, M., Shulman, G.L., 2011. Spatial neglect and attention networks. Annu. Rev. Neurosci. 34, 569–599.

Corbetta, M., Patel, G., Shulman, G.L., 2008. The reorienting system of the human brain: from environment to theory of mind. Neuron 58, 306–324.

Courchesne, E., 2004. Brain development in autism: early overgrowth followed by premature arrest of growth. Ment. Retard. Dev. Disabil. Res. Rev. 10, 106–111.

Courchesne, E., Campbell, K., Solso, S., 2011a. Brain growth across the life span in autism: age-specific changes in anatomical pathology. Brain Res. 1380, 138–145.

Courchesne, E., Mouton, P.R., Calhoun, M.E., Semendeferi, K., Ahrens-Barbeau, C., et al., 2011b. Neuron number and size in prefrontal cortex of children with autism. JAMA 306, 2001–2010.

Couturier, J.L., Speechley, K.N., Steele, M., Norman, R., Stringer, B., et al., 2005. Parental perception of sleep problems in children of normal intelligence with pervasive developmental disorders: prevalence, severity, and pattern. J. Am. Acad. Child Adolesc. Psychiatry 44, 815–822.

Crane, L., Goddard, L., Pring, L., 2009. Sensory processing in adults with autism spectrum disorders. Autism Int. J. Res. Pract. 13, 215–228.

Ćurin, J.M., Terzić, J., Petković, Z.B., Zekan, L., Terzić, I.M., et al., 2003. Lower cortisol and higher ACTH levels in individuals with autism. J. Autism Dev. Disord. 33, 443–448.

Dahlgren, S.O., Gillberg, C., 1989. Symptoms in the first two years of life. A preliminary population study of infantile autism. Eur. Arch. Psychiatry Neurol. Sci. 238, 169–174.

Daluwatte, C., Miles, J.H., Christ, S.E., Beversdorf, D.Q., Takahashi, T.N., et al., 2013. Atypical pupillary light reflex and heart rate variability in children with autism spectrum disorder. J. Autism Dev. Disord. 43, 1910–1925.

Damasio, A.R., Maurer, R.G., 1978. A neurological model for childhood autism. Arch. Neurol. 35, 777–786.

Davidson, P.W., Myers, G.J., Weiss, B., 2004. Mercury exposure and child development outcomes. Pediatrics 113, 1023–1029.

Dawson, G., Meltzoff, A.N., Osterling, J., Rinaldi, J., Brown, E., 1998. Children with autism fail to orient to naturally occurring social stimuli. J. Autism Dev. Disord. 28, 479–485.

Dawson, G., Jones, E.J.H., Merkle, K., Venema, K., Lowy, R., et al., 2012. Early behavioral intervention is associated with normalized brain activity in young children with autism. J. Am. Acad. Child Adolesc. Psychiatry 51, 1150–1159.

de Lecea, L., Carter, M., Adamantidis, A., 2012. Shining light on wakefulness and arousal. Biol. Psychiatry 71, 1046–1052.

Değerliyurt, A., Yalnizoğlu, D., Bakar, E.E., Topçu, M., Turanli, G., 2015. Electrical status epilepticus during sleep: a study of 22 patients. Brain and Development 37, 250–264.

Di Martino, A., Kelly, C., Grzadzinski, R., Zuo, X.-N., Mennes, M., et al., 2011. Aberrant striatal functional connectivity in children with autism. Biol. Psychiatry 69, 847–856.

Diomedi, M., Curatolo, P., Scalise, A., Placidi, F., Caretto, F., et al., 1999. Sleep abnormalities in mentally retarded autistic subjects: Down's syndrome with mental retardation and normal subjects. Brain and Development 21, 548–553.

Dringenberg, H.C., Vanderwolf, C.H., 1998. Involvement of direct and indirect pathways in electrocorticographic activation. Neurosci. Biobehav. Rev. 22, 243–257.

Durkin, M.S., Maenner, M.J., Newschaffer, C.J., Lee, L.-C., Cunniff, C.M., et al., 2008. Advanced parental age and the risk of autism spectrum disorder. Am. J. Epidemiol. 168, 1268–1276.

Dykens, E.M., Sutcliffe, J.S., Levitt, P., 2004. Autism and 15q11-q13 disorders: behavioral, genetic, and pathophysiological issues. Ment. Retard. Dev. Disabil. Res. Rev. 10, 284–291.

Ecker, C., Shahidiani, A., Feng, Y., Daly, E., Murphy, C., et al., 2014. The effect of age, diagnosis, and their interaction on vertex-based measures of cortical thickness and surface area in autism spectrum disorder. J. Neural Transm. Vienna Austria 1996 121, 1157–1170.

Elia, M., Ferri, R., Musumeci, S.A., Del Gracco, S., Bottitta, M., et al., 2000. Sleep in subjects with autistic disorder: a neurophysiological and psychological study. Brain and Development 22, 88–92.

Elison, J.T., Paterson, S.J., Wolff, J.J., Reznick, J.S., Sasson, N.J., et al., 2013. White matter microstructure and atypical visual orienting in 7-month-olds at risk for autism. Am. J. Psychiatry 170, 899–908.

Elsabbagh, M., Divan, G., Koh, Y.-J., Kim, Y.S., Kauchali, S., et al., 2012. Global prevalence of autism and other pervasive developmental disorders. Autism Res. Off. J. Int. Soc. Autism Res. 5, 160–179.

Elsabbagh, M., Fernandes, J., Webb, S.J., Dawson, G., Charman, T., et al., 2013. Disengagement of visual attention in infancy is associated with emerging autism in toddlerhood. Biol. Psychiatry 74, 189–194.

Ernst, M., Nelson, E.E., McClure, E.B., Monk, C.S., Munson, S., et al., 2004. Choice selection and reward anticipation: an fMRI study. Neuropsychologia 42, 1585–1597.

Erwin, R., Buchwald, J.S., 1986a. Midlatency auditory evoked responses: differential effects of sleep in the human. Electroencephalogr. Clin. Neurophysiol. Potentials Sect. 65, 383–392.

Erwin, R.J., Buchwald, J.S., 1986b. Midlatency auditory evoked responses: differential recovery cycle characteristics. Electroencephalogr. Clin. Neurophysiol. 64, 417–423.

Estes, A., Shaw, D.W.W., Sparks, B.F., Friedman, S., Giedd, J.N., et al., 2011. Basal ganglia morphometry and repetitive behavior in young children with autism spectrum disorder. Autism Res. Off. J. Int. Soc. Autism Res. 4, 212–220.

Farley, M.A., McMahon, W.M., Fombonne, E., Jenson, W.R., Miller, J., et al., 2009. Twenty-year outcome for individuals with autism and average or near-average cognitive abilities. Autism Res. Off. J. Int. Soc. Autism Res. 2, 109–118.

Fombonne, E., 2009. Epidemiology of pervasive developmental disorders. Pediatr. Res. 65, 591–598.

Freedman, R., Adler, L.E., Waldo, M.C., Pachtman, E., Franks, R.D., 1983. Neurophysiological evidence for a defect in inhibitory pathways in schizophrenia: comparison of medicated and drug-free patients. Biol. Psychiatry 18, 537–551.

Freeman, B.J., 1993. The syndrome of autism: Update and guidelines for diagnosis. Infants Young Child 6, 1.

Fuchs, R.A., Eaddy, J.L., Su, Z.-I., Bell, G.H., 2007. Interactions of the basolateral amygdala with the dorsal hippocampus and dorsomedial prefrontal cortex regulate drug context-induced reinstatement of cocaine-seeking in rats. Eur. J. Neurosci. 26, 487–498.

Gadow, K.D., DeVincent, C.J., Pomeroy, J., Azizian, A., 2004. Psychiatric symptoms in preschool children with PDD and clinic and comparison samples. J. Autism Dev. Disord. 34, 379–393.

Gail, W.P., Sears, L.L., Allard, A., 2004. Sleep problems in children with autism. J. Sleep Res. 13, 265–268.

Garcia-Rill, E., 1997. Disorders of the reticular activating system. Med. Hypotheses 49, 379–387.

Garcia-Rill, E., 2015. Waking and the Reticular Activating System in Health and Disease. Academic Press.

Garcia-Rill, E., Simon, C., Smith, K., Kezunovic, N., Hyde, J., 2011. The pedunculopontine tegmental nucleus: from basic neuroscience to neurosurgical applications. J. Neural Transm. Vienna Austria 1996 118, 1397–1407.

Geschwind, D.H., 2011. Genetics of autism spectrum disorders. Trends Cogn. Sci. 15, 409–416.

Geschwind, D.H., Levitt, P., 2007. Autism spectrum disorders: developmental disconnection syndromes. Curr. Opin. Neurobiol. 17, 103–111.

Geva, R., Feldman, R., 2008. A neurobiological model for the effects of early brainstem functioning on the development of behavior and emotion regulation in infants: implications for prenatal and perinatal risk. J. Child Psychol. Psychiatry 49, 1031–1041.

Gharani, N., Benayed, R., Mancuso, V., Brzustowicz, L.M., Millonig, J.H., 2004. Association of the homeobox transcription factor, ENGRAILED 2, 3, with autism spectrum disorder. Mol. Psychiatry 9, 474–484.

Gillette, G.M., Skinner, R.D., Rasco, L.M., Fielstein, E.M., Davis, D.H., et al., 1997. Combat veterans with posttraumatic stress disorder exhibit decreased habituation of the P1 midlatency auditory evoked potential. Life Sci. 61, 1421–1434.

Gillott, A., Furniss, F., Walter, A., 2001. Anxiety in high-functioning children with autism. Autism 5, 277–286.

Godbout, R., Bergeron, C., Limoges, E., Stip, E., Mottron, L., 2000. A laboratory study of sleep in Asperger's syndrome. Neuroreport 11, 127–130.

Gozal, D., 1998. Sleep-disordered breathing and school performance in children. Pediatrics 102, 616–620.

Gozal, D., 2001. Morbidity of obstructive sleep apnea in children: facts and theory. Sleep Breath. 05, 035–042.

Grandin, T., Sacks, O., 2006. Thinking in Pictures, Expanded Edition: My Life With Autism. Expanded ed. edition, Vintage, New York.

Grandin, T., Scariano, M.M., 1996. Emergence: Labeled Autistic. Reissue edition, Warner Books, New York.

Grayton, H.M., Fernandes, C., Rujescu, D., Collier, D.A., 2012. Copy number variations in neurodevelopmental disorders. Prog. Neurobiol. 99, 81–91.

Green, S.A., Ben-Sasson, A., 2010. Anxiety disorders and sensory over-responsivity in children with autism spectrum disorders: is there a causal relationship? J. Autism Dev. Disord. 40, 1495–1504.

Gringras, P., Gamble, C., Jones, A.P., Wiggs, L., Williamson, P.R., et al., 2012. Melatonin for sleep problems in children with neurodevelopmental disorders: randomized double masked placebo controlled trial. BMJ 345, e6664.

Haller, J., Tóth, M., Halász, J., 2005. The activation of raphe serotonergic neurons in normal and hypoarousal-driven aggression: a double labeling study in rats. Behav. Brain Res. 161, 88–94.

Hallmayer, J., Cleveland, S., Torres, A., Phillips, J., Cohen, B., et al., 2011. Genetic heritability and shared environmental factors among twin pairs with autism. Arch. Gen. Psychiatry 68, 1095–1102.

Hamani, C., Pilitsis, J., Rughani, A.I., Rosenow, J.M., Patil, P.G., et al., 2014. Deep brain stimulation for obsessive-compulsive disorder: systematic review and evidence-based guideline sponsored by the American Society for Stereotactic and Functional Neurosurgery and the Congress of Neurological Surgeons (CNS) and endorsed by the CNS and American Association of Neurological Surgeons. Neurosurgery 75, 327–333 (quiz 333).

Hardan, A.Y., Jou, R.J., Keshavan, M.S., Varma, R., Minshew, N.J., 2004. Increased frontal cortical folding in autism: a preliminary MRI study. Psychiatry Res. 131, 263–268.

Harris, N.S., Courchesne, E., Townsend, J., Carper, R.A., Lord, C., 1999. Neuroanatomic contributions to slowed orienting of attention in children with autism. Brain Res. Cogn. Brain Res. 8, 61–71.

Harrison, J., Hare, D.J., 2004. Brief report: assessment of sensory abnormalities in people with autistic spectrum disorders. J. Autism Dev. Disord. 34, 727–730.

Hashimoto, T., Tayama, M., Murakawa, K., Yoshimoto, T., Miyazaki, M., et al., 1995. Development of the brainstem and cerebellum in autistic patients. J. Autism Dev. Disord. 25, 1–18.

Herbert, M.R., Ziegler, D.A., Makris, N., Filipek, P.A., Kemper, T.L., et al., 2004. Localization of white matter volume increase in autism and developmental language disorder. Ann. Neurol. 55, 530–540.

Herman, G.E., Henninger, N., Ratliff-Schaub, K., Pastore, M., Fitzgerald, S., et al., 2007. Genetic testing in autism: how much is enough? Genet. Med. Off. J. Am. Coll. Med. Genet. 9, 268–274.

Hermelin, B., O'Connor, N., 1970. Psychological Experiments With Autistic Children, first ed. Pergamon Press, Oxford, New York.

Hirsch, L.E., Pringsheim, T., 2016. Aripiprazole for autism spectrum disorders (ASD). Cochrane Database Syst. Rev. CD009043.

Hirstein, W., Iversen, P., Ramachandran, V.S., 2001. Autonomic responses of autistic children to people and objects. Proc. Biol. Sci. 268, 1883–1888.

Hobson, J.A., 1995. Sleep. W. H. Freeman, New York.

Hobson, J.A., Pace-Schott, E.F., 2002. The cognitive neuroscience of sleep: neuronal systems, consciousness and learning. Nat. Rev. Neurosci. 3, 679–693.

Hohmann, C.F., Berger-Sweeney, J., 1998. Cholinergic regulation of cortical development and plasticity. New twists to an old story. Perspect. Dev. Neurobiol. 5, 401–425.

Hoon, A.H., Reiss, A.L., 1992. The mesial-temporal lobe and autism: case report and review. Dev. Med. Child Neurol. 34, 252–259.

Horne, J.A., 1988. Sleep loss and "divergent" thinking ability. Sleep 11, 528–536.

Howard, M.A., Cowell, P.E., Boucher, J., Broks, P., Mayes, A., et al., 2000. Convergent neuroanatomical and behavioural evidence of an amygdala hypothesis of autism. Neuroreport 11, 2931–2935.

Hua, X., Thompson, P.M., Leow, A.D., Madsen, S.K., Caplan, R., et al., 2013. Brain growth rate abnormalities visualized in adolescents with autism. Hum. Brain Mapp. 34, 425–436.

Hutt, C., Hutt, S.J., Lee, D., Ounsted, C., 1964. Arousal and childhood autism. Nature 204, 908–909.

Janecka, M., Mill, J., Basson, M.A., Goriely, A., Spiers, H., et al., 2017. Advanced paternal age effects in neurodevelopmental disorders—review of potential underlying mechanisms. Transl. Psychiatry 7, e1019.

Jones, B.E., 2003. Arousal systems. Front. Biosci. J. Virtual Libr. 8, s438–s451.

Joosten, A.V., Bundy, A.C., 2010. Sensory processing and stereotypical and repetitive behaviour in children with autism and intellectual disability. Aust. Occup. Ther. J. 57, 366–372.

Kanner, L., 1943. "Autistic disturbances of affective contact" (1943), by Leo Kanner. Nerv. Child 2, 217–250.

Karmel, B.Z., Gardner, J.M., Meade, L.S., Cohen, I.L., London, E., et al., 2010. Early medical and behavioral characteristics of NICU infants later classified with ASD. Pediatrics 126, 457–467.

Keehn, B., Lincoln, A.J., Müller, R.-A., Townsend, J., 2010. Attentional networks in children and adolescents with autism spectrum disorder. J. Child Psychol. Psychiatry 51, 1251–1259.

Kemner, O.B., van Engeland, H., 2002. Normal P50gating in children with autism [CME]. J. Clin. Psychiatry 63, 214–217.

Kemper, T.L., Bauman, M., 1998. Neuropathology of infantile autism. J. Neuropathol. Exp. Neurol. 57, 645–652.

Kern, J.K., Jones, A.M., 2006. Evidence of toxicity, oxidative stress, and neuronal insult in autism. J. Toxicol. Environ. Health B Crit. Rev. 9, 485–499.

Kevanishvili, Z.S., Von Specht, H., 1979. Human slow auditory evoked potentials during natural and drug-induced sleep. Electroencephalogr. Clin. Neurophysiol. 47, 280–288.

Kim, Y.S., Leventhal, B.L., Koh, Y.-J., Fombonne, E., Laska, E., et al., 2011. Prevalence of autism spectrum disorders in a total population sample. Am. J. Psychiatry 168, 904–912.

Kinsbourne, M., 1987. Cerebral-brainstem relations in infantile autism. In: Schopler, E., Mesibov, G.B. (Eds.), Neurobiol. Issues Autism. Springer US, Boston, MA, pp. 107–125.

Kirino, E., 2014. Efficacy and tolerability of pharmacotherapy options for the treatment of irritability in autistic children. Clin. Med. Insights Pediatr. 8, 17–30.

Kirov, R., Kinkelbur, J., Banaschewski, T., Rothenberger, A., 2007. Sleep patterns in children with attention-deficit/hyperactivity disorder, tic disorder, and comorbidity. J. Child Psychol. Psychiatry 48, 561–570.

Klin, A., 1993. Auditory brainstem responses in autism: brainstem dysfunction or peripheral hearing loss? J. Autism Dev. Disord. 23, 15–35.

Knutson, B., Adams, C.M., Fong, G.W., Hommer, D., 2001. Anticipation of increasing monetary reward selectively recruits nucleus accumbens. J. Neurosci. Off. J. Soc. Neurosci. 21, RC159.

Kolevzon, A., Gross, R., Reichenberg, A., 2007. Prenatal and perinatal risk factors for autism: a review and integration of findings. Arch. Pediatr. Adolesc. Med. 161, 326–333.

Krakowiak, P., Goodlin-Jones, B., Hertz-Picciotto, I., Croen, L.A., Hansen, R.L., 2008. Sleep problems in children with autism spectrum disorders, developmental delays, and typical development: a population-based study. J. Sleep Res. 17, 197–206.

Kryger, M., Roth, T., Dement, W., 2011. Principles and Practice of Sleep Medicine, fifth ed. Saunders.

Kupchik, Y.M., Brown, R.M., Heinsbroek, J.A., Lobo, M.K., Schwartz, D.J., et al., 2015. Coding the direct/indirect pathways by D1 and D2 receptors is not valid for accumbens projections. Nat. Neurosci. 18, 1230–1232.

Lane, A.E., Young, R.L., Baker, A.E.Z., Angley, M.T., 2010. Sensory processing subtypes in autism: association with adaptive behavior. J. Autism Dev. Disord. 40, 112–122.

Langen, M., Durston, S., Staal, W.G., Palmen, S.J.M.C., van Engeland, H., 2007. Caudate nucleus is enlarged in high-functioning medication-naive subjects with autism. Biol. Psychiatry 62, 262–266.

Le, J.F., Lohr, W.D., 2012. Aggression and self-injury in a patient with severe autism. Pediatr. Ann. 41, e207–e209.

Leblois, A., Wendel, B.J., Perkel, D.J., 2010. Striatal dopamine modulates basal ganglia output and regulates social context-dependent behavioral variability through D1 receptors. J. Neurosci. 30, 5730–5743.

Leekam, S.R., Nieto, C., Libby, S.J., Wing, L., Gould, J., 2007. Describing the sensory abnormalities of children and adults with autism. J. Autism Dev. Disord. 37, 894–910.

Lerner, T.N., Shilyansky, C., Davidson, T.J., Evans, K.E., Beier, K.T., et al., 2015. Intact-brain analyses reveal distinct information carried by SNc dopamine subcircuits. Cell 162, 635–647.

Levitt, J.G., Blanton, R.E., Smalley, S., Thompson, P.M., Guthrie, D., et al., 2003. Cortical sulcal maps in autism. Cereb. Cortex N. Y. N 1991 13, 728–735.

Limoges, É., Bolduc, C., Berthiaume, C., Mottron, L., Godbout, R., 2013. Relationship between poor sleep and daytime cognitive performance in young adults with autism. Res. Dev. Disabil. 34, 1322–1335.

Liss, M., Saulnier, C., Fein, D., Kinsbourne, M., 2006. Sensory and attention abnormalities in autistic spectrum disorders. Autism Int. J. Res. Pract. 10, 155–172.

London, E.A., 2000. The environment as an etiologic factor in autism: a new direction for research. Environ. Health Perspect. 108 (Suppl. 3), 401–404.

Lopez, J., Roffwarg, H.P., Dreher, A., Bissette, G., Karolewicz, B., et al., 2008. Rapid eye movement sleep deprivation decreases long-term potentiation stability and affects some glutamatergic signaling proteins during hippocampal development. Neuroscience 153, 44–53.

Luster, B.R., Urbano, F.J., Garcia-Rill, E., 2016. Intracellular mechanisms modulating gamma band activity in the pedunculopontine nucleus (PPN). Physiol. Rep. 4, e12787.

Madsen, G.F., Bilenberg, N., Jepsen, J.R., Glenthøj, B., Cantio, C., et al., 2015. Normal P50 gating in children with autism, yet attenuated p50 amplitude in the asperger subcategory. Autism Res. 8, 371–378.

Magnée, M.J.C.M., Oranje, B., van Engeland, H., Kahn, R.S., Kemner, C., 2009. Cross-sensory gating in schizophrenia and autism spectrum disorder: EEG evidence for impaired brain connectivity? Neuropsychologia 47, 1728–1732.

Malow, B.A., 2004. Sleep disorders, epilepsy, and autism. Ment. Retard. Dev. Disabil. Res. Rev. 10, 122–125.

Malow, B.A., Marzec, M.L., McGrew, S.G., Wang, L., Henderson, L.M., et al., 2006. Characterizing sleep in children with autism spectrum disorders: a multidimensional approach. Sleep 29, 1563–1571.

Malow, B., Adkins, K.W., McGrew, S.G., Wang, L., Goldman, S.E., et al., 2012a. Melatonin for sleep in children with autism: a controlled trial examining dose, tolerability, and outcomes. J. Autism Dev. Disord. 42, 1729–1737 (author reply 1738).

Malow, B.A., Byars, K., Johnson, K., Weiss, S., Bernal, P., et al., 2012b. A practice pathway for the identification, evaluation, and management of insomnia in children and adolescents with autism spectrum disorders. Pediatrics 130, S106–S124.

Maquet, P., 2001. The role of sleep in learning and memory. Science 294, 1048–1052.

Mashhoon, Y., Wells, A.M., Kantak, K.M., 2010. Interaction of the rostral basolateral amygdala and prelimbic prefrontal cortex in regulating reinstatement of cocaine-seeking behavior. Pharmacol. Biochem. Behav. 96, 347–353.

Mason, T.B.A., Pack, A.I., 2007. Pediatric parasomnias. Sleep 30, 141–151.

Mattila, M.-L., Kielinen, M., Linna, S.-L., Jussila, K., Ebeling, H., et al., 2011. Autism spectrum disorders according to DSM-IV-TR and comparison with DSM-5 draft criteria: an epidemiological study. J. Am. Acad. Child Adolesc. Psychiatry 50, 583–592.e11.

Maurer, R.G., Damasio, A.R., 1982. Childhood autism from the point of view of behavioral neurology. J. Autism Dev. Disord. 12, 195–205.

Mazurek, M.O., Vasa, R.A., Kalb, L.G., Kanne, S.M., Rosenberg, D., et al., 2013. Anxiety, sensory over-responsivity, and gastrointestinal problems in children with autism spectrum disorders. J. Abnorm. Child Psychol. 41, 165–176.

McCarley, R.W., 2007. Neurobiology of REM and NREM sleep. Sleep Med. 8, 302–330.

McGrath, M.J., Cohen, D.B., 1978. REM sleep facilitation of adaptive waking behavior: a review of the literature. Psychol. Bull. 85, 24–57.

McPheeters, M.L., Warren, Z., Sathe, N., Bruzek, J.L., Krishnaswami, S., et al., 2011. A systematic review of medical treatments for children with autism spectrum disorders. Pediatrics 127, e1312–e1321.

Megens, A.A., Awouters, F.H., Schotte, A., Meert, T.F., Dugovic, C., et al., 1994. Survey on the pharmacodynamics of the new antipsychotic risperidone. Psychopharmacology 114, 9–23.

Milad, M.R., Rauch, S.L., 2012. Obsessive-compulsive disorder: beyond segregated cortico-striatal pathways. Trends Cogn. Sci. 16, 43–51.

Miles, J.H., 2011. Autism spectrum disorders—a genetics review. Genet. Med. Off. J. Am. Coll. Med. Genet. 13, 278–294.

Miller, L.J., Anzalone, M.E., Lane, S.J., Cermak, S.A., Osten, E.T., 2007. Concept evolution in sensory integration: a proposed nosology for diagnosis. Am. J. Occup. Ther. 61, 135–140.

Ming, X., Sun, Y.-M., Nachajon, R.V., Brimacombe, M., Walters, A.S., 2009. Prevalence of parasomnia in autistic children with sleep disorders. Clin. Med. Pediatr. 3, 1–10.

Minshew, N.J., Sweeney, J., Luna, B., 2002. Autism as a selective disorder of complex information processing and underdevelopment of neocortical systems. Mol. Psychiatry 7, S14–S15.

Mirmiran, M., van den Dungen, H., Uylings, H.B., 1982. Sleep patterns during rearing under different environmental conditions in juvenile rats. Brain Res. 233, 287–298.

Miyake, K., Hirasawa, T., Koide, T., Kubota, T., 2012. Epigenetics in autism and other neurodevelopmental diseases. Adv. Exp. Med. Biol. 724, 91–98.

Montgomery, P., Stores, G., Wiggs, L., 2004. The relative efficacy of two brief treatments for sleep problems in young learning disabled (mentally retarded) children: a randomized controlled trial. Arch. Dis. Child. 89, 125–130.

Mundy, P., Jarrold, W., 2010. Infant joint attention, neural networks and social cognition. Neural Netw. Off. J. Int. Neural Netw. Soc. 23, 985–997.

Mundy, P., Rebecca, N.A., 2000. Neural plasticity, joint attention, and a transactional social-orienting model of autism. Int. Rev. Res. Ment. Retard. 23, 139–168. Academic Press.

Muris, P., Steerneman, P., Merckelbach, H., Holdrinet, I., Meesters, C., 1998. Comorbid anxiety symptoms in children with pervasive developmental disorders. J. Anxiety Disord. 12, 387–393.

Nagamoto, H.T., Adler, L.E., Waldo, M.C., Freedman, R., 1989. Sensory gating in schizophrenics and normal controls: effects of changing stimulation interval. Biol. Psychiatry 25, 549–561.

Newschaffer, C.J., Croen, L.A., Daniels, J., Giarelli, E., Grether, J.K., et al., 2007. The epidemiology of autism spectrum disorders. Annu. Rev. Public Health 28, 235–258.

O'Neill, M., Jones, R.S.P., 1997. Sensory-perceptual abnormalities in autism: a case for more research? J. Autism Dev. Disord. 27, 283–293.

Ohnishi, T., Matsuda, H., Hashimoto, T., Kunihiro, T., Nishikawa, M., et al., 2000. Abnormal regional cerebral blood flow in childhood autism. Brain J. Neurol. 123 (Pt 9), 1838–1844.

Orekhova, E.V., Stroganova, T.A., Prokofyev, A.O., Nygren, G., Gillberg, C., et al., 2008. Sensory gating in young children with autism: relation to age, IQ, and EEG gamma oscillations. Neurosci. Lett. 434, 218–223.

Ornitz, E.M., 1974. The modulation of sensory input and motor output in autistic children. J. Autism Child. Schizophr. 4, 197–215.

Ornitz, E.M., 1988. Autism: a disorder of directed attention. Brain Dysfunct. 1, 309–322.

Owens, J.A., Spirito, A., McGuinn, M., Nobile, C., 2000. Sleep habits and sleep disturbance in elementary school-aged children. J. Dev. Behav. Pediatr. JDBP 21, 27–36.

Øyane, N.M.F., Bjorvatn, B., 2005. Sleep disturbances in adolescents and young adults with autism and Asperger syndrome. Autism 9, 83–94.

Panksepp, J., 2005. Affective consciousness: core emotional feelings in animals and humans. Conscious. Cogn. 14, 30–80.

Park, H.R., Kim, I.H., Kang, H., Lee, D.S., Kim, B.-N., et al., 2017. Nucleus accumbens deep brain stimulation for a patient with self-injurious behavior and autism spectrum disorder: functional and structural changes of the brain: report of a case and review of literature. Acta Neurochir. 159, 137–143.

Patel, D.R., Greydanus, D.E., 2012. Autism Spectrum Disorders: Practical Overview for Pediatricians, An Issue of Pediatric Clinics. Elsevier Health Sciences.

Pfaff, D.W., Kieffer, B.L., 2008. Preface. Ann. N. Y. Acad. Sci. 1129, xi.

Posey, D.J., McDougle, C.J., 2001. Pharmacotherapeutic management of autism. Expert. Opin. Pharmacother. 2, 587–600.

Preuss, T.M., 1995. Do rats have prefrontal cortex? The rose-woolsey-akert program reconsidered. J. Cogn. Neurosci. 7, 1–24.

Quine, L., 1991. Sleep problems in children with mental handicap. J. Ment. Defic. Res. 35 (Pt 4), 269–290.

Quirk, G.J., Beer, J.S., 2006. Prefrontal involvement in the regulation of emotion: convergence of rat and human studies. Curr. Opin. Neurobiol. 16, 723–727.

Quirk, G.J., Mueller, D., 2008. Neural mechanisms of extinction learning and retrieval. Neuropsychopharmacol. Off. Publ. Am. Coll. Neuropsychopharmacol. 33, 56–72.

Rapin, I., Katzman, R., 1998. Neurobiology of autism. Ann. Neurol. 43, 7–14.

Rasmussen, S.A., Eisen, J.L., 1992. The epidemiology and clinical features of obsessive compulsive disorder. Psychiatr. Clin. 15, 743–758.

Reese, N.B., Garcia-Rill, E., Skinner, R.D., 1995. The pedunculopontine nucleus—auditory input, arousal and pathophysiology. Prog. Neurobiol. 47, 105–133.

Reynolds, S., Lane, S.J., 2008. Diagnostic validity of sensory over-responsivity: a review of the literature and case reports. J. Autism Dev. Disord. 38, 516–529.

Richdale, A.L., 1999. Sleep problems in autism: prevalence, cause, and intervention. Dev. Med. Child Neurol. 41, 60–66.

Richdale, A.L., Prior, M.R., 1992. Urinary cortisol circadian rhythm in a group of high-functioning children with autism. J. Autism Dev. Disord. 22, 433–447.

Richdale, A.L., Prior, M.R., 1995. The sleep/wake rhythm in children with autism. Eur. Child Adolesc. Psychiatry 4, 175–186.

Ring, H.A., Baron-Cohen, S., Wheelwright, S., Williams, S.C.R., Brammer, M., et al., 1999. Cerebral correlates of preserved cognitive skills in autism: a functional MRI study of embedded figures task performance. Brain 122, 1305–1315.

Ritvo, E.R., Ritvo, R., Yuwiler, A., Brothers, A., Freeman, B.J., et al., 1993. Elevated daytime melatonin concentrations in autism: a pilot study. Eur. Child Adolesc. Psychiatry 2, 75–78.

Robbins, T.W., 1997. Arousal systems and attentional processes. Biol. Psychol. 45, 57–71.

Robbins, T.W., Arnsten, A.F.T., 2009. The neuropsychopharmacology of fronto-executive function: monoaminergic modulation. Annu. Rev. Neurosci. 32, 267–287.

Roberts, E.M., English, P.B., Grether, J.K., Windham, G.C., Somberg, L., et al., 2007. Maternal residence near agricultural pesticide applications and autism spectrum disorders among children in the California Central Valley. Environ. Health Perspect. 115, 1482–1489.

Robinson, A.A., Malow, B.A., 2013. Gabapentin shows promise in treating refractory insomnia in children. J. Child Neurol. 28, 1618–1621.

Rogers, S.J., Ozonoff, S., 2005. Annotation: what do we know about sensory dysfunction in autism? A critical review of the empirical evidence. J. Child Psychol. Psychiatry 46, 1255–1268.

Rosenberg, R.E., Law, J.K., Yenokyan, G., McGready, J., Kaufmann, W.E., et al., 2009. Characteristics and concordance of autism spectrum disorders among 277 twin pairs. Arch. Pediatr. Adolesc. Med. 163, 907–914.

Rubenstein, J.L.R., Merzenich, M.M., 2003. Model of autism: increased ratio of excitation/inhibition in key neural systems. Genes Brain Behav. 2, 255–267.

Saemundsen, E., Ludvigsson, P., Rafnsson, V., 2007. Autism spectrum disorders in children with a history of infantile spasms: a population-based study. J. Child Neurol. 22, 1102–1107.

Sah, P., Faber, E.S.L., Lopez De Armentia, M., Power, J., 2003. The amygdaloid complex: anatomy and physiology. Physiol. Rev. 83, 803–834.

Sajdyk, T.J., Shekhar, A., 1997. Excitatory amino acid receptor antagonists block the cardiovascular and anxiety responses elicited by gamma-aminobutyric acid A receptor blockade in the basolateral amygdala of rats. J. Pharmacol. Exp. Ther. 283, 969–977.

Sajdyk, T.J., Shekhar, A., 2000. Sodium lactate elicits anxiety in rats after repeated GABA receptor blockade in the basolateral amygdala. Eur. J. Pharmacol. 394, 265–273.

Salgado, S., Kaplitt, M.G., 2015. The nucleus accumbens: a comprehensive review. Stereotact. Funct. Neurosurg. 93, 75–93.

Saxena, S., Rauch, S.L., 2000. Functional neuroimaging and the neuroanatomy of obsessive-compulsive disorder. Psychiatr. Clin. North Am. 23, 563–586.

Schaaf, R.C., Toth-Cohen, S., Johnson, S.L., Outten, G., Benevides, T.W., 2011. The everyday routines of families of children with autism: examining the impact of sensory processing difficulties on the family. Autism 15, 373–389.

Schrock, L.E., Mink, J.W., Woods, D.W., Porta, M., Servello, D., et al., 2015. Tourette syndrome deep brain stimulation: a review and updated recommendations. Mov. Disord. Off. J. Mov. Disord. Soc. 30, 448–471.

Schultz, W., 2007. Multiple dopamine functions at different time courses. Annu. Rev. Neurosci. 30, 259–288.

Schumann, C.M., Bloss, C.S., Barnes, C.C., Wideman, G.M., Carper, R.A., et al., 2010. Longitudinal MRI study of cortical development through early childhood in autism. J. Neurosci. 30, 4419–4427.

Scott-Van Zeeland, A.A., Dapretto, M., Ghahremani, D.G., Poldrack, R.A., Bookheimer, S.Y., 2010. Reward processing in autism. Autism Res. Off. J. Int. Soc. Autism Res. 3, 53–67.

Sears, L.L., Vest, C., Mohamed, S., Bailey, J., Ranson, B.J., et al., 1999. An MRI study of the basal ganglia in autism. Prog. Neuro-Psychopharmacol. Biol. Psychiatry 23, 613–624.

Skinner, R.D., Rasco, L.M., Fitzgerald, J., Karson, C.N., Matthew, M., et al., 1999. Reduced sensory gating of the P1 potential in rape victims and combat veterans with posttraumatic stress disorder. Depress. Anxiety 9, 122–130.

Skinner, R.D., Homma, Y., Garcia-Rill, E., 2004. Arousal mechanisms related to posture and locomotion: 2. Ascending modulation. Prog. Brain Res. 143, 291–298.

Solso, S., Xu, R., Proudfoot, J., Hagler, D.J., Campbell, K., et al., 2016. Diffusion tensor imaging provides evidence of possible axonal overconnectivity in frontal lobes in autism spectrum disorder toddlers. Biol. Psychiatry 79, 676–684.

Sotres-Bayon, F., Quirk, G.J., 2010. Prefrontal control of fear: more than just extinction. Curr. Opin. Neurobiol. 20, 231–235.

Souders, M.C., Mason, T.B.A., Valladares, O., Bucan, M., Levy, S.E., et al., 2009. Sleep behaviors and sleep quality in children with autism spectrum disorders. Sleep 32, 1566–1578.

Sparks, B.F., Friedman, S.D., Shaw, D.W., Aylward, E.H., Echelard, D., et al., 2002. Brain structural abnormalities in young children with autism spectrum disorder. Neurology 59, 184–192.

Stein, M.A., Mendelsohn, J., Obermeyer, W.H., Amromin, J., Benca, R., 2001. Sleep and behavior problems in school-aged children. Pediatrics 107, E60.

Steriade, M., 2000. Corticothalamic resonance, states of vigilance and mentation. Neuroscience 101, 243–276.

Steriade, M., 2006. Grouping of brain rhythms in corticothalamic systems. Neuroscience 137, 1087–1106.

Steriade, M., McCarley, R.W., 1990. Neuronal control of the sleep—wake states. In: Steriade, M., McCarley, R.W. (Eds.), Brainstem Control Wakefulness Sleep. Springer US, Boston, MA, pp. 325–361.

Stickgold, R., 2005. Sleep-dependent memory consolidation. Nature 437, 1272–1278.

Stigler, K.A., McDonald, B.C., Anand, A., Saykin, A.J., McDougle, C.J., 2011. Structural and functional magnetic resonance imaging of autism spectrum disorders. Brain Res. 1380, 146–161.

Stocco, A., Baizabal-Carvallo, J.F., 2014. Deep brain stimulation for severe secondary stereotypies. Parkinsonism Relat. Disord. 20, 1035–1036.

Stoner, R., Chow, M.L., Boyle, M.P., Sunkin, S.M., Mouton, P.R., et al., 2014. Patches of disorganization in the neocortex of children with autism. N. Engl. J. Med. 370, 1209–1219.

Strömland, K., Nordin, V., Miller, M., Akerström, B., Gillberg, C., 1994. Autism in thalidomide embryopathy: a population study. Dev. Med. Child Neurol. 36, 351–356.

Sturm, W., Willmes, K., 2001. On the functional neuroanatomy of intrinsic and phasic alertness. NeuroImage 14, S76–S84.

Sturm, V., Fricke, O., Bührle, C.P., Lenartz, D., Maarouf, M., et al., 2012. DBS in the basolateral amygdala improves symptoms of autism and related self-injurious behavior: a case report and hypothesis on the pathogenesis of the disorder. Front. Hum. Neurosci. 6, 341.

Stuss, D.T., Knight, R.T., 2012. Principles of Frontal Lobe Function, second ed. Oxford University Press, Oxford, New York.

Taghva, A.S., Malone, D.A., Rezai, A.R., 2013. Deep brain stimulation for treatment-resistant depression. World Neurosurg. 80 . S27.e17–24.

Tanguay, P.E., Ornitz, E.M., Forsythe, A.B., Ritvo, E.R., 1976. Rapid eye movement (REM) activity in normal and autistic children during REM sleep. J. Autism Child. Schizophr. 6, 275–288.

Tani, P., Lindberg, N., Nieminen-von, W.T., von Wendt, L., Alanko, L., et al., 2003. Insomnia is a frequent finding in adults with Asperger syndrome. BMC Psychiatry 3, 12.

Taylor, D.C., Neville, B.G., Cross, J.H., 1999. Autistic spectrum disorders in childhood epilepsy surgery candidates. Eur. Child Adolesc. Psychiatry 8, 189–192.

Thirumalai, S.S., Shubin, R.A., Robinson, R., 2002. Rapid eye movement sleep behavior disorder in children with autism. J. Child Neurol. 17, 173–178.

Townsend, J., Courchesne, E., Egaas, B., 1996. Slowed orienting of covert visual-spatial attention in autism: specific deficits associated with cerebellar and parietal abnormality. Dev. Psychopathol. 8, 563–584.

Tuchman, R., Rapin, I., 2002. Epilepsy in autism. Lancet Neurol. 1, 352–358.

Tucker, D.M., Luu, P., Derryberry, D., 2005. Love hurts: the evolution of empathic concern through the encephalization of nociceptive capacity. Dev. Psychopathol. 17, 699–713.

Tudor, M.E., Hoffman, C.D., Sweeney, D.P., 2012. Children with autism: sleep problems and symptom severity. Focus Autism Dev. Disabil. 27, 254–262.

Turner, K.C., Frost, L., Linsenbardt, D., McIlroy, J.R., Müller, R.-A., 2006. Atypically diffuse functional connectivity between caudate nuclei and cerebral cortex in autism. Behav. Brain Funct. BBF 2, 34.

Veatch, O.J., Maxwell-Horn, A.C., Malow, B.A., 2015. Sleep in autism spectrum disorders. Curr. Sleep Med. Rep. 1, 131–140.

Voineagu, I., Wang, X., Johnston, P., Lowe, J.K., Tian, Y., et al., 2011. Transcriptomic analysis of autistic brain reveals convergent molecular pathology. Nature 474, 380–384.

Volkmar, F.R., Nelson, D.S., 1990. Seizure Disorders in Autism. J. Am. Acad. Child Adolesc. Psychiatry 29, 127–129.

Volkmar, F., Siegel, M., Woodbury-Smith, M., King, B., McCracken, J., et al., 2014. Practice parameter for the assessment and treatment of children and adolescents with autism spectrum disorder. J. Am. Acad. Child Adolesc. Psychiatry 53, 237–257.

Wacker, J., Dillon, D.G., Pizzagalli, D.A., 2009. The role of the nucleus accumbens and rostral anterior cingulate cortex in anhedonia: integration of resting EEG, fMRI, and volumetric techniques. NeuroImage 46, 327–337.

Walsh, J.J., Christoffel, D.J., Heifets, B.D., Ben-Dor, G.A., Selimbeyoglu, A., et al., 2018. 5-HT release in nucleus accumbens rescues social deficits in mouse autism model. Nature 560, 589–594.

Wassink, T.H., Brzustowicz, L.M., Bartlett, C.W., Szatmari, P., 2004. The search for autism disease genes. Ment. Retard. Dev. Disabil. Res. Rev. 10, 272–283.

Weiss, L.A., Shen, Y., Korn, J.M., Arking, D.E., Miller, D.T., et al., 2008. Association between microdeletion and microduplication at 16p11.2 and autism. N. Engl. J. Med. 358, 667–675.

Wichniak, A., Wierzbicka, A., Walęcka, M., Jernajczyk, W., 2017. Effects of antidepressants on sleep. Curr. Psychiatry Rep. 19, 63–69.

Wiggs, L., Stores, G., 2004. Sleep patterns and sleep disorders in children with autistic spectrum disorders: insights using parent report and actigraphy. Dev. Med. Child Neurol. 46, 372–380.

Wood, J.N., Grafman, J., 2003. Human prefrontal cortex: processing and representational perspectives. Nat. Rev. Neurosci. 4, 139–147.

Young, A.W., Hellawell, D.J., Van De Wal, C., Johnson, M., 1996. Facial expression processing after amygdalotomy. Neuropsychologia 34, 31–39.

Zielinski, B.A., Prigge, M.B.D., Nielsen, J.A., Froehlich, A.L., Abildskov, T.J., et al., 2014. Longitudinal changes in cortical thickness in autism and typical development. Brain J. Neurol. 137, 1799–1812.

Zwaigenbaum, L., Bryson, S., Rogers, T., Roberts, W., Brian, J., et al., 2005. Behavioral manifestations of autism in the first year of life. Int. J. Dev. Neurosci. Off. J. Int. Soc. Dev. Neurosci. 23, 143–152.

Chapter 7

Arousal and drug abuse

Francisco J. Urbano, Veronica Bisagno
Department of Physiology, Molecular and Cellular Biology "Dr. Héctor Maldonado "; CONICET, Institute of Physiology, Molecular Biology and Neurosciences (IFIBYNE) and CONICET, Institute of Pharmacological Research (ININFA), University of Buenos Aires, Buenos Aires, Argentina

Introduction

Wakefulness is one of the daily recurring states of arousal, during which neural networks actively acquire external information, continuously sense and logically process them, and continuously control voluntary movements. The reticular activating system (RAS) serves to modulate the oscillating rhythms between thalamocortical networks, directly controlling transitions in EEG activity during sleep-wake states (Garcia-Rill, 2009, 2015), and it is also required for vigilance and consciousness (Gottesmann, 1988). Indeed, transection of the brain stem (i.e., *encéphale isolé*) while leaving untouched the RAS produced EEG patterns normally observed during waking and REM sleep (Kerkhofs and Lavie, 2000; Garcia-Rill, 2009, 2015). The RAS includes the pedunculopontine (PPN, cholinergic), locus coeruleus (LC, noradrenergic), and raphe nuclei (RN) (Garcia-Rill, 2009; Urbano et al., 2015). The RAS relays sensory "arousal" information to the cortex through the "nonspecific" intralaminar thalamic nuclei. This system functions in parallel with the "specific" sensory input through the primary sensory thalamic nuclei that are also relayed to the cortex (Garcia-Rill, 2009, 2015). The temporal summation ("binding") of "specific" and "nonspecific" afferents to the cortex is thought to provide the content and context of sensory experience that permits binding of perceptions, respectively (Llinas et al., 1994, 2002; Llinás et al., 1999). Alterations in thalamocortical processing have been suggested to mediate the transition from recreational to a more compulsive, dysregulated drug-seeking behavior resulting in the well-documented changes associated with addiction (Balleine et al., 2015). Therefore the disturbance in the ability of the RAS to modulate thalamocortical oscillations can lead to disturbances in perception and awareness and promote abnormal drug-seeking behavior (Garcia-Rill, 2009; Urbano et al., 2018).

Arousal in Neurological and Psychiatric Diseases. https://doi.org/10.1016/B978-0-12-817992-5.00007-6

Thalamocortical dysrhythmia, a potential mechanism for drug-seeking behavior

Upon waking, sensory information onto PPN neurons induces arousal, representing a continuous flow of afferent input (Garcia-Rill, 2009, 2015). The role of gamma band activity in PPN neurons has been proposed to stabilize coherence related to arousal, providing a stable state during waking (Garcia-Rill et al., 2013). The PPN relays gamma band activity to cells in the intralaminar "nonspecific" thalamus that projects to the cortex to elicit arousal. Coincidence detection by coactivation of the "specific" (sensory relay) and the "nonspecific" (intralaminar) thalamic nuclei at gamma band frequencies has been proposed as the basis for the temporal conjunction that supports cognitive binding in the brain (Llinas et al., 1994, 2002; Llinás et al., 1999). During awake states, thalamocortical neurons manifest a tonic firing mode when their resting membrane potential levels are depolarized enough to activate sodium- ("specific") and P/Q-type ("nonspecific") channel-mediated gamma oscillations. However, when thalamocortical neurons manifest more hyperpolarized membrane potentials (e.g., during enhanced GABA receptor activation or lower glutamate receptor activation), a burst firing pattern has been described due to enhanced T-type calcium channel activation (Jahnsen and Llinás, 1984a, b; McCormick and Feeser, 1990; Pedroarena and Llinás, 1997).

Transitions between these modes of thalamocortical activity depend on slow membrane depolarizations that modulate calcium channels to be open/closed/inactivated according to their voltage dependence of activation (Jahnsen and Llinás, 1984a, b; McCormick and Feeser, 1990; Pedroarena and Llinás, 1997). It is important to emphasize that thalamocortical low-frequency bursting blunted effective sensory information (i.e., distorted perception/arousal), whereas high-frequency gamma oscillation thalamocortical activity accurately relayed sensory information (McCormick and Feeser, 1990). Abnormal low-frequency oscillatory activity of thalamocortical neurons has been related to altered thalamocortical dynamics described as the basis for several types of neurological and neuropsychiatric conditions, collectively named *thalamocortical dysrhythmia syndrome* (Llinás et al., 1999, 2005). That is, the occurrence of decreased thalamic activity levels shifting from high-frequency tonic firing patterns to low-frequency bursting patterns represents conditions that lead to *thalamocortical dysrhythmia*. Furthermore, gamma oscillation deficits have been described as a *pathophysiological marker* of a number of brain diseases (Ribary et al., 1991; Stam et al., 2002; Uhlhaas and Singer, 2013). The presence of long-lasting, protracted activation of such low-voltage-activated (LVA) T-type calcium currents at the thalamic level would in turn be relayed to the cortical mantle. Low- and high-frequency interaction between nearby cortical columns generated asymmetrical cortical inhibition called the *edge effect* (Llinás et al., 2005). This contrast between mismatched inputs is thought to produce errors in perception, which do not occur when the inputs are matched.

Cocaine abusers show GABAergic disruption in the striatum, thalamus, and parietal cortex (Volkow et al., 1998). In a visuospatial attention task, using functional magnetic resonance imaging (fMRI), cocaine abusers also showed lower thalamic activation, higher occipital and prefrontal cortex activation, and higher deactivation of parietal regions than controls (Tomasi et al., 2007). In rats, systemically administered acute cocaine is also able to modulate somatosensory thalamocortical processing (Rutter et al., 1998, 2005), akin to *thalamocortical dysrhythmia*, characterized at the thalamic level as an enhancement of synaptic GABAergic release on ventrobasal (VB) neurons and higher T-type calcium and hyperpolarization-activated cyclic nucleotide-gated (h-current) density in VB neurons (Urbano et al., 2009; Bisagno et al., 2010). These conditions would reduce thalamocortical activity from high frequencies to low frequencies, leading to a lack of coincidence or mismatching between "specific" and "nonspecific" inputs to the cortex, producing a *thalamocortical dysrhythmia*, leading to perceptual disturbances.

Impaired *thalamocortical-based* arousal brain states have been described during psychostimulant withdrawal, and insomnia-like symptoms are present in addicted users of cocaine, methamphetamine, and other psychostimulants (Hasler et al., 2014; Cadet and Bisagno, 2016). Basically, disturbed sleep has been described as a risk factor that *predicts* relapse to stimulant abuse (Brower and Perron, 2010). Therefore, rebalancing arousal, wake/sleep homeostasis would be key to design a treatment response to addicted users. In fact the absence of *thalamocortical dysrhythmia syndrome*-like symptoms would signal successful alleviation of symptoms associated with abnormal sensory processing.

Symptoms and etiology of arousal alterations mediated by stimulants

A number of arousal-related problems have been associated with the consumption of stimulants (Krasnova and Cadet, 2009; Zhang et al., 2014; Urbano et al., 2015; Cadet and Bisagno, 2016). Heavy exposure to cocaine in utero has been proposed to reduce the infant's capacity for modulation of arousal during the early months of life (Bendersky and Lewis, 1998). Similar findings were found in rodents (Morgan et al., 2002). Repetitive administration of cocaine in rats resulted in lower total sleep time plus alterations in sleep-wake rhythms that were attenuated by systemic (intraperitoneal, *i.p.*) administration of a serotonin type 2 (5-HT_2) receptor antagonist (Dugovic et al., 1992). Furthermore the long-term effects of cocaine on arousal have been described to produce alterations in gene expression (Black et al., 2006; Zhou et al., 2008). Methamphetamine abuse has also been associated with deleterious effects on arousal (Edgar and Seidel, 1997; Bisagno et al., 2002; Krasnova and Cadet, 2009; González et al., 2018). Interestingly, modafinil is a stimulant that acts by blocking dopamine transporters at high concentration (Kim et al., 2014) and increasing electric coupling, especially in GABAergic neurons (Garcia-Rill et al., 2007; Urbano et al., 2007), that can improve the methamphetamine-mediated impairments described in the prefrontal cortex (González et al., 2014).

The underlying mechanisms mediating the potent arousal-enhancing actions of stimulants involve several monoaminergic neurotransmitter systems. Cocaine promotes elevated synaptic levels of dopamine (DA) neurotransmission (Wilson et al., 1976; Wise and Bozarth, 1987; Volkow et al., 1997), followed by increases in both norepinephrine (NE) and 5-HT neurotransmission related to monoamine reuptake inhibition (Glowinski and Axelrod, 1966; Ross and Renyi, 1969). While rapid increases in brain concentrations of cocaine are able to block all dopamine (DAT), norepinephrine (NET), and serotonin (SERT) transporters, methylphenidate mainly inhibits DAT and NET (Kuczenski and Segal, 1997; Segal and Kuczenski, 1999; Han and Gu, 2006). Higher levels of monoaminergic neurotransmission have been linked to the direct modulation of intrinsic properties of thalamic (Goitia et al., 2016; Rozas et al., 2017) and cortical neurons (Bisagno et al., 2016; González et al., 2016) induced by stimulants. Indeed, monoaminergic synaptic receptors act through the activation of G-protein-coupled receptors (GPCR, Fig. 7.1) (Roth et al., 1998; Johnson and Lovinger, 2016; Borroto-Escuela et al., 2017; Derouiche and Massotte, 2018) that mediate the activation/inhibition of voltage-gated ionic channels responsible for neuronal excitability, endowing them with autorhythmic membrane oscillatory capabilities and their capability of enhancing oscillations throughout arousal-controlling brain stem-thalamocortical networks (Llinás, 1988; Garcia-Rill et al., 2013). In addition, neurotrophic factors that bind to tropomyosin-related kinase (Trk) receptors (Fig. 7.1) are functionally coupled to monoaminergic pathways (McGinty et al., 2008, 2010). Trophic factor brain-derived neurotrophic factor (BDNF) induced long-lasting potentiation of cocaine seeking during abstinence (Lu et al., 2004), while BDNF immune neutralization attenuated addictive cocaine behavioral effects (Graham et al., 2007).

Alterations in plasma membrane expression of several voltage-gated ion channels (and intracellular pathways coupled to their activity) have been proposed as the etiology of arousal alterations mediated by stimulants. For example, cocaine was found to mediate alterations in T-type calcium channel expression at thalamocortical and striatal networks (Urbano et al., 2009; Bisagno et al., 2010; Gangarossa et al., 2014). Higher T-type channel and h-current activation, plus higher synaptic GABAergic activity induced by cocaine administration in mice, was suggested to "lock" the whole thalamocortical system into low frequencies in the EEG (Urbano et al., 2009). In addition, methamphetamine reduced voltage-gated calcium and increased I_H channels that led to lower glutamate synaptic release in deep-layer pyramidal mPFC neurons (Bisagno et al., 2016; González et al., 2016). Moreover, modafinil enhanced gap junction opening in PPN and thalamocortical neurons, activating an intracellular pathway involving Ca^{2+}/calmodulin protein kinase II (Garcia-Rill et al., 2007; Urbano et al., 2007). Furthermore, level of arousal mediated by PPN activation has been suggested to be waking and is modulated by the Ca^{2+}-calmodulin-dependent

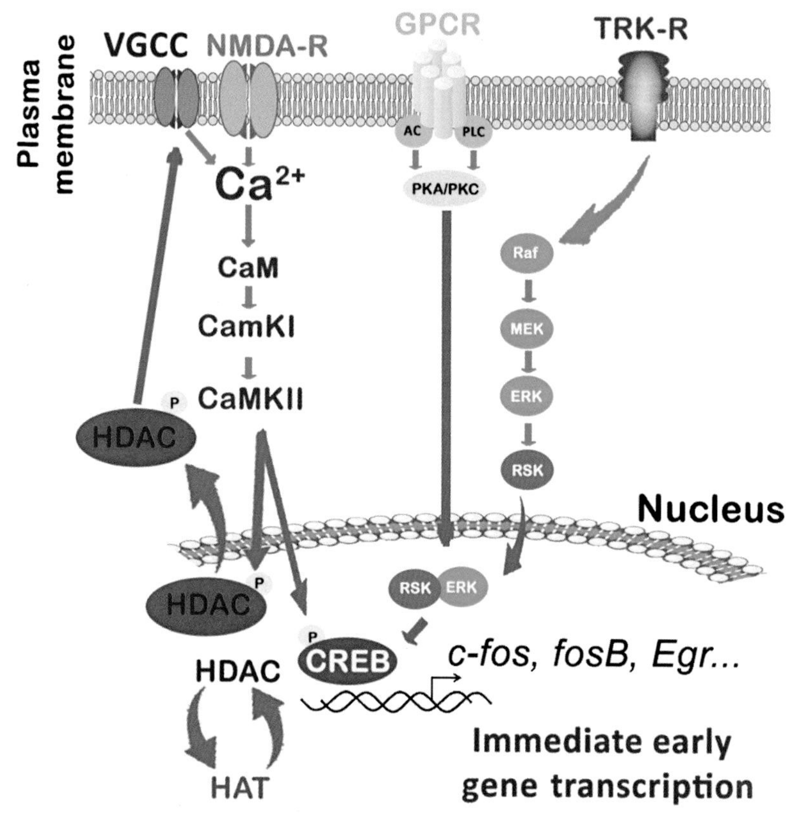

FIG. 7.1 Overview of intracellular signaling pathways that transform plasma membrane effects of several monoamine and synaptic receptors in nuclear neuroadaptations underlying arousal alterations. At least three pathways have been described to positively activate the CREB signaling pathway via CREB phosphorylation (pCREB), which has been found to increase early transcription genes: (1) Ca^{2+}-dependent pathways triggered by the activation of voltage-gated calcium channels (P/Q-type voltage-gated calcium channel—VGCC) or glutamatergic NMDA receptor. *Calcium-dependent pathway*: Intracellular $[Ca^{2+}]$ increase activates CAM-dependent pathways resulting in the upregulation of CaM kinases (CaMKI and CaMKII). A CaMKII-dependent phosphorylation is required for the action of modafinil on thalamocortical (Urbano et al., 2007) and pedunculopontine nuclei (Garcia-Rill et al., 2007). Furthermore, CaMKII is also necessary for class II histone deacetylase (HDAC) translocation from the nucleus to the cytoplasm in order to control the normal activation of P/Q-type calcium channels (Urbano et al., 2018). (2) PKA/PKC-mediated pathway that requires the activation of G-protein-coupled receptors (GPCR). Either homomeric (Roth et al., 1998; Johnson and Lovinger, 2016) or heteromeric (Borroto-Escuela et al., 2017; Derouiche and Massotte, 2018) GPCRs mediate the effects of several stimulants and are key elements to understanding long-lasting drug abuse. (3) A signal-regulated kinase (ERK), ribosomal S6 kinase (RSK), and mitogen- and stress-activated kinase (MSK) pathway mediated by the activation of tyrosine kinase-coupled receptors (TRK) that are known to activate in response to growth factors, neurotransmitters, and other stimuli (Hauge and Frödin, 2006). MSK has only one well-established function: regulation of gene expression by phosphorylation of transcription factors and chromatin-associated proteins (Hauge and Frödin, 2006). Intracellular ERK-mediated pathways induced dendrite complexity via sustained signaling and CREB-mediated signaling (Ha and Redmond, 2008).

protein kinase II (CaMKII) pathway (Datta et al., 2011), while REM sleep is modulated by the cAMP/PKA pathway (Datta and Desarnaud, 2010) (Fig. 7.1).

Downstream from the plasma membrane, stimulant-mediated activation of intracellular pathways (Fig. 7.1) exerts epigenetic effects (González et al., 2018, 2019; Jayanthi et al., 2018; Krasnova et al., 2013; Sadri-Vakili, 2015; Schmidt et al., 2013; Urbano et al., 2018). Epigenetics refers to nongenetic mechanisms (not explained by the DNA sequence) that regulate gene expression (Fig. 7.1). This regulation can be short or long term, and for this regulation, there are three groups of epigenetic "markers": modifications on histones, on DNA, and on nonmessenger RNA. Regarding applications to pharmacology, it is interesting to note that these "markers" can be reversible: that is, drugs targeting certain enzymes (e.g., histone deacetylases or HDACs) could be designed to try to reverse alterations in transcriptional networks. Importantly, histone deacetylases have been proposed to finely modulate, within 15–30 min, the activity of voltage-gated calcium channels expressed at the plasma membrane level (Urbano et al., 2018) and to modulate immediate early genes like *c-fos* and *c-jun* (Fig. 7.1; Clayton et al., 2000; Khan and Davie, 2013a, b).

Modafinil and methamphetamine have been described to have differential effects on epigenetic and transcriptional markers. Methamphetamine showed drastic epigenetic alterations that have been correlated with the long-term cognitive decline effects of the drug on prefrontal cortex function (González et al., 2018, 2019). Modafinil and methamphetamine repetitive systemic treatment differentially enriched levels of acetylated histones 3 (H3ac) and 4 (H4ac) at promoters of the following arousal-dependent receptors: DA receptor 1, alpha-adrenergic (α1) receptors, orexin receptors, histamine receptors 1 and 3, and glutamate α-amino-3-hydroxy-5-methyl-4-isoxazolepropionic acid (AMPA) and NMDA receptors (González et al., 2018). Single-dose systemic administration of modafinil or methamphetamine also presented distinct effects on acetylated histone 3 (González et al., 2019). The lack of similar epigenetic effects of chronic modafinil administration suggested a possible mechanism for its neuroprotective effects against methamphetamine-induced toxicity (González et al., 2014; Raineri et al., 2011, 2012, 2015). Indeed, modafinil prevented methamphetamine-induced toxic effects at the striatum including DA depletion and reductions in tyrosine hydroxylase and DA transporter levels (Raineri et al., 2011). In addition, previous administration of modafinil blunted methamphetamine-induced hyperthermia, activation of astroglia/microglia, and proapoptotic protein expression in the mouse striatum (Raineri et al., 2012, 2015).

Repetitive methamphetamine treatment in mice showed in the medial prefrontal cortex (a) decreased levels of total histones 3 and 4 tail acetylation and DNA cytosine methylation and (b) decreased histone 3 tail acetylation enrichment at promoters of dopaminergic, orexinergic, histaminergic, and NMDA receptors, while it increased histone 4 tail enrichment at dopaminergic, histaminergic, and NMDA receptors. Therefore modafinil-treated mice shared none of these methamphetamine-induced effects (González et al., 2018).

However, single-dose methamphetamine administration, but not modafinil, altered mRNA levels in medial prefrontal cortex of genes representing histone acetylation that mediated higher expression levels of several receptors (dopaminergic, adrenergic, orexinergic, and histaminergic) while decreased expression of NMDA receptors (González et al., 2019). Thus acute methamphetamine and modafinil increased DA neurotransmission in the medial prefrontal cortex but showing contrasting epigenetic and transcriptional consequences that may account for their divergent clinical effects.

Symptoms and etiology of arousal alterations mediated by sedative-hypnotics and anesthetics

Precise modulation of thalamocortical rhythmicity is required during arousal transitions after anesthesia (Choi et al., 2015). Alterations in arousal mediated by general anesthetics, like the GABA-A agonist propofol (Fig. 7.2), have been suggested to be due to disrupting activity within the thalamocortical system (Balleine et al., 2015; Urbano et al., 2015) and its brain stem afferent modulation (Garcia-Rill, 2009; Leung et al., 2014; Scheib, 2017). Coherent low-frequency oscillations between thalamic and cortical networks are correlated with the loss of arousal levels (Flores et al., 2017). General anesthetics have been used as illicit drugs (Jansen and Darracot-Cankovic, 2001). In addition, ketamine, a dissociative anesthetic pharmacologically classified as an NMDA antagonist with high abuse liability (Liu et al., 2016), induced abnormal thalamocortical connectivity of resting state in the brains of chronic abusers (Liao et al., 2016).

Abuse of benzodiazepines, a group of sedative-hypnotics that act as agonists of GABA-A receptors and represent the most commonly prescribed medications in the world (Guina and Merrill, 2018), has been linked to several arousal alterations such as the disruption of sleep architecture (Michelini et al., 1996; Guina and Merrill, 2018). Adverse effects of benzodiazepines on arousal can be exacerbated when they are abused in combination with other anesthetics (De la Peña and Cheong, 2016). Therefore mechanisms underlying the effects of sedative-hypnotics and anesthetics on arousal have been related to the enhancement of inhibitory levels (either by enhancing GABA-A receptor activation or by blocking NMDA receptors) of thalamocortical networks in a widespread manner, that is, producing conditions akin to *thalamocortical dysrhythmia and its cortical correlation*, "the edge effect" (Fig. 7.2). Anesthetics were able to alter epigenetic markers in several neuronal nuclei. Significantly, epigenetic regulation of histone deacetylase in the cortical-accumbens network caused by the anesthetic ketamine has been described as a key molecular event that was necessary to reverse specific stress-induced behavior (Reus et al., 2013).

Fig. 7.2 illustrates how GABA-A anesthetics would exacerbate the overactivation at low frequencies of the entire ThCo system in the awake state (which should be characterized by sensory activation at high frequencies). GABA release from thalamic reticular nucleus (TRN) and cortical

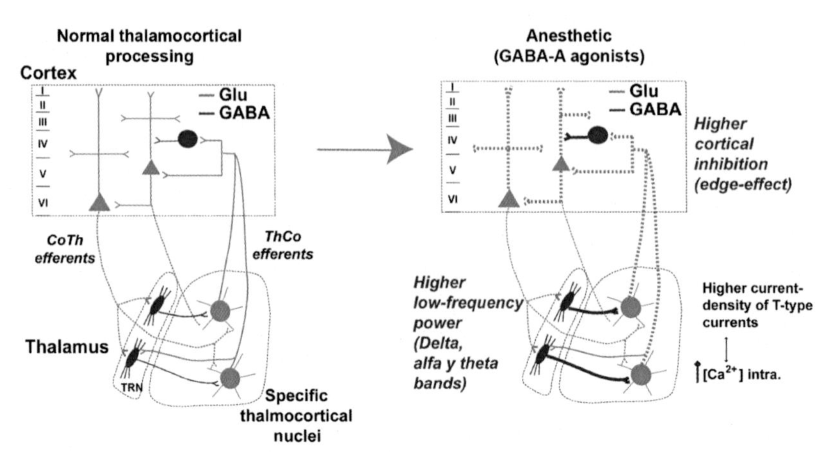

FIG. 7.2 Overview of thalamocortical dysrhythmia-like effects mediated by anesthetics and sedative-hypnotics. During normal sensory processing, the thalamus and cortex together with neostriatal areas are massively interconnected (Jones, 1985). The sensory thalamus is populated by two distinct sets of neurons: (1) cortically projecting relay neurons that subserve the thalamocortical loop via glutamatergic synapses to the central layers of the somatosensory cortex and (2) thalamic reticular nucleus (TRN) GABAergic neurons projecting to sensory neurons. Cortical layer V and VI pyramidal neurons subserve the corticothalamic loop via glutamatergic synapses directly on both TRN and sensory thalamic nuclei. The thalamic reticular nucleus (TRN) surrounds the sensory thalamus and acts to inhibit the activation of the sensory thalamus. At the neuronal level, low and high frequencies are segregated in the thalamocortical system due to the fact that both specific and reticular thalamic neurons can display two mutually exclusive electrophysiological modes (Steriade and Llinas, 1988; Llinas and Pare, 1991; Llinas and Ribary, 1993): (1) a tonic firing mode in which thalamocortical neurons are partly depolarized and (2) a burst firing mode when thalamocortical neurons are hyperpolarized (Jahnsen and Llinás, 1984a, 1984b; McCormick and Feeser, 1990). Thalamic relay neurons in tonic firing mode discharge in direct relation to afferent sensory input, whereas in burst firing mode, sensory information is not transmitted. The activation of GABA receptors on thalamic relay neurons by GABA-agonist-like anesthetics (right panel) released by reticular thalamic terminals will trigger prolonged membrane hyperpolarization. Then, T-type calcium channels will deactivate, allowing thalamocortical neurons to respond with bursts of high-amplitude low-threshold spikes generating abnormally higher power of low-frequency bands that mediate a *thalamocortical dysrhythmia and its cortical correlation*, "the edge effect" (Llinás et al., 1999, 2005; Urbano et al., 2009).

interneurons (red dots and thick lines) would prolong the hyperpolarization of specific ThCo neurons throughout the increment of GABA-A receptors open time (Guina and Merrill, 2018). This would cause the abnormal increment in bursting mediated by T channels of the NRT and VB neurons (Urbano et al., 2009), resulting in alterations of sensory input processing and abrupt changes in sleep-wake cycles (Llinas et al., 1994, 2002; Llinás et al., 1999). Furthermore, higher cortical inhibition mediated by sustained low-frequency ThCo afferents (Fig. 7.2, right panel) would blunt cortical processing of sensory information, a condition previously defined as *the edge effect* (Llinás et al., 2005).

Clinical manifestations of arousal alterations by drug abuse and its social and economic impact

Substance abuse is often initiated during adolescence, while brain areas related to reward circuits are undergoing developmental changes (Schramm-Sapyta et al., 2004) and are prone to an increasing number of vulnerable conditions (Gerra et al., 1998; Compas et al., 1995). Indeed, epidemiological data suggest that the use of various kinds of psychoactive agents is widespread during this period, ranging from 11 to 18 years of age (Compas et al., 1995; Mathias, 1996; Witt, 1994). Indeed, adolescents are likely to start with tobacco and/or alcohol, followed by marijuana, and eventually use psychostimulants and/or opiates (Yamaguchi and Kandel, 1984).

Clinically, cocaine-induced delirium can sometimes be associated with such perceptual distortions as hallucinations, suggesting abnormal thalamocortical processing (Behrendt, 2006; Urbano et al., 2009). For example, cocaine abusers showed lower thalamic activation, higher occipital and prefrontal cortex activation, and higher deactivation of parietal regions, compared with controls (Tomasi et al., 2007), and several cognitive deficits (Cadet and Bisagno, 2016). As mentioned earlier, abnormal coherent low-frequency alteration at the thalamocortical level has been associated with the lower arousal levels induced by general anesthetics (Flores et al., 2017). Given the fact that general anesthesia is a widely used manipulation of the brain and central nervous system in medicine (Brown et al., 2010), understanding and further establishing the role of thalamocortical networks on anesthesia-induced unconsciousness and altered arousal would greatly reduce neurocognitive side effects of these drugs (Ching and Brown, 2014; Akeju and Brown, 2017). Perhaps blunted thalamocortical rhythmicity might underlie deficits in cognitive and emotional processes that contribute to dysregulated ketamine-seeking behavior (Balleine et al., 2015).

Concluding remarks

In conclusion the pursuit of better descriptions of the effects on acetylation/deacetylation and methylation balance as key regulatory mechanisms of neuronal intrinsic properties is essential. Modafinil and methamphetamine effects on prefrontal cortex function have been suggested to be mediated by their differential epigenetic modulation of the expression of key proteins to maintain arousal like DA receptor 1, alpha-adrenergic ($\alpha 1$) receptors, orexin receptors, histamine receptors 1 and 3, and glutamate AMPA and NMDA receptors. These epigenetic findings point to a larger number of potentially fruitful areas of research that need to be pursued. In particular, further understanding of thalamocortical physiology mediated by epigenetic alterations in patients after the abuse of stimulant or sedative-hypnotics/anesthetics might be crucial to develop future treatments.

Acknowledgments

Supported by NIH award P30 GM110702 from the IDeA program at NIGMS to the CTN. In addition, this work was supported by grants from FONCYT-ANPCyT and Préstamo BID 1728 OC.AR. PICT-2016-1728 (to Dr. Urbano) and PICT 2015-2594 (to Dr. Bisagno).

References

Akeju, O., Brown, E.N., 2017. Neural oscillations demonstrate that general anesthesia and sedative states are neurophysiologically distinct from sleep. Curr. Opin. Neurobiol. 44, 178–185.

Balleine, B.W., Morris, R.W., Leung, B.K., 2015. Thalamocortical integration of instrumental learning and performance and their disintegration in addiction. Brain Res. 1628, 104–116.

Bendersky, M., Lewis, M., 1998. Arousal modulation in cocaine-exposed infants. Dev. Psychol. 34 (3), 555–564.

Behrendt, R.P., 2006. Dysregulation of thalamic sensory "transmission" in schizophrenia: neurochemical vulnerability to hallucinations. J. Psychopharmacol. 20 (3), 356–372.

Bisagno, V., Ferguson, D., Luine, V.N., 2002. Short toxic methamphetamine schedule impairs object recognition task in male rats. Brain Res. 940 (1–2), 95–101.

Bisagno, V., González, B., Urbano, F.J., 2016. Cognitive enhancers versus addictive psychostimulants: the good and bad side of dopamine on prefrontal cortical circuits. Pharmacol. Res. 109, 108–118.

Bisagno, V., Raineri, M., Peskin, V., Wikinski, S.I., Uchitel, O.D., Llinás, R.R., Urbano, F.J., 2010. Effects of T-type calcium channel blockers on cocaine-induced hyperlocomotion and thalamocortical GABAergic abnormalities in mice. Psychopharmacology (Berlin) 212 (2), 205–214.

Black, Y.D., Maclaren, F.R., Naydenov, A.V., Carlezon Jr., W.A., Baxter, M.G., Konradi, C., 2006. Altered attention and prefrontal cortex gene expression in rats after binge-like exposure to cocaine during adolescence. J. Neurosci. 26 (38), 9656–9665.

Borroto-Escuela, D.O., Carlsson, J., Ambrogini, P., Narváez, M., Wydra, K., Tarakanov, A.O., Li, X., Millón, C., Ferraro, L., Cuppini, R., Tanganelli, S., Liu, F., Filip, M., Diaz-Cabiale, Z., Fuxe, K., 2017. Understanding the role of GPCR heteroreceptor complexes in modulating the brain networks in health and disease. Front. Cell Neurosci. 11, 37.

Brower, K.J., Perron, B.E., 2010. Sleep disturbance as a universal risk factor for relapse in addictions to psychoactive substances. Med. Hypotheses 74 (5), 928–933.

Brown, E.N., Lydic, R., Schiff, N.D., 2010. General anesthesia, sleep, and coma. New Eng. J. Med. 363, 2638–2650.

Cadet, J.L., Bisagno, V., 2016. Neuropsychological consequences of chronic drug use: relevance to treatment approaches. Front. Psychiatry. 6, 189.

Ching, S., Brown, E.N., 2014. Modeling the dynamical effects of anesthesia on brain circuits. Curr. Opin. Neurobiol. 25, 116–122.

Choi, S., Yu, E., Lee, S., Llinás, R.R., 2015. Altered thalamocortical rhythmicity and connectivity in mice lacking CaV3.1 T-type Ca^{2+} channels in unconsciousness. Proc. Natl. Acad. Sci. U. S. A. 112 (25), 7839–7844.

Clayton, A.L., Rose, S., Barratt, M.J., Mahadevan, L.C., 2000. Phosphoacetylation of histone H3 on c-fos- and c-jun-associated nucleosomes upon gene activation. EMBO J. 19, 3714–3726.

Compas, B.E., Hinden, B.R., Gerhardt, C.A., 1995. Adolescent development: pathways and processes of risk and resilience. Ann. Rev. Psychol. 46, 265–293.

Datta, S., Desarnaud, F., 2010. Protein kinase A in the pedunculopontine tegmental nucleus of rat contributes to regulation of rapid eye movement sleep. J. Neurosci. 30, 12263–12273.

Datta, S., O'Malley, M., Patterson, E.H., 2011. Calcium/Calmodulin Kinase II in the pedunculopontine tegmental nucleus modulates the initiation and maintenance of wakefulness. J. Neurosci. 31, 17007–17016.

De la Peña, J.B., Cheong, J.H., 2016. The abuse liability of the NMDA receptor antagonist-benzodiazepine (tiletamine-zolazepam) combination: evidence from clinical case reports and preclinical studies. Drug Test. Anal. 8 (8), 760–767.

Derouiche, L., Massotte, D., 2018. G protein-coupled receptor heteromers are key players in substance use disorder. Neurosci. Biobehav. Rev. pii: S0149-7634(18), 30158-1.

Dugovic, C., Meert, T.F., Ashton, D., Clincke, G.H.C., 1992. Effects of ritanserin and chlordiazepoxide on sleep-wakefulness alterations in rats following chronic cocaine treatment. Psychopharmacology 108 (3), 263–270.

Edgar, D.M., Seidel, W.F., 1997. Modafinil induces wakefulness without intensifying motor activity or subsequent rebound hypersomnolence in the rat. J. Pharmacol. Exp. Ther. 283 (2), 757–769.

Flores, F.J., Hartnack, K.E., Fath, A.B., Kim, S.E., Wilson, M.A., Brown, E.N., Purdon, P.L., 2017. Thalamocortical synchronization during induction and emergence from propofol-induced unconsciousness. Proc. Natl. Acad. Sci. U. S. A. 114 (32), E6660–E6668.

Garcia-Rill, E., 2009. Reticular activating system. In: Stickgold, R., Walker, M. (Eds.), The Neuroscience of Sleep. Elsevier, Oxford, pp. 133–139.

Garcia-Rill, E., 2015. Waking and the Reticular Activating System. Academic Press, p.330.

Garcia-Rill, E., Heister, D.S., Ye, M., Charlesworth, A., Hayar, A., 2007. Electrical coupling: novel mechanism for sleep-wake control. Sleep 30 (11), 1405–1414.

Garcia-Rill, E., Kezunovic, N., Hyde, J., Simon, C., Beck, P., Urbano, F.J., 2013. Coherence and frequency in the reticular activating system (RAS). Sleep Med. Rev. 17 (3), 227–238.

Gangarossa, G., Laffray, S., Bourinet, E., Valjent, E., 2014. T-type calcium channel Cav3.2 deficient mice show elevated anxiety, impaired memory and reduced sensitivity to psychostimulants. Front. Behav. Neurosci. 8, 92.

Gerra, G., Zaimovic, A., Giucastro, G., Maestri, D., Monica, C., Sartori, R., Caccavari, R., Delsignore, R., 1998. Serotonergic function after ± 3,4-methylene-dioxymethamphetamine (ecstasy) in humans. Int. Clin. Psychopharmacol. 13, 1–9.

Guina, J., Merrill, B., 2018. Benzodiazepines I: upping the care on downers: the evidence of risks, benefits and alternatives. J. Clin. Med. 7 (2). pii: E17.

Goitia, B., Rivero-Echeto, M.C., Weisstaub, N.V., Gingrich, J.A., Garcia-Rill, E., Bisagno, V., Urbano, F.J., 2016. Modulation of GABA release from the thalamic reticular nucleus by cocaine and caffeine: role of serotonin receptors. J. Neurochem. 136 (3), 526–535.

González, B., Jayanthi, S., Gomez, N., Torres, O.V., Sosa, M.H., Bernardi, A., Urbano, F.J., García-Rill, E., Cadet, J.L., Bisagno, V., 2018. Repeated methamphetamine and modafinil induce differential cognitive effects and specific histone acetylation and DNA methylation profiles in the mouse medial prefrontal cortex. Prog. Neuropsychopharmacol. Biol. Psychiatry. 82, 1–11.

González, B., Raineri, M., Cadet, J.L., García-Rill, E., Urbano, F.J., Bisagno, V., 2014. Modafinil improves methamphetamine-induced object recognition deficits and restores prefrontal cortex ERK signaling in mice. Neuropharmacology 87, 188–197.

González, B., Rivero-Echeto, C., Muñiz, J.A., Cadet, J.L., García-Rill, E., Urbano, F.J., Bisagno, V., 2016. Methamphetamine blunts Ca^{2+} currents and excitatory synaptic transmission through D1/5 receptor-mediated mechanisms in the mouse medial prefrontal cortex. Addict. Biol. 21 (3), 589–602.

González, B., Torres, O.V., Jayanthi, S., Gomez, N., Sosa, M.H., Bernardi, A., Urbano, F.J., García-Rill, E., Cadet, J.L., Bisagno, V., 2019. The effects of single-dose injections of modafinil and methamphetamine on epigenetic and functional markers in the mouse medial prefrontal cortex: potential role of dopamine receptors. Prog. Neuro-Psychopharmacol. Biol. Psychiatry 88, 222–234.

Gottesmann, C., 1988. What the cerveau isole preparation tells us nowadays about sleep-wake mechanisms? Neurosci. Biobehav. Rev. 12, 39–48.

Glowinski, J., Axelrod, J., 1966. Effects of drugs on the disposition of H-3-norepinephrine in the rat brain. Pharmacol. Rev. 18 (1), 775–785.

Graham, D.L., Edwards, S., Bachtell, R.K., DiLeone, R.J., Rios, M., Self, D.W., 2007. Dynamic BDNF activity in nucleus accumbens with cocaine use increases self-administration and relapse. Nat. Neurosci. 10, 1029–1037.

Ha, S., Redmond, L., 2008. ERK mediates activity dependent neuronal complexity via sustained activity and CREB-mediated signaling. Dev. Neurobiol. 68 (14), 1565–1579.

Han, D.D., Gu, H.H., 2006. Comparison of the monoamine transporters from human and mouse in their sensitivities to psychostimulant drugs. BMC Pharmacol. 6, 6.

Hasler, B.P., Soehner, A.M., Clark, D.B., 2014. Circadian rhythms and risk for substance use disorders in adolescence. Curr. Opin. Psychiatry 27 (6), 460–466.

Hauge, C., Frödin, M., 2006. RSK and MSK in MAP kinase signaling. J. Cell Sci. 119, 3021–3023.

Jansen, K.L., Darracot-Cankovic, R., 2001. The nonmedical use of ketamine, part two: a review of problem use and dependence. J. Psychoactive Drugs 33 (2), 151–158.

Jahnsen, H., Llinás, R., 1984a. Electrophysiological properties of guinea-pig thalamic neurones: an in vitro study. J. Physiol. (Lond.) 349, 205–226.

Jahnsen, H., Llinás, R., 1984b. Ionic basis for the electro-responsiveness and oscillatory properties of guinea-pig thalamic neurones in vitro. J. Physiol. (Lond.) 349, 227–247.

Jayanthi, S., Gonzalez, B., McCoy, M.T., Ladenheim, B., Bisagno, V., Cadet, J.L., 2018. Methamphetamine induces TET1- and TET3-dependent DNA hydroxymethylation of Crh and Avp genes in the rat nucleus accumbens. Mol. Neurobiol. 55 (6), 5154–5166.

Johnson, K.A., Lovinger, D.M., 2016. Presynaptic G protein-coupled receptors: gatekeepers of addiction? Front. Cell. Neurosci. 10, 264.

Jones, E.G., 1985. The Thalamus. Plenum Press, New York.

Khan, D.H., Davie, J.R., 2013a. HDAC inhibitors prevent the induction of the immediate-early gene FOSL1, but do not alter the nucleosome response. FEBS Lett. 587, 1510–1517.

Khan, D.H., Davie, J.R., 2013b. HDAC inhibitors prevent the induction of the immediate-early gene FOSL1, but do not alter the nucleosome response. FEBS Lett. 587, 1510–1517.

Kim, W., Tateno, A., Arakawa, R., Sakayori, T., Ikeda, Y., Suzuki, H., Okubo, Y., 2014. In vivo activity of modafinil on dopamine transporter measured with positron emission tomography and [^{18}F]FE-PE2I. Int. J. Neuropsychopharmacol. 17 (5), 697–703.

Krasnova, I.N., Cadet, J.L., 2009. Methamphetamine toxicity and messengers of death. Brain Res. Rev. 60 (2), 379–407.

Krasnova, I.N., Chiflikyan, M., Justinova, Z., McCoy, M.T., Ladenheim, B., Jayanthi, S., Quintero, C., Brannock, C., Barnes, C., Adair, J.E., Lehrmann, E., Kobeissy, F.H., Gold, M.S., Becker, K.G., Goldberg, S.R., Cadet, J.L., 2013. CREB phosphorylation regulates striatal transcriptional responses in the self-administration model of methamphetamine addiction in the rat. Neurobiol. Dis. 58, 132–143.

Kerkhofs, M., Lavie, P., 2000. Frederic Bremer 1892–1982: a pioneer in sleep research. Sleep Med. Rev. 4, 505–514.

Kuczenski, R., Segal, D.S., 1997. Effects of methylphenidate on extracellular dopamine, serotonin, and norepinephrine: comparison with amphetamine. J. Neurochem. 68 (5), 2032–2037.

Leung, L.S., Luo, T., Ma, J., Herrick, I., 2014. Brain areas that influence general anesthesia. Prog. Neurobiol. 122, 24–44.

Liao, Y., Tang, J., Liu, J., Xie, A., Yang, M., Johnson, M., Wang, X., Deng, Q., Chen, H., Xiang, X., Liu, T., Chen, X., Song, M., Hao, W., 2016. Decreased thalamocortical connectivity in chronic ketamine users. PLoS One 11 (12), e0167381.

Liu, Y., Lin, D., Wu, B., Zhou, W., 2016. Ketamine abuse potential and use disorder. Brain Res. Bull. 126 (Pt 1), 68–73.

Llinás, R.R., 1988. The intrinsic electrophysiological properties of mammalian neurons: insights into central nervous system function. Science 242 (4886), 1654–1664.

Llinas, R.R., Leznik, E., Urbano, F.J., 2002. Temporal binding via cortical coincidence detection of specific and nonspecific thalamocortical inputs: a voltage-dependent dye-imaging study in mouse brain slices. Proc. Natl. Acad. Sci. U. S. A. 99, 449–454.

Llinas, R.R., Pare, D., 1991. Of dreaming and wakefulness. Neuroscience 44 (3), 521–535.

Llinas, R.R., Ribary, U., 1993. Coherent 40-Hz oscillation characterizes dream state in humans. Proc. Natl. Acad. Sci. U. S. A. 90, 2078–2081.

Llinas, R., Ribary, U., Joliot, M., Wang, X.J., 1994. Content and context in temporal thalamocortical binding. In: Buzsaki, G. (Ed.), Temporal Coding in the Brain. Springer, Berlin, pp. 251–272.

Llinás, R.R., Ribary, U., Jeanmonod, D., Kronberg, E., Mitra, P.P., 1999. Thalamocortical dysrhythmia: a neurological and neuropsychiatric syndrome characterized by magnetoencephalography. Proc. Natl. Acad. Sci. U. S. A. 96 (26), 15222–15227.

Llinás, R., Urbano, F.J., Leznik, E., Ramírez, R.R., van Marle, H.J., 2005. Rhythmic and dysrhythmic thalamocortical dynamics: GABA systems and the edge effect. Trends Neurosci. 28 (6), 325–333.

Lu, L., Dempsey, J., Liu, S.Y., Bossert, J.M., Shaham, Y., 2004. A single infusion of brain-derived neurotrophic factor into the ventral tegmental area induces long-lasting potentiation of cocaine seeking after withdrawal. J. Neurosci. 24, 1604–1611.

Mathias, R., 1996. Students' use of marijuana, other illicit drugs and cigarettes continued to rise in 1995. NIDA Notes 11, 8–9.

McCormick, D.A., Feeser, H.R., 1990. Functional implications of burst firing and single spike activity in lateral geniculate relay neurons. Neuroscience 39, 103–113.

McGinty, J.F., Shi, X.D., Schwendt, M., Saylor, A., Toda, S., 2008. Regulation of psychostimulant induced signaling and gene expression in the striatum. J. Neurochem. 104, 1440–1449.

McGinty, J.F., Whitfield, T.W., Berglind, W.J., 2010. Brain-derived neurotrophic factor and cocaine addiction. Brain Res. 1314, 183–193.

Michelini, S., Cassano, G.B., Frare, F., Perugi, G., 1996. Long-term use of benzodiazepines: tolerance, dependence and clinical problems in anxiety and mood disorders. Pharmacopsychiatry 29 (4), 127–134.

Morgan, R.E., Garavan, H.P., Mactutus, C.F., Levitsky, D.A., Booze, R.M., Strupp, B.J., 2002. Enduring effects of prenatal cocaine exposure on attention and reaction to errors. Behav. Neurosci. 116 (4), 624–633.

Pedroarena, C., Llinás, R., 1997. Dendritic calcium conductances generate high-frequency oscillation in thalamocortical neurons. Proc. Natl. Acad. Sci. U. S. A. 94 (2), 724–728.

Raineri, M., Peskin, V., Goitia, B., Taravini, I.R., Giorgeri, S., Urbano, F.J., Bisagno, V., 2011. Attenuated methamphetamine induced neurotoxicity by modafinil administration in mice. Synapse 65 (10), 1087–1098.

Raineri, M., Gonzalez, B., Goitia, B., Garcia-Rill, E., Krasnova, I.N., Cadet, J.L., Urbano, F.J., Bisagno, V., 2012. Modafinil abrogates methamphetamine-induced neuroinflammation and apoptotic effects in the mouse striatum. PLoS One 7 (10), e46599.

Raineri, M., González, B., Rivero-Echeto, C., Muñiz, J.A., Gutiérrez, M.L., Ghanem, C.I., Cadet, J.L., García-Rill, E., Urbano, F.J., Bisagno, V., 2015. Differential effects of environment-induced changes in body temperature on modafinil's actions against methamphetamine-induced striatal toxicity in mice. Neurotox. Res. 27 (1), 71–83.

Reus, G.Z., Abelaira, H.M., dos Santos, M.A., Carlessi, A.S., Tomaz, D.B., Neotti, M.V., Liranço, J.L., Gubert, C., Barth, M., Kapczinski, F., Quevedo, J., 2013. Ketamine and imipramine in the nucleus accumbens regulate histone deacetylation induced by maternal deprivation and are critical for associated behaviors. Behav. Brain Res. 256, 451–456.

Ribary, U., Ioannides, A.A., Singh, K.D., Hasson, R., Bolton, J.P., Lado, F., Mogliner, A., Llinas, R.R., 1991. Magnetic field tomography of coherent thalamocortical 40-Hz oscillations in humans. Proc. Natl. Acad. Sci. U. S. A. 88, 11037–11041.

Ross, S.B., Renyi, A.L., 1969. Inhibition of the uptake of tritiated 5-hydroxytryptamine in brain tissue. Eur. J. Pharmacol. 7 (3), 270–277.

Roth, B.L., Willins, D.L., Kroeze, W.K., 1998. G protein-coupled receptor (GPCR) trafficking in the central nervous system: relevance for drugs of abuse. Drug Alcohol Depend. 51 (1–2), 73–85.

Rozas, J.L., Goitia, B., Bisagno, V., Urbano, F.J., 2017. Differential alterations of intracellular [Ca2+] dynamics induced by cocaine and methylphenidate in thalamocortical ventrobasal neurons. Transl. Brain Rhythm. 2 (1), 1–15.

Rutter, J.J., Baumann, M.H., Waterhouse, B.D., 1998. Systemically administered cocaine alters stimulus-evoked responses of thalamic somatosensory neurons to perithreshold vibrissae stimulation. Brain Res. 798 (1–2), 7–17.

Rutter, J.J., Devilbiss, D.M., Waterhouse, B.D., 2005. Effects of systemically administered cocaine on sensory responses to peri-threshold vibrissae stimulation: individual cells, ensemble activity, and animal behaviour. Eur. J. Neurosci. 22 (12), 3205–3216.

Sadri-Vakili, G., 2015. Cocaine triggers epigenetic alterations in the corticostriatal circuit. Brain Res. 1628 (Pt A), 50–59.

Scheib, C.M., 2017. Brainstem influence on thalamocortical oscillations during anesthesia emergence. Front. Syst. Neurosci. 11, 66.

Schramm-Sapyta, N.L., Pratt, A.R., Winder, D.G., 2004. Effect of periadolescent versus adult cocaine exposure on cocaine conditioned place preference and motor sensitization in mice. Psychopharmacology 173, 41–48.

Schmidt, H.D., McGinty, J.F., West, A.E., Sadri-Vakili, G., 2013. Epigenetics and psychostimulant addiction. Cold Spring Harbor Perspect. Med. 3 (3), a012047.

Segal, D.S., Kuczenski, R., 1999. Escalating dose-binge treatment with methylphenidate: role of serotonin in the emergent behavioral profile. J. Pharmacol. Exp. Ther. 291 (1), 19–30.

Stam, C.J., van Cappellen van Walsum, A.M., Pijnenburg, Y.A., Berendse, H.W., de Munck, J.C., Scheltens, P., van Dijk, B.W., 2002. Generalized synchronization of MEG recordings in Alzheimer's disease: evidence for involvement of the gamma band. J. Clin. Neurophysiol. 19, 562–574.

Steriade, M., Llinas, R.R., 1988. The functional states of the thalamus and the associated neuronal interplay. Physiol. Rev. 68 (3), 649–742.

Tomasi, D., Goldstein, R.Z., Telang, F., Maloney, T., Alia-Klein, N., Caparelli, E.C., Volkow, N.D., 2007. Thalamo-cortical dysfunction in cocaine abusers: implication in attention and perception. Psychiatry Res. Neuroimaging 155, 189–201.

Uhlhaas, P.J., Singer, W., 2013. High-frequency oscillations and the neurobiology of schizophrenia. Dialog. Clin. Neurosci. 15, 301–313.

Urbano, F.J., Bisagno, V., González, B., Rivero-Echeto, C.M., Muñiz, J.A., Luster, B., D'Onofrio, S., Mahaffey, S., Garcia-Rill, E., 2015. Pedunculopontine arousal system physiology-effects of psychostimulant abuse. Sleep Sci. 8 (3), 162–168.

Urbano, F.J., Bisagno, V., Mahaffey, S., Lee, S.H., Garcia-Rill, E., 2018. Class II histone deacetylases require P/Q-type Ca^{2+} channels and CaMKII to maintain gamma oscillations in the pedunculopontine nucleus. Sci. Rep. 8 (1), 13156.

Urbano, F.J., Bisagno, V., Wikinski, S.I., Uchitel, O.D., Llinas, R.R., 2009. Cocaine acute "Binge" administration results in altered thalamocortical interactions in mice. Biol. Psychiatry 66, 769–776.

Urbano, F.J., Leznik, E., Llinás, R.R., 2007. Modafinil enhances thalamocortical activity by increasing neuronal electrotonic coupling. Proc. Natl. Acad. Sci. U. S. A. 104 (30), 12554–12559.

Volkow, N.D., Gur, R.C., Wang, G.J., Fowler, J.S., Moberg, P.J., Ding, Y.S., Hitzemann, R., Smith, G., Logan, J., 1998. Association between decline in brain dopamine activity with age and cognitive and motor impairment in healthy individuals. Am. J. Psychiatry 155 (3), 344–349.

Volkow, N.D., Wang, G.J., Fischman, M.W., Foltin, R.W., Fowler, J.S., Abumrad, N.N., Vitkun, S., Logan, J., Gatley, S.J., Pappas, N., Hitzemann, R., Shea, C.E., 1997. Relationship between subjective effects of cocaine and dopamine transporter occupancy. Nature 386 (6627), 827–830.

Witt, E.D., 1994. Mechanisms of alcohol abuse and alcoholism: a case for developing animal models. Behav. Neural Biol. 62, 168–177.

Wilson, M.C., Bedford, J.A., Buelke, J., Kibbe, A.H., 1976. Acute pharmacological activity of intravenous cocaine in the rhesus monkey. Psychopharmacol. Commun. 2, 251–261.

Wise, R.A., Bozarth, M.A., 1987. A psychomotor stimulant theory of addiction. Psychol. Rev. 94, 469–492.

Yamaguchi, K., Kandel, D.B., 1984. Patterns of drug use from adolescent to young adulthood. III. Predictors of progression. Am. J. Publ. Health 74, 673–681.

Zhang, S., Hu, S., Bednarski, S.R., Erdman, E., Li, C.S., 2014. Error-related functional connectivity of the thalamus in cocaine dependence. Neuroimage Clin. 4, 585–592.

Zhou, Y., Cui, C.L., Schlussman, S.D., Choi, J.C., Ho, A., Han, J.S., Kreek, M.J., 2008. Effects of cocaine place conditioning, chronic escalating-dose "binge" pattern cocaine administration and acute withdrawal on orexin/hypocretin and preprodynorphin gene expressions in lateral hypothalamus of Fischer and Sprague-Dawley rats. Neuroscience 153 (4), 1225–1234.

Further reading

Biederman, J., Wilens, T., Mick, E., Spencer, T., Faraone, S.V., 1999. Pharmacotherapy of attention-deficit/hyperactivity disorder reduces risk for substance use disorder. Pediatrics 104 (2), e20.

Urbano, F.J., Bisagno, V., Garcia-Rill, E., 2017. Arousal and drug abuse. Behav. Brain Res. 333, 276–281.

Chapter 8

Arousal and the Alzheimer disease

Edgar Garcia-Rill
Center for Translational Neuroscience, Department of Neurobiology and Developmental Sciences, University of Arkansas for Medical Sciences, Little Rock, AR, United States

Introduction

AD involves dementia (~70% of cases of dementia are due to AD), memory loss, and other cognitive deficits. Dementia itself involves significant impairments in memory, language, attention, reasoning, and perception. These symptoms get worse over time, that is, dementia also is progressive. AD not only is present in many persons over 65 years of age and is progressive but also may have an early onset in the forties. While survival may range from 5 to 20 years after diagnosis, average AD patients live ~8 years after diagnosis. The diagnosis is tentative until postmortem verification of the presence of plaques and tangles in the brain is made. Plaques are deposits of beta amyloid protein in the extracellular space, and tangles refer to *tau* protein fibers that build up inside cells. In AD, there is a progression of plaques and tangles that begins in the temporal lobe (explaining the earlier onset of memory problems) and spreads to include occipital and frontal lobes in later stages (explaining the deficits in vision and executive function). Early onset may be present as much as 20 years before diagnosis (and can be detected with neuropsychological testing), with mild AD lasting 3–10 years and severe AD lasting 2–5 years.

Symptoms

AD involves difficulty in remembering events and decreases in short-term memory. The formation of new memories is more affected than episodic memory of past events and implicit memory of learned movements. As the disease progresses, there is confusion and long-term memory loss, along with decrements in executive function and attention. The lack of interaction with the external world leads to pronounced apathy, until patients disregard even themselves, failing to perform even the most basic hygienic functions. In some (20%–45%) patients, there is sundowning agitation, in which a range of symptoms including

Arousal in Neurological and Psychiatric Diseases. https://doi.org/10.1016/B978-0-12-817992-5.00008-8

confusion, fatigue, restlessness, and anger increase in the evening, suggestive of circadian dysregulation. Sundowning agitation symptoms in addition include agitation, confusion, anxiety, and aggressiveness that occur in the late afternoon or evening (Khachiyants et al., 2011). Sometimes other clinical features such as mood swings, paranoia, and hallucinations are present in the late afternoon or evening. In some cases, ataxia and postural collapse are observed.

Given the previous discussion (see Chapter 2) on the role of the RAS in controlling sleep and waking and the primary function of the pedunculopontine nucleus (PPN) in maintaining waking and rapid eye movement (REM) sleep, it is clear that this region is disturbed in AD. For example, in Chapter 3, the manifestation of hallucinations was attributed to REM sleep intrusion into waking, suggesting that REM sleep drive is increased in AD, just as it is in schizophrenia (Chapter 3), depression and bipolar disorder (chapter 4), and PTSD and anxiety disorders (Chapter 5).

Etiology

AD is now present in over 30 million people and will soon affect 1 in 100 persons worldwide. Less than half of cases have a strictly genetic etiology, most cases involving combined environmental and genetic risk factors. The most common genetic risk factor is inherited apolipoprotein E 4 allele, which may contribute to amyloid protein accumulation. *Tau* protein abnormalities may lead to microtubule breakdown and accumulation of neurofibrillary tangles. Head trauma and high cholesterol are additional risk factors. The cholinergic hypothesis (cell loss of cholinergic basal forebrain neurons) was developed from animal models showing memory deficits after inactivation of cholinergic basal forebrain neurons. This theory is less accepted today because of the ineffectiveness of medications that treat acetylcholine deficiency in alleviating memory dysfunction. The amyloid hypothesis became the theory of choice based on the findings on patients with extra gene copies of amyloid precursor protein and animal studies showing memory deficits in mice with extra copies of amyloid precursor protein. This was later linked with an inflammatory process that became the preferred theory. However, inflammation appears to be a consequence of some event, not the original cause of the cell damage. Moreover, new drugs that inhibit the formation of beta amyloid protein do not affect the cognitive and functional decline in AD (Egan et al., 2018). The emphasis on amyloid as a causal factor in AD is waning.

There is an abnormal aggregation of *tau* protein in AD and is thus considered a *tau*opathy (Iqbal et al., 2005). The internal support structure of neurons is made up in part of microtubules that are cemented by *tau* phosphorylation. In AD, there is excessive phosphorylation of *tau* protein, which leads to neurofibrillary tangles that impair the transport of nutrients and molecules along microtubules. This process could also result in inflammation and subsequent cellular damage. There is renewed interest in *tau* protein as a causal factor in AD.

The histopathology in the brain of AD patients results in such significant cell loss that the cortex appears atrophied (Wenk, 2003). In addition, there is cell loss in the locus coeruleus (LC), part of the RAS (Braak and Del Tredici, 2012). In the succeeding text, we discuss the symptoms such damage might induce in those patients with cortical and LC degeneration.

EEG, reflexes, and P50 potential

In general, the EEG findings in AD suggest an increase in lower frequencies such as delta and a decrease in higher frequencies such as beta and gamma (Horie et al., 1990; Kim et al., 2012; Koponen et al., 1989; Park et al., 2012). On the contrary, some studies point to increased gamma band EEG activity in some patients with AD (Koenig et al., 2005; Van Deusen et al., 2008). The blink reflex and startle response are delayed and/or exaggerated in AD (Bonanni et al., 2007; Green et al., 1997; Ueki et al., 2006), indicative of decreased sensory gating, at least in some patients. The P50 potential was found to be reduced in amplitude and to show a decreased in habituation (Buchwald et al., 1989). This profile is similar to the P50 findings in autism (Buchwald et al., 1992) discussed in Chapter 6. On the one hand, these findings suggest that the PPN is underactive in AD that would be reflected in lower amplitude of the P50 potential. This could account for the decreased REM sleep duration (Bliwise, 2004) and decreased high frequencies in the EEG. The decreased habituation of the P50 potential may be explained by decreased descending cortical modulation of the RAS. Therefore both ascending RAS output and descending cortical output are reduced, making it very difficult to reestablish appropriate levels of vigilance. On the other hand, the increased REM sleep drive and increased gamma power observed in some studies suggest overactivity in the RAS. This potential paradox will be discussed in the succeeding text.

Treatment

There is no cure for AD. Treatment is directed at specific symptoms, especially in later stages to address behavioral problems with the use of antidepressants and antipsychotics. Life expectancy after diagnosis is less than 10 years. Considering that there seems to be a decrease in arousal marked by apathy, it is surprising that stimulants, including modafinil, failed to induce improvements in apathy (Rea et al., 2014). One possibility is that cortical and subcortical cell loss is so advanced that there is not enough tissue to reactivate efficiently. In the case of AD, the cell loss may be so extreme that the RAS, even when stimulated pharmacologically, has little to activate. This devastating disease is still frustratingly difficult to treat effectively.

There is a potentially effective new treatment being tested. A drug developed in Japan named T-817MA is basically an inhibitor of *tau* protein phosphorylation, and its effects include restoration of transmitter function (Moreno et al.,

2011) and cessation of cognitive decline in animals treated with amyloid protein (Kimura et al., 2009). This agent appears to be neuroprotective in *tau* transgenic animals and reduces the motor and cognitive impairments in these animals (Fukushima et al., 2011). Clinical trials on this agent are ongoing.

Mechanisms

The exact mechanism that leads to cell death in AD is controversial, with one suggestion being disruption of intracellular calcium levels (Khachaturian, 1994; Mattson et al., 2000), with mutations in the presenilin genes of mice leading to increased intracellular calcium release from the endoplasmic reticulum (Stutzmann et al., 2006). As mentioned earlier, neurofibrillary tangles from *tau* hyperphosphorylation induce a structural obstacle to the transport of proteins and molecules. This kind of damage could be cumulative in the same or different cells. However, in order to understand and better treat AD, these brain-wide disturbances may not explain the diversity of symptoms manifested by AD patients. As far as arousal symptoms are concerned, we can conclude that there may be two populations of AD patients, one with decreased gamma activity and one with increased gamma activity. EEG recordings may help differentiate between these subpopulations. If gamma activity is decreased, these patients may be amenable to treatment with the atypical stimulant modafinil, but if gamma activity is increased, this agent would be counterproductive. Clinical trials would need to effectively identify those that could versus those that would not respond.

In patients manifesting degeneration in the LC, it may be indicative of an increase in arousal levels. As we saw in Chapter 2, the LC normally inhibits the PPN, and its degeneration is a hallmark of anxiety disorders. Cell loss in the LC would disinhibit or release the PPN, increasing arousal and hyperresponsiveness to sensory inputs. In Chapter 5, we considered PTSD and anxiety disorders that are generated by decreased inhibition of PPN by LC. In some proportion of patients with AD, a decrease in LC cell number similar to that in anxiety disorders could easily account for the anxiety and similar symptoms, including sundowning agitation. Future studies on AD need to establish if sundowning is present mainly in patients with LC cell loss, a result that would indicate treatment with anxiolytics or antidepressants.

In AD patients with increased REM sleep drive, there may be a differential activation of N-type calcium channel activity (Chapter 2 describes the modulation of gamma oscillations in REM sleep is preferentially mediated by N-type calcium channels) or the inhibition of P/Q-type calcium channels (modulation of gamma oscillations in waking is preferentially modulated by P/Q-type calcium channels). In these patients, the manifestation of hallucinations (considered to be due to REM sleep intrusion into waking) may be caused by preferential activation of REM sleep mechanisms. Moreover, the decrease in waking drive could account for some of the lack of arousal.

Using the P50 potential, those patients with decreased amplitude response to the first stimulus of a pair would be indicative of decreased levels of arousal. That is, only some patients may manifest such decreased arousal, while other may have normal amplitude responses to the first stimulus. If instead, there is decreased habituation of the P50 potential, in which the response to the second stimulus is higher than normal due to the lack of descending modulation, this may indicate a sensory gating deficit as in anxiety disorder but without a decrement in arousal. Such patients may also be the ones manifesting increased LC cell loss and perhaps sundowning agitation reminiscent of an anxiety disorder. In addition, sleep studies in AD could determine which subpopulation of patients exhibits hallucinations, permitting the design of therapies to dampen REM sleep drive. That is, the use of EEG and P50 potential recordings in a large sample could help differentiate potential subpopulations of AD patients. Another possibility is that it is the stage of the disease that dictates the dysregulation observed in arousal and REM sleep drive. All of these questions could be answered by such investigations.

We carried out a small survey of AD patients and found that there was a trend toward two distinct populations based on P50 potential criteria. There was a group of ~40% of AD patients with very low P50 potential amplitude in the response to the first stimulus of a pair. This is indicative of low levels of arousal. The other group of ~60% of AD patients manifested normal amplitude in the response to the first stimulus of a pair, indicative of normal arousal levels. Fig. 8.1 is a bar graph of the mean ± SE of the amplitude of controls compared with all AD subjects, who showed slightly but not significantly reduced amplitude. When the AD group was divided into low-amplitude and high-amplitude groups, there was a significant difference between the subgroups of AD subjects. This suggested the presence of two groups of AD patients, one group with low arousal levels who might have attentional and memory deficits and a second group with normal arousal levels.

However, the subgroup of AD subjects who had high-amplitude responses also showed decreased habituation of the response to the second stimulus of a pair. This is indicative of a sensory gating deficit. Fig. 8.2 shows the percent habituation of the P50 potential response to the second stimulus as a ratio of the response to the first stimulus. When pooled, the habituation of all AD subjects was not different from controls, but when divided into low- versus high-amplitude subgroups, the high-amplitude subgroup showed decreased habituation at the 250 ms interstimulus interval, indicative of decreased habituation or sensory gating deficit. When we compared the performance of neuropsychological tests, the patients with low-amplitude P50 potentials tested worse on the Word List 1 memory task, Trails B, and Stroop 1 tests of executive function and the LM 1 and LM 2 for delayed memory. On the other hand, the patients with normal amplitude but decreased habituation tested well on these tasks but had significantly higher scores on the Geriatric Depression and Anxiety Scale. These results suggest the presence of at least two populations of AD patients, the larger

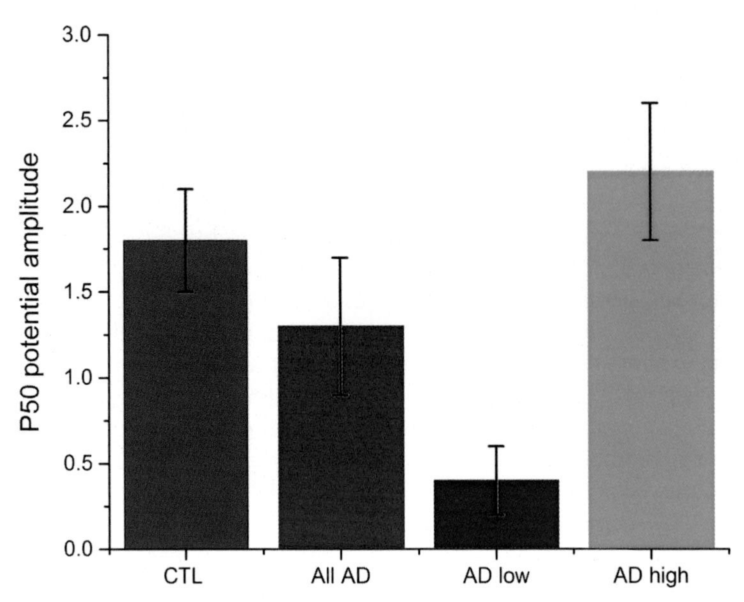

FIG. 8.1 P50 potential amplitude in AD. The mean and standard error of the mean of the amplitude following the first stimulus of a pair in a group of control subjects (black bar) was similar to the group of AD subjects as a whole *(red bar)*. However, when the amplitudes were divided into low-amplitude versus high-amplitude responses, there was a subgroup with low amplitude *(blue bar)* and a subgroup with high-amplitude (green bar) P50 potential that were significantly different from each other but not to the other groups.

one with a tendency toward depression and anxiety and the smaller one with lower arousal and attention and worse memory. Additional studies are needed to expand these observations.

Recently, studies in both humans and animals have reported that prolonged episodes of waking increase amyloid aggregation and that such accumulation of amyloid is related to fragmented sleep (Vanderhayden et al., 2018). As mentioned earlier, sleep disturbances are a characteristic of AD and insomnia affects at least one-third of patients (Moran et al., 2005). A mechanism for insomnia based on overexpression of P/Q-type calcium channels, which are involved in modulating waking in the PPN, has been proposed (Garcia-Rill et al., 2015). Although this may only be one of several insomnia mechanisms driving excessive waking, there are certainly others. However, the link between amyloid deposition in the cortex and major changes in sleep-wake control seems remote. Rather than amyloid deposition causing such changes the disturbances in sleep-wake cycles that are observed quite early in the disease may be a sign of disturbances in this homeostatic process. After all, brain stem centers such as the RAS drive cortical EEG rhythms that manifest gamma coherence between distant sites in the cortex, but not during REM sleep (Castro et al., 2013; Cavelli et al., 2015). This suggests that ascending projections dictate the level of EEG coherence between states, so that it is unlikely that amyloid deposition in the

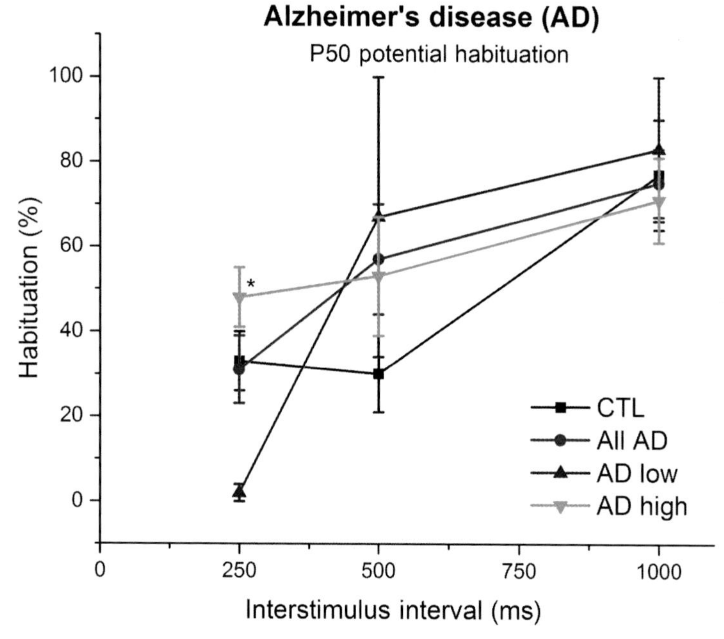

FIG. 8.2 P50 potential habituation in AD. The mean and SE of the mean percent habituation of the ratio of the second to the first response as a percent is shown for three interstimulus intervals (ISI). The control group *(black squares)* showed similar habituation as the overall AD group *(red circles)* at all three ISIs. However, when the AD population was divided into low amplitude *(blue triangles)* and high amplitude *(green inverted triangles)*, they were significantly different from each other at the 250 ms ISI, suggestive of a sensory gating deficits in the high-amplitude subgroup.

cortex could drive the sleep-wake disturbances, at least early in the disease. The commonality of sleep-wake disturbances between AD and schizophrenia has been proposed to lead to the development of neuropsychiatric disturbances that arise later in the disease process (Winsky-Sommerer et al., 2018). In a number of neurological and psychiatric disorders, as repeatedly pointed out throughout this book, sleep-wake abnormalities presage the onset of the disorder. Such consistent observations suggest that the RAS and its control of waking and sleep are first responder to the disease process, manifesting disturbed sleep-wake cycles indicative of a serious disease process that has dysregulated a critical homeostatic system. This concept is discussed further in subsequent chapters on the Parkinson disease, REM sleep behavior disorder, and narcolepsy.

Cortical mechanisms

Accounting for sleep and arousal symptoms would seem straightforward given the knowledge we possess regarding the RAS and the PPN in particular. However, it is less clear how the deficits in perceptual and cognitive symptoms in AD can be explained, which may be of cortical origin. One useful concept for consciousness and perception has been described in previous chapters, especially in Chapter 7, and that is the idea of recurrent thalamocortical oscillations. Briefly, sensory afferent information activates the "specific" sensory pathways such as the primary visual, auditory, and somatosensory projections. This information terminates in the "specific" sensory thalamus, and thalamocortical neurons relay this information to layer IV of the cortex, providing the "content" of sensory experience. The same sensory afferents in parallel activate the RAS and its projections to the "nonspecific" intralaminar thalamus. Intralaminar thalamocortical neurons then relay this information to layer I of the cortex, providing the "context" of sensory experience (Garcia-Rill, 2009, 2015). The temporal summation of "specific" and "nonspecific" afferents to the cortex is thought to provide the "content" and "context" of sensory experience that permits binding of perceptions, respectively (Llinas et al., 1994, 2002; Llinás et al., 1999). When the two inputs summate at the level of the cortex, cortical output neurons are activated to begin corticothalamic reentrant signaling and promoting thalamocortical oscillations. The recurrent signaling from repeated oscillations helps bind perception and provides a mechanism for consciousness. Fig. 8.3 outlines the mechanism but does not show how cortical columns, arranged radially to the cortical surface, inhibit each other through lateral inhibition. Such lateral inhibition is essential to detection and perception since the critical role of the cortex is to provide differentiation of sensory information via specific and localized activation of columns, for example, directional selectivity and orientational selectivity.

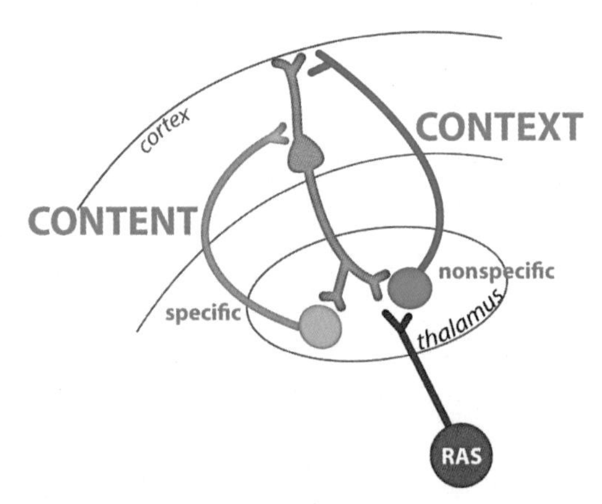

FIG. 8.3 Mechanism for consciousness and perception. According to Llinas et al. (1994, 2002) and Llinás et al. (1999), sensory afferent information is relayed to "specific" thalamic nuclei that project to layer IV of the cortex *(green cell)*, providing the "content" of sensory experience. In parallel, this sensory input activates the RAS projections *(magenta cell)* to the intralaminar or "nonspecific" thalamus that projects to later I of the cortex *(blue cell)*, providing the "context" of sensory experience. When the two cortical inputs coincide, cortical output is triggered to begin corticothalamic oscillations that reinforce the sensation. If the two inputs do not coincide, the sensory event does not produce a response from the cortex and is not perceived or misperceived.

If the lateral inhibition between cortical columns is lost, there is no specificity to the column activated; therefore perception does not occur. This is what happens during slow-wave sleep, when large groups of columns fire together without providing lateral inhibition and, therefore, lack sensory specificity. Upon waking, lateral inhibition returns to cortical columns, and sensory specificity once again is the result. In AD, there is massive cell loss across cortical columns, such that, as the disease progresses, specific sensation is degraded and perception is impaired. This unspecific degeneration of cells downgrades the perceptual process, so that the coincident activation of "specific" and "nonspecific" sensory inputs to the cortex does not occur. Moreover, the capacity for forming a conscious individual is impaired. If the cell loss is drastic enough, there is insufficient cortical mass to provide the percept of the self. More on the topic of the formation of the self appears in Chapter 13. The disturbance in this process may lead to the absence of self-control and personal disregard.

References

Bliwise, D.L., 2004. Sleep disorders in Alzheimer's disease and other dementias. Clin. Cornerst. 6, S16–S28.

Bonanni, L., Anzellotti, F., Varanese, S., Thomas, A., Manzoli, L., et al., 2007. Delayed blink reflex in dementia with Lewy bodies. J. Neurol. Neurosurg. Psychiatr. 78, 1137–1139.

Braak, H., Del Tredici, K., 2012. Where, when, and in what form does sporadic Alzheimer's disease begin? Curr. Opin. Neurol. 25, 708–714.

Buchwald, J.S., Erwin, R., Read, S., Van Lancker, D., Cummings, J.L., 1989. Midlatency auditory evoked responses: differential abnormality of P1 in Alzheimer's disease. Electroenceph. Clin. Neurophyiol. 74, 378–384.

Buchwald, J.S., Erwin, R., Van Lancker, D., Guthrie, D., Schwafel, J., et al., 1992. Midlatency auditory evoked responses: P1 abnormalities in adult autistic subjects. Electroenceph. Clin. Neurophyiol. 84, 164–171.

Castro, S., Falconi, A., Chase, M., Torterolo, P., 2013. Coherent neocortical 40-Hz oscillations are not present during REM sleep. Eur. J. Neurosci. 37, 1330–1339.

Cavelli, M., Castro, S., Schwartzkopf, N., Chase, M., Falconi, A., et al., 2015. Coherent cortical oscillations decrease during REM sleep in the rat. Behav. Brain Res. 281, 318–325.

Egan, M.F., Kost, J., Tariot, P.N., Aisen, P.S., Cummings, J.L., et al., 2018. Randomized trial of verubecestat for mild-to-moderate Alzheimer's disease. New. Engl. J. Med. 378, 1691–1703.

Fukushima, T., Nakamura, A., Iwakami, N., Nakada, Y., Hattori, H., et al., 2011. T-817MA, a neuroprotective agent, attenuates the motor and cognitive impairments associated with neuronal degeneration in P301L tau transgenic mice. Biochim. Biophys. Acta 407, 730–734.

Garcia-Rill, E., 2009. Reticular activating system. In: Stickgold, R., Walker, M. (Eds.), The Neuroscience of Sleep. Elsevier, Oxford, pp. 133–139.

Garcia-Rill, E., 2015. Waking and the Reticular Activating System. Academic Press, p. 330.

Garcia-Rill, E., Luster, B., Mahaffey, S., Bisagno, V., Urbano, F.J., 2015. Pedunculopontine arousal system physiology—implications for insomnia. Sleep Sci. 8, 92–99.

Green, J.B., Burba, A., Freed, D.M., Elder, W.W., Xu, W., 1997. The P1 component of the middle latency auditory potential may differentiate a brainstem subgroup of Alzheimer's disease. Alz. Dis. Assoc. Disord. 11, 153–157.

Horie, T., Koshino, Y., Murata, T., Omori, M., Isaki, K., 1990. EEG analysis in patients with senile dementia and Alzheimer's disease. Jpn. J. Psychiat. Neurol. 44, 91–98.

Iqbal, K., Alonso, A.C., Chen, S., Chohan, M.O., El-Akkad, E., et al., 2005. Tau pathology in Alzheimer disease and other tauopathies. Biochim. Biophys. Acta 1739, 198–210.

Khachaturian, Z.S., 1994. Calcium hypothesis of Alzheimer's disease and brain aging. Ann. N. Y. Acad. Sci. 747, 1–11.

Khachiyants, N., Trinkle, D., Son, S.J., Kim, K.Y., 2011. Sundown syndrome in persons with dementia: an update. Psychiatr. Invest. 8, 275–287.

Kim, J.S., Lee, S.H., Park, G., Kim, S., Bae, S.M., et al., 2012. Clinical implications of quantitative electroencephalography and current source density in patients with Alzheimer's disease. Brain Topog. 25, 461–474.

Kimura, T., Nguyen, P.T.H., Ho, S.A., Tran, A.H., Ono, T., et al., 2009. T-817MA, a neurotrophic agent, ameliorates the deficits in adult neurogenesis and spatial memory in rats infused i.c.v. with amyloid-β peptide. Br. J. Pharmacol. 157, 451–463.

Koenig, T., Prichep, L., Dierks, T., Hubl, D., Wahlund, L.O., et al., 2005. Decreased EEG synchronization in Alzheimer's disease and mild cognitive impairment. Neurobiol. Aging 26, 165–171.

Koponen, H., Partanen, J., Paakonen, A., Mattila, E., Riekkinen, P.J., 1989. EEG spectral analysis in delirium. J. Neurol. Neurosurg. Psychiatr. 52, 980–985.

Llinas, R.R., Leznik, E., Urbano, F.J., 2002. Temporal binding via cortical coincidence detection of specific and nonspecific thalamocortical inputs: a voltage-dependent dye-imaging study in mouse brain slices. Proc. Natl. Acad. Sci. U. S. A. 99, 449–454.

Llinás, R.R., Ribary, U., Jeanmonod, D., Kronberg, E., Mitra, P.P., 1999. Thalamocortical dysrhythmia: a neurological and neuropsychiatric syndrome characterized by magnetoencephalography. Proc. Natl. Acad. Sci. U. S. A. 96 (26), 15222–15227.

Llinas, R., Ribary, U., Joliot, M., Wang, X.J., 1994. Content and context in temporal thalamocortical binding. In: Buzsaki, G. (Ed.), Temporal Coding in the Brain. Springer, Berlin, pp. 251–272.

Mattson, M., LaFerla, F., Chan, S., Leissring, M., Shepel, P., et al., 2000. Calcium signaling in the ER: its role in neuronal plasticity and neurodegenerative disorders. Trends Neurosci. 23, 222–229.

Moran, M., Lynch, C.A., Walsh, C., Coen, R., Coakley, D., et al., 2005. Sleep disturbance in mild to moderate Alzheimer's disease. Sleep Med. 6, 347–352.

Moreno, H., Choi, S., Yu, E., Brusco, J., Avila, J., et al., 2011. Blocking effects of human tau on squid giant synapse transmission and its prevention by T-817MA. Front. Synap. Neurosci. 3 (3), 1–8.

Park, J.Y., Lee, S.K., An, S.K., Lee, S.J., Kim, J.J., et al., 2012. Gamma oscillatory activity in relation to memory ability in older adults. Int. J. Psychophysiol. 86, 58–65.

Rea, R., Carotenuto, A., Fasanaro, A.M., Traini, E., Amenta, F., 2014. Apathy in Alzheimer's disease: any effective treatment? Sci. World J. 2. 421385 (1-9).

Stutzmann, G.E., Smith, I., Caccamo, A., Oddo, S., LaFerla, F.M., et al., 2006. Enhanced ryanodine receptor recruitment contributes to Ca^{2+} disruptions in young, adult, and aged Alzheimer's disease mice. J. Neurosci. 26, 5180–5189.

Ueki, A., Goto, K., Sato, N., Iso, H., Morita, Y., 2006. Prepulse inhibition of acoustic startle response in mild cognitive impairment and mild dementia of Alzheimer type. Psychiatr. Clin. Neurosci. 60, 55–62.

Vanderhayden, W.M., Lim, M.N., Musiek, E.S., Gerstner, J.R., 2018. Alzheimer's disease and sleep-wake disturbances: amyloid, astrocytes, and animal models. J. Neurosci. 38, 2901–2910.

Van Deusen, J.A., Vuurman, E.F.M.P., Verhey, F.R.J., van Kranen-Mastenbroek, V.H.J.M., Riedel, W.J., 2008. Increased EEG gamma band activity in Alzheimer's disease and mild cognitive impairment. J. Neural Transm. 115, 1301–1311.

Wenk, G.L., 2003. Neuropathologic changes in Alzheimer's disease. J. Clin. Psychiatr. 64, 7–10.
Winsky-Sommerer, R., de Oliveira, P., Loomis, S., Wafford, K., Dijk, D.-J., et al., 2018. Disturbances in sleep quality, timing and structure and their relationship with other neuropsychiatric symptoms in Alzheimer's disease and schizophrenia: insights from studies in patient populations and animal models. Neurosci. Biobehav. Rev. In press.

Chapter 9

Deep brain stimulation of the pedunculopontine tegmental nucleus and arousal in Parkinson's disease

Paolo Mazzone*, Giacomo Della Marca†, Eugenio Scarnati‡

**Stereotactic and Functional Neurosurgery, Centro Chirurgico Toscano, Arezzo, Italy, †Institute of Neurology, Catholic University, Rome, Italy, ‡Department of Biotechnological and Applied Clinical Sciences, University of L'Aquila, L'Aquila, Italy*

Introduction

The PPTg is part of the mesencephalic locomotor region and together with the locus coeruleus (LC) and the dorsal raphe nucleus is a major constituent of the RAS. As such, it is involved in different aspects of motor and nonmotor functions. In the last two decades, an impressive number of morphological, functional, and clinical investigations have addressed the role of the PPTg in such functions, and the results have been the object of many exhaustive reviews in which numerous details can be found (Garcia-Rill, 1991; Goetz et al., 2016; Gut and Winn, 2016; Manaye et al., 1999; Mena-Segovia and Bolam, 2017; Sebille et al., 2017; Vitale et al., 2018; Wang and Morales, 2009).

Briefly, the PPTg is composed of neurons located in a region bordered by the substantia nigra (SN) and the retrorubral field rostrally, the ascending limb of the superior cerebral peduncle medially, the lateral lemniscus and related nuclei laterally, the cuneiform nucleus dorsally, the pontine tegmental field ventrally, and the parabrachial nucleus caudally.

The neuronal population forming the PPTg is rather heterogeneous, and different subpopulations have been identified that utilize either acetylcholine (ACh), glutamate (GLU), or ɣ-aminobutiric acid (GABA). Some of these neurotransmitters may be also colocalized in the same neuron even together with other neuroactive substances. The ACh neurons are mainly localized in the dorsal part of the nucleus, that is, in the pars compacta, while those expressing other neurotransmitters are localized in the ventrolateral regions that form the pars dissipata.

Arousal in Neurological and Psychiatric Diseases. https://doi.org/10.1016/B978-0-12-817992-5.00009-X

The PPTg output fibers directed rostrally give rise to a dense and highly ordered innervation of the basal ganglia and the thalamus. The SN, the subthalamic nucleus (STN), and the inner segment of the globus pallidus (GPi) are the most densely innervated structures among the basal ganglia nuclei. Thalamic nuclei (anterior, reticular, mediodorsal, centromedian, centrolateral, posterior, lateral geniculate, and habenular) are also densely innervated, while other structures (the hypothalamus, external segment of the globus pallidus, ventral tegmental area, amygdala, septum, tectum, and frontal cortical areas) receive sparse projections.

In turn the PPTg is reached by SN, STN, and GPi fibers, thus forming with these structures three reciprocating circuits. It also receives fibers from the motor cortices and cerebellum and from limbic hypothalamic and raphe nuclei.

In addition to ascending fibers the PPTg is a source of descending axons that innervate mainly the gigantocellular reticular nucleus, cranial nerve nuclei, and pontine reticular nuclei, thus establishing a route to influence spinal cord motor mechanisms and autonomic functions. Other targets for descending fibers include pontine, vestibular, and cerebellar nuclei; the spinal cord; and the rostral ventrolateral medulla.

Loss of PPTg neurons has been reported in Parkinson's Disease (PD) and in atypical parkinsonisms, and it has been proposed that this contributes to gait and axial deficits and to sleep disorders that develop in these pathologies (Garcia-Rill, 2015; Jellinger, 1988; Lee et al., 2000; Obeso et al., 1997; Pahapill and Lozano, 2000; Wichmann and Delong, 2006; Zweig et al., 1989).

Deep brain stimulation (DBS) has been consequently applied to the PPTg in PD patients unresponsive to medications or DBS STN in the attempt to improve their motor deficits. Benefits in freezing of gait and axial instability have been reported in various studies (Ferraye et al., 2010; Mazzone et al., 2014, 2016, 2018; Moro et al., 2010; Perera et al., 2018; Thevathasan et al., 2011).

Less is known about the effects of PPTg DBS on cortical activity during wake-sleep states. Experimental data provided evidence that the PPTg participates in the induction and maintenance of REM sleep and gamma activity in the cerebral cortex (Datta and Siwek, 1997; Deurveilher and Hennevin, 2001; Urbano et al., 2012). Only a few studies have been devoted to this aspect in humans (Arnulf et al., 2010; Lim et al., 2009; Romigi et al., 2008), but the results are difficult to compare given the heterogeneity of patients, different electrode positioning, stimulation parameters, and time elapsed after PPTg implantation. However, taken together, such investigations show that PPTg DBS may induce REM sleep and maintenance of cortical arousal irrespective of the nature of disease (i.e., PD or atypical parkinsonisms). In a patient in which recordings were taken from the PPTg, REM sleep episodes could be preceded or not by a P wave originating in the pons, which was followed 20–140 ms later by a cortical potential (Lim et al., 2007). This wave was considered similar to the ponto-geniculo-occipital (PGO) waves that are characteristic of the transition from slow-wave sleep (SWS) to REM sleep in animal studies (Callaway et al., 1987; Datta, 1997; Steriade et al., 1990a, 1990b).

The PPTg cholinergic neurons projecting to thalamic midline and intralaminar nuclei that innervate the cerebral cortex (Capozzo et al., 2003; Datta and Siwek, 1997; Deurveilher and Hennevin, 2001; Mena-Segovia et al., 2008; Mesulam et al., 1989; Steriade et al., 1990a, b; Urbano et al., 2012) should be primarily involved in modulating the PPTg DBS-mediated cortical activity.

The implantation of electrodes for DBS in the PPTg, STN, and GPi in different basal ganglia nuclei in PD patients gave us the opportunity to explore by means of intracerebral recordings the activity of these nuclei across wake-sleep states. Given the purpose of this book and the location of the PPTg in the RAS, in this chapter, we describe the results concerning the influence of PPTg on cortical arousal activity.

Therefore the aims of the present study were (1) to record the spontaneous PPTg electric activity during waking-sleep states and (2) to evaluate the modifications of functional connectivity between the PPTg and the cortex during sleep, using coherence analysis.

Patients and methods

Patients

Eleven patients aged 61.2 ± 7.2 years (mean ± SD) were enrolled in the study. They were implanted in the PPTg in association or not with implantation of the STN or GPI as detailed in Table 9.1

The surgical procedures for PPTg targeting and electrode implantation have been described elsewhere (Mazzone et al., 2011, 2013, 2016). Surgery was performed in the Regional Center for Stereotactic Neurosurgery and DBS, CTO Hospital ASL Roma 2, in Rome. The study was approved by the local Ethics Committee, and the patients gave written informed consent.

Methods

Polysomnographic recordings

The patients underwent a full-night laboratory nocturnal polysomnography (PSG) performed in a quiet room 2–5 days post surgery. The four contacts of the stimulating electrode (mod. 3389 or 3387, Medtronic Neurological Division, Minneapolis, the United States) were connected to the recording device, and DBS was applied once the patient had accommodated to bed and all the procedures around were completed. EEG was recorded using the Brain Quick System 98 digital polygraph (Micromed® SpA, Mogliano Veneto, TV, Italy). The recording setup included one EEG scalp electrode at C4 (contralateral to the side of PPTg implantation), reference electrodes applied to the left and right mastoids, two electrooculographic electrodes applied to the canthus of each eye, two surface EMG electrodes applied to submental muscles, and standard electrodes for EKG.

TABLE 9.1 Patients and electrode implantations

Patient	Age	Gender	Diagnosis	PPTg		STN		GPi	
				Monolateral	Bilateral	Monolateral	Bilateral	Monolateral	Bilateral
1	62	M	PD		+		+		
2	61	M	PD		+		+		
3	67	M	PD		+		+		
4	66	M	PD		+		+		
5	62	M	PD		+		+		
6	69	M	PD		+		+		
7	66	F	PD dystonic	+				+	
8	56	M	PD dystonic		+				+
9	49	M	PSP	+					
10	48	M	PD dystonic	+					
11	67	M	PD	+					

PSP, progressive supranuclear palsy.

The effects of PPTg DBS were investigated switching on stimulation medially for 15 min when the patients entered into waking, NREM sleep, and REM sleep. Stimulation was delivered at parameters that we commonly use for PPTg DBS, that is, 20–40 Hz, pulse width 60 μs, and voltage 1–3 V.

Recordings from the PPTg were carried out through the four contacts of the DBS electrode, referred to the external mastoid, and in a bipolar configuration. Sleep recordings lasted from 11:00 p.m. to 7:00 a.m. in the morning. Impedances were kept below 5 kΩ before starting the recording and checked again at the end of the recording session. Sampling frequency was 512 Hz with A/D conversion at 16 bit. Preamplifier amplitude range was ± 3200 μV, and prefilters were set at 0.15 Hz.

Sleep analysis

Sleep recordings were analyzed on computer monitors, while sleep stages were visually classified according to criteria proposed by the American Academy of Sleep Medicine (Iber et al., 2007). The sleep structure was evaluated by means of arousal detection and analysis of the cyclic alternating pattern. Fast-frequency EEG arousals were visually detected and quantified in accordance to the rules of the American Sleep Disorders Association (Bonnet et al., 2007). The arousal index (number of arousals per hour of sleep) was calculated for the whole sleep period and for non-REM (NREM) and REM sleep.

Frequency analysis

The EEG frequency analysis was performed before the connectivity analysis on epochs of EEG tracings lasting 30 s that were selected for the wake-sleep stages (NREM and REM). Each 30 s epoch was visually screened for artifacts (EMG, temporary disconnect spikes, sweating, and body movements), and epochs with artifacts were removed from further analysis. The remaining data were extracted from the scored sleep data file by a software program (Rembrandt SleepView and Medcare) and stored in a separate binary data file. Fast Fourier transform was performed with a 1 sec interval on the EEG signal both from the scalp electrode and from the PPTg contacts. Power spectra bands were computed per 30 s epochs for each EEG electrode for the entire recording period. The following frequency bands were considered: delta (0.5–4 Hz), theta (4.5–7.5 Hz), alpha (8–12.5 Hz), beta (13–30 Hz), and gamma (30.5–256 Hz). To compensate for variability among subjects and across the night in EEG power, the spectra were normalized, that is, value in each frequency bin was divided by the total power. These spectra were matched with the scored sleep data by stage and converted to data files for statistical analysis.

Connectivity analysis

The connectivity analysis was performed using the ASA4 software (Advanced Source Analysis 4.0, ANT Software BV® Colosseum 2). The EEG coherence analysis was performed in blocks of recordings lasting 3 min and selected for wake-sleep stages (NREM and REM). The connectivity analysis was performed in the waking, NREM sleep, and REM sleep stages excluding the deep NREM N3 stage since it was not clearly recognizable in all patients.

Artifact rejection was performed visually. Coherence values were computed for each frequency band (delta, theta, alpha, beta, and gamma) in the frequency range 0–256 Hz. Connectivity analysis was performed by the computation of both instantaneous (I-Coh) and lagged coherence (L-Coh). This approach allowed better separation of synchronization from true connectivity.

PPTg-to-scalp instantaneous and lagged coherence were computed between signals recorded from the scalp and signals recorded from the PPTg derivations. Coherence was computed for each of the five EEG frequency bands, for each subject and for each condition (waking, NREM sleep, and REM sleep).

The definition for the complex value of coherence between time series x and y in the frequency band ω was

$$r_{xy\omega} = \frac{\operatorname{Re}\operatorname{cov}(x,y) + i\operatorname{Im}\operatorname{cov}(x,y)}{\sqrt{Var(x) \times Var(y)}}$$

This definition is based on the cross spectrum given by the covariance and variances of the signals and where i is the imaginary unit ($\sqrt{-1}$) (Lehmann et al., 2012).

The squared modulus of the coherence was

$$r_{xy\omega}^2 = \frac{[\operatorname{Re}\operatorname{cov}(x,y)]^2 + [\operatorname{Im}\operatorname{cov}(x,y)]^2}{Var(x) \times Var(y)}$$

and the lagged coherence was

$$LagR_{xy\omega}^2 = \frac{[\operatorname{Im}\operatorname{cov}(x,y)]^2}{Var(x) \times Var(y) - [\operatorname{Re}\operatorname{cov}(x,y)]^2}$$

The instantaneous EEG coherence was calculated according to the following formula:

$$Rx_1x_2(\infty) = \frac{ComplexSpectrum(x_1(t)) \times ComplexSpectrum(x_2(t))}{\sqrt{PowerSpectrum(x_1(t)) \times PowerSpectrum(x_2(t))}}$$

Statistical analysis

To analyze power changes a three-way analysis of variance (ANOVA) was carried out with the following repeated measures: band (frequency bands) and stage (waking, NREM sleep, and REM sleep) as repeated measures and site (scalp and PPTg) as independent variables. The same design was adopted for the comparison of instantaneous (I-Coh) and lagged coherence (L-Coh). Correction for degrees of freedom was made with the Greenhouse-Geisser procedure, and the Duncan test was used for post hoc comparisons. Statistics were performed by means of dedicated analysis using the package SYSTAT 12 ver. 12.02.00 for Windows (SYSTAT Software Inc. 2007).

Results

PPTg electrode position

Postoperative MRI showed that the DBS electrodes were correctly positioned in the planned targets. In particular, the active contacts of the PPTg electrode were positioned, although with unavoidable stochastic variations, in the ventrolateral pontine tegmentum just below the pontomesencephalic junction, that is, in a region that includes mostly the caudal PPTg (Fig. 9.1). In our experience, this stimulating site has given the best outcome (Mazzone et al., 2013, 2016).

Sleep staging

The analytic details of wake-sleep staging in each patient are reported in Table 9.2.

Power spectra analysis

The mean values of relative power spectra (± SD) for each sleep stage recorded from the scalp and PPTg electrodes are listed in Table 9.3 (power spectra). The time course of spectra recorded from the PPTg during wake-sleep stages closely resembled those recorded from the scalp. No significant differences were observed in relative power spectra values comparing PPTg and scalp recordings for each of the frequency bands considered in any of the sleep stages. A representative example of a power spectra curve is shown in Fig. 9.2, while the profiles of power spectra are shown in Fig. 9.3.

Connectivity analysis: I-Coh and L-Coh

The profiles of I-Coh and L-Coh are reported in Fig. 9.4. In the multivariate approach, the condition "stage" (waking, NREM sleep, and REM sleep) was associated with a significant effect on I-Coh ($F = 6.294$ and $P \leq 0.002$) and on L-Coh ($F = 5.300$ and $P \leq 0.006$) as well. In the successive univariate analysis, significant changes were observed in I-Coh in the beta ($F = 10.513$ and $P \leq 0.001$) and gamma bands ($F = 11.080$ and $P \leq 0.001$). Conversely, no significant modifications in I-Coh were recorded in the low-frequency delta, theta, and alpha bands. As concerns L-Coh, no significant increase in any of the specific frequency bands was recorded across wake-sleep stages, but a trend toward increased L-Coh in REM sleep was recorded in all frequency bands. This suggests

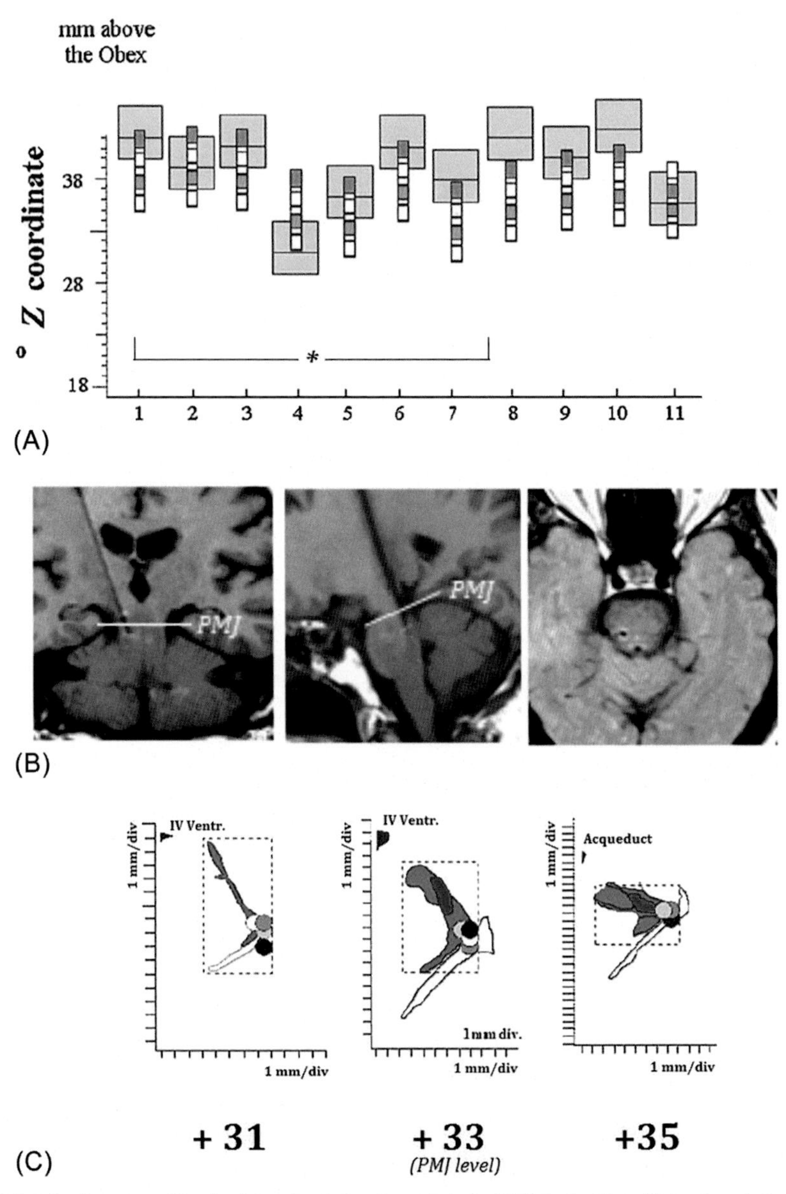

FIG. 9.1 Location of the tip of the stimulating electrode in the PPTg in the 11 studied patients inferred from postoperative MRI. (A) The position of the electrode is represented with respect to the position of the PPTg (gray vertical rectangles) in each patient. The horizontal line in each rectangle corresponds to the pontomesencephalic junction (PMJ). The two active contacts through which stimulation was applied are represented in red (positive) and green (negative). The asterisk refers to the patients who underwent a bilateral PPTg implantation. (B) Representative example of electrode position in coronal (left), sagittal (center), and axial view (right) on MRI. (C) Location of electrode contacts as reconstructed in axial representations of the PPTg in slides +31, +33, and +35 in the Paxinos and Huang's Atlas (1995). White dot, 0 contact; light gray, contact 1; dark gray, contact 2; black dot, contact 3. The pars dissipata of the PPTg is in blue; the pars compacta is in red. The medial lemniscus is shown in white.

TABLE 9.2 Polysomnographic scores

	Mean ± SD
Total time in bed (h)	413.2 ± 54.2
Total sleep time (TST) (h)	312.5 ± 55.3
Sleep efficiency index	87.5 ± 15.5
N1 (% of TST)	9.3 ± 4.2
N2 (% of TST)	56.2 ± 21.3
N3 (% of TST)	7.4 ± 2.2
REM (% of TST)	12.8 ± 3.6
Arousal index in sleep	16.8 ± 5.2
Arousal index in NREM	18.2 ± 3.6
Arousal index in REM	14.6 ± 2.1

The amount of sleep stages (N1, N2, and N3) is expressed as percent of the total sleep time recorded in the cohort of patients. The arousal indexes are expressed as number of arousal per hour of sleep.

that L-Coh was overall modified during REM sleep although this effect could not be detected in the detailed analysis of discrete frequency bands.

Discussion

The brainstem reticular formation is critical for the induction and maintenance of REM sleep and cortical activity required for vigilance and consciousness. Among the various nuclei comprising the brainstem reticular formation the PPTg, in addition to participating in motor control, has a central role in inducing fast EEG rhythms and REM sleep. Much of what we know about these aspects stems from animal studies. In the present study, stimulating electrodes implanted in the PPTg in patients suffering from PD allowed us to investigate the functional relationships between the PPTg and cortex during wake-sleep states. We focused on investigating the EEG activity recorded from the PPTg during the wake-sleep cycle and the correlational relationships occurring between PPTg activity and scalp EEG. To our knowledge, this is the first report in humans concerning such functions investigated by means of power spectrum and coherence analysis.

During waking, the power frequency spectrum of the EEG recorded from the PPTg showed a main peak in the delta band and a peak in the beta band, while the spectrum of EEG coherence showed a peak in the alpha band. This latter result is in accordance with data from a previous study in which recordings from the PPTg were performed in awake patients (Androulidakis et al., 2008).

TABLE 9.3 Power spectra and connectivity analysis

Condition	Frequency band	Power spectra		Connectivity	
		PPTg	Scalp	I-Coh	L-Coh
Wake	Delta	69.1	76.4	0.17	12.2
	Theta	6.4	3.7	0.34	21.3
	Alpha	7.6	6.6	0.35	32.5
	Beta	10.3	9.4	0.55	11.3
	Gamma	5.6	3.9	0.63	2.0
NREM	Delta	67.8	51.7	0.30	20.2
	Theta	12.5	21.3	0.34	38.7
	Alpha	7.2	6.7	0.34	55.0
	Beta	8.8	10.1	0.25	28.9
	Gamma	3,7	10.2	0.25	10.3
REM	Delta	73.2	60.7	0.28	27.9
	Theta	4.8	11.4	0.24	56.8
	Alpha	83.0	7.9	0.59	59.8
	Beta	9.3	11.8	0.53	47.1
	Gamma	5.4	8.2	0.63	15.3

The values of power spectra are expressed as percent of total power.

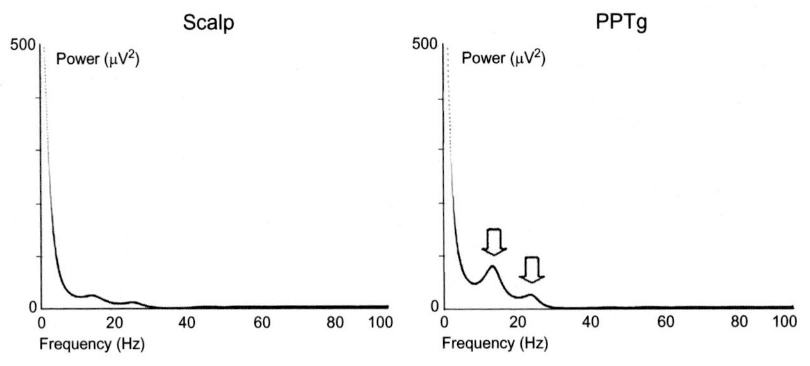

FIG. 9.2 Representative power spectra from scalp (left) and PPTg (right) recordings showing peaks of high-frequency activity in the beta and gamma bands (arrows) in the PPTg during waking or REM sleep.

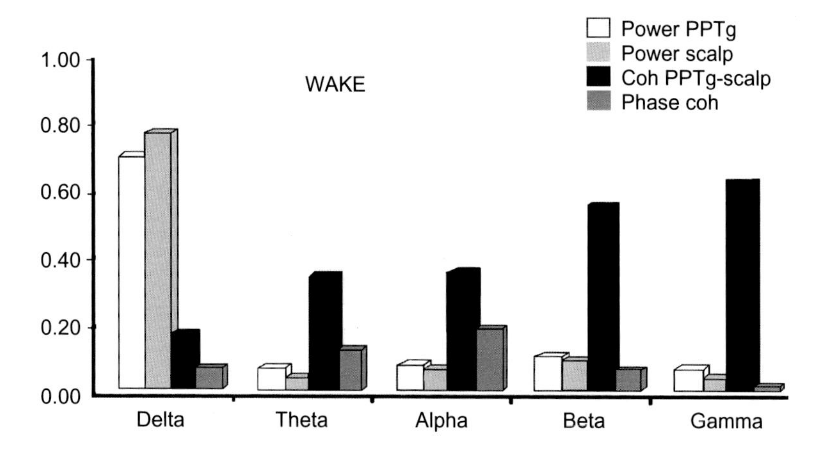

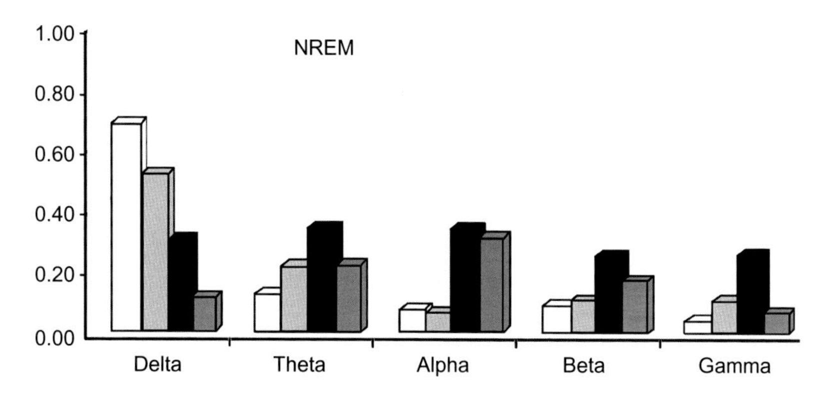

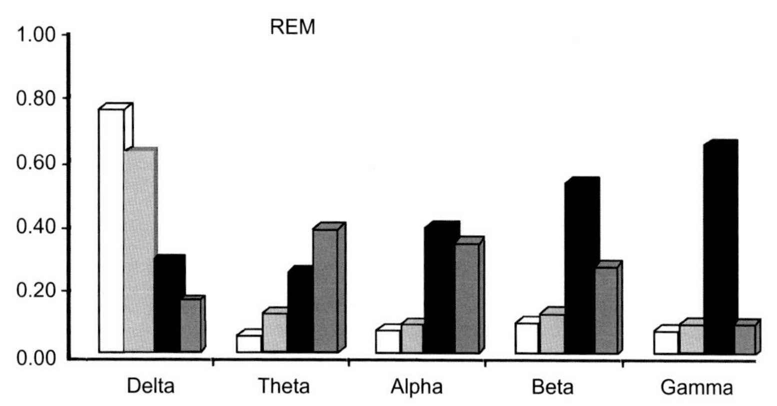

FIG. 9.3 Power spectra for each frequency band (delta, theta, alpha, beta, and gamma) during waking, NREM sleep, and REM sleep

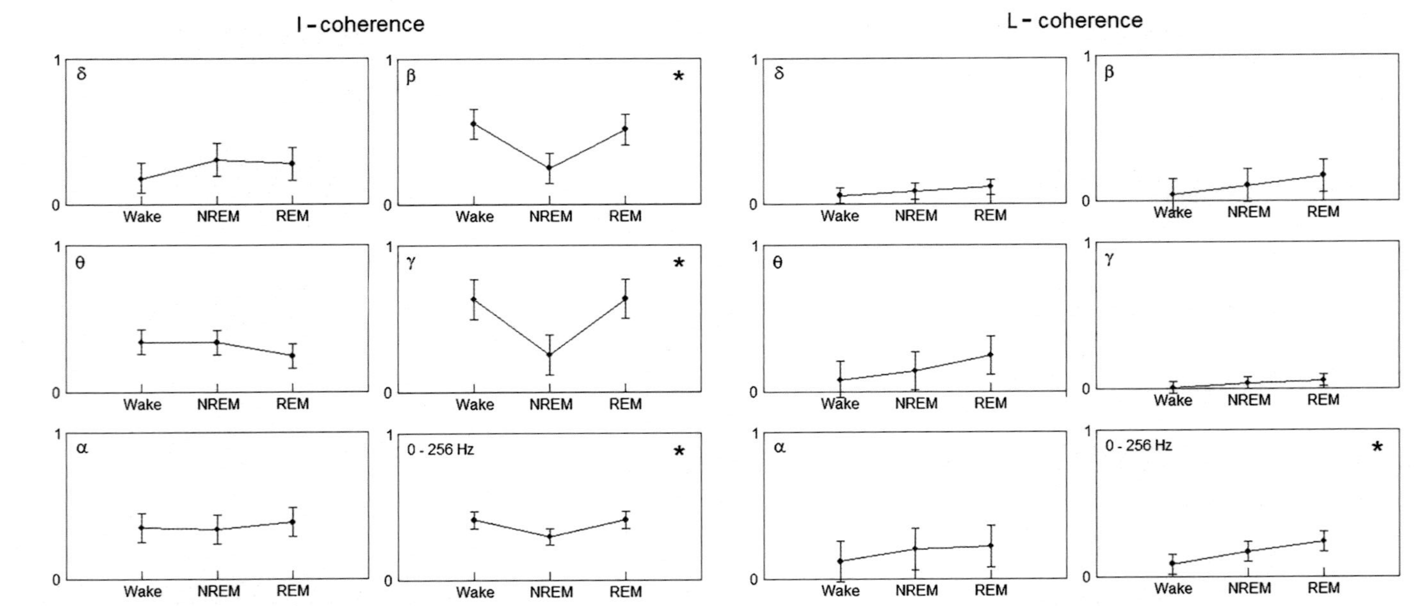

FIG. 9.4 Instantaneous and lagged coherence between PPTg and scalp recordings (mean ± SD) for delta, theta, alpha, beta, and gamma bands and for global range (0–256 Hz) in the 11 investigated patients. I-Coh significantly increased in the beta ($P \leq 0.001$) and gamma bands ($P \leq 0.001$) during waking and REM sleep (asterisk panels 4 and 5) and when measured in the global range (panel 6) ($P \leq 0.002$). In contrast, L-Coh showed a trend to increase during REM sleep in all the frequency bands but never reached levels of significance except in REM when measured in the global range (asterisk panel 12) ($P \leq 0.006$).

Frequency

Although no significant changes in the power spectrum were measured across wake-sleep stages, waking and REM sleep showed increased fast-frequency rhythms (beta and gamma), whereas NREM sleep was characterized by a relative increase in the amount of theta band. This finding is in agreement with results from other coherence studies (Garcia-Rill et al., 2013; Valencia et al., 2014).

As concerns the changes of functional connections between the PPTg and the scalp EEG, we quantified the coupling between electric activity of the PPTg and scalp. The results suggest that EEG synchronization (measured in the whole 0–256 Hz spectrum) between the PPTg and scalp was high during waking and REM sleep and lower during NREM sleep. When coherence was measured in the frequency domain, the effects of waking and REM sleep were frequency-specific since the fast-frequency oscillations (beta and gamma, 30–256 Hz) were predominantly involved. This finding is fully consistent with results from animal models, which showed that gamma band activity was present in regions of the RAS and that neurons in the PPTg, parafascicular, and pontine subcoeruleus dorsalis nuclei fired in the beta/gamma band range when maximally activated (Kezunovic et al., 2012; Urbano et al., 2012). Accordingly, it has been also reported that neurons within the PPTg fired at gamma band frequency when subjected to depolarizing currents (Simon et al., 2011) and that PPTg stimulation potentiated fast oscillations (20–40 Hz) in thalamocortical neurons and in cortical EEG (Steriade et al., 1991).

Coherence

The different results that we have obtained when measuring I-Coh and L-Coh suggest that different neural activities were explored.

I-Coh increased during waking and REM sleep, and this effect could be seen selectively in the high-frequency bands (beta and gamma). I-Coh is a measure of synchronization that reflects the simultaneous activation of different structures driven by a same pacemaker. As concern with our recordings, two hypotheses may be proposed: (1) The synchronization between PPTg and cortex could be due to the simultaneous activation of the RAS and an arousal-related structure projecting to the cortex. This view would be in line with the activation of reticular neurons following PPTg stimulation (Garcia-Rill et al., 2001) and with the occurrence of electric coupling in RAS nuclei, a mechanism that allows groups of neurons to fire together and that is crucial to wake-sleep control (Garcia-Rill et al., 2007; Urbano et al., 2012). Electric coupling was found to be enhanced by the administration of modafinil, a CNS stimulant, which increases coupling via gap junctions and is capable of driving coherence at higher frequencies to induce arousal (Beck et al., 2008). (2) Alternatively, the increase in cortical-PPTg coherence could be due to a synchronization process mediated by intralaminar thalamic nuclei, that is, a process in which PPTg neurons are also involved as they innervate mainly thalamic neurons, which project to the cortex. High-frequency synchronous oscillations within this circuitry could therefore explain the increase in EEG coherence between PPTg and scalp.

L-Coh increased in REM sleep, compared with waking and NREM sleep, but this effect was state-dependent and not frequency-dependent. In fact, it could only be measured in the analysis of the whole EEG spectrum (0–256 Hz) and not in separate or discrete frequency bands. L-Coh does not depend on EEG synchronization, and it is considered a measure of "true" connectivity. Therefore the increase in L-Coh expressed the activation of a neural pathway that would connect, directly or indirectly, the pontine tegmentum with the cortex.

The communication between PPTg and cortex measured by L-Coh has revealed two peculiarities: (1) it was minimal during waking, increased during NREM sleep, and reached its maximal value in REM sleep and (2) it was not selectively active in a specific frequency band.

The profile of L-Coh during the wake-sleep cycle is closely similar to that of ACh release in the cerebral cortex and basal forebrain where ACh release has been found to be high during both waking and REM sleep compared with NREM sleep (Marrosu et al., 1995; Vazquez and Baghdoyan, 2001).

From the present data, it cannot be excluded that at least part of the EEG frequency modifications, and consequently the I-Coh and L-Coh, was affected by recruitment of the LC, which is also located in the upper pons. Indeed, LC neurons have been found to discharge during waking, decrease their activity during SWS, and become nearly quiescent or discharging in the form of rapid bursts during REM sleep (Aston-Jones et al., 1991; Chu and Bloom, 1973; Hobson et al., 1975). In such a way, LC neurons, if coactivated by PPTg stimulation, might have contributed to the reported effects.

Limitations of the study

DBS provided the opportunity to record spontaneous brainstem activity in humans, but data must be considered keeping in mind some limitations that were unavoidable in the experimental protocol. First, we studied a cohort of patients, affected by degeneration of a region that directly involved the structure explored (this is the reason why they underwent DBS at that site). Second, another possible bias comes from the pharmacological treatment administered to patients. All the patients included in the study consumed at the time of PSG recordings, various combinations of L-dopa and dopamine agonists, but in no case anticholinergic drugs. The possible effects of such unavoidable drug treatments must be taken into consideration when evaluating the results of this study.

Acknowledgments

The authors are grateful to Prof. Edgar Garcia-Rill for his critical reading of the manuscript.

References

Androulidakis, A.G., Mazzone, P., Litvak, V., Penny, W., Dileone, M., et al., 2008. Oscillatory activity in the pedunculopontine area of patients with Parkinson's disease. Exp. Neurol. 211, 59–66.

Arnulf, I., Ferraye, M., Fraix, V., Benabid, A.L., Chabardes, S., et al., 2010. Sleep induced by stimulation in the human pedunculopontine nucleus area. Ann. Neurol. 67, 546–549.

Aston-Jones, G., Chiang, C., Alexinsky, T., 1991. Discharge of noradrenergic locus coeruleus neurons in behaving rats and monkeys suggests a role in vigilance. Prog. Brain Res. 88, 501–520.

Beck, P., Odle, A., Wallace-Huitt, T., Skinner, R.D., Garcia-Rill, E., 2008. Modafinil increases arousal determined by P13 potential amplitude: an effect blocked by gap junction antagonists. Sleep 31, 1647–1654.

Bonnet, M.H., Doghramji, K., Roehrs, T., Stepanski, E.J., Sheldon, S.H., et al., 2007. The scoring of arousal in sleep: reliability, validity, and alternatives. J. Clin. Sleep Med. 3, 133–145.

Callaway, C.W., Lydic, R., Baghdoyan, H.A., Hobson, J.A., 1987. Pontogeniculooccipital waves: spontaneous visual system activity during rapid eye movement sleep. Cell Mol. Neurobiol. 7, 105–149.

Capozzo, A., Florio, T., Cellini, R., Moriconi, U., Scarnati, E., 2003. The pedunculopontine nucleus projection to the parafascicular nucleus of the thalamus: an electrophysiological investigation in the rat. J. Neural. Transm. 110, 733–747.

Chu, N., Bloom, F.E., 1973. Norepinephrine-containing neurons: changes in spontaneous discharge patterns during sleeping and waking. Science 179, 908–910.

Datta, S., 1997. Cellular basis of pontine ponto-geniculo-occipital wave generation and modulation. Cell Mol. Neurobiol. 17, 341–365.

Datta, S., Siwek, D.F., 1997. Excitation of the brain stem pedunculopontine tegmentum cholinergic cells induces wakefulness and REM sleep. J. Neurophysiol. 77, 2975–2988.

Deurveilher, S., Hennevin, E., 2001. Lesions of the pedunculopontine tegmental nucleus reduce paradoxical sleep (PS) propensity: evidence from a short-term PS deprivation study in rats. Eur. J. Neurosci. 13, 1963–1976.

Ferraye, M.U., Debu, B., Fraix, V., Goetz, L., Ardouin, C., et al., 2010. Effects of pedunculopontine nucleus area stimulation on gait disorders in Parkinson's disease. Brain 133, 205–214.

Garcia-Rill, E., 1991. The pedunculopontine nucleus. Prog. Neurobiol. 36, 363–389.

Garcia-Rill, E., 2015. Waking and the Reticular Activating System. Elsevier Academic Press, Amsterdam.

Garcia-Rill, E., Heister, D.S., Ye, M., Charlesworth, A., Hayar, A., 2007. Electrical coupling: novel mechanism for sleep-wake control. Sleep 30, 1405–1414.

Garcia-Rill, E., Kezunovic, N., Hyde, J., Simon, C., Beck, P., et al., 2013. Coherence and frequency in the reticular activating system (RAS). Sleep Med. Rev. 17, 227–238.

Garcia-Rill, E., Skinner, R.D., Miyazato, H., Homma, Y., 2001. Pedunculopontine stimulation induces prolonged activation of pontine reticular neurons. Neuroscience 104, 455–465.

Goetz, L., Piallat, B., Bhattacharjee, M., Mathieu, H., David, O., et al., 2016. The primate pedunculopontine nucleus region: towards a dual role in locomotion and waking state. J. Neural. Transm..

Gut, N.K., Winn, P., 2016. The pedunculopontine tegmental nucleus—a functional hypothesis from the comparative literature. Mov. Disord. 31, 615–624.

Hobson, J.A., McCarley, R.W., Wyzinski, P.W., 1975. Sleep cycle oscillation: reciprocal discharge by two brainstem neuronal groups. Science 189, 55–58.

Iber, C., Ancoli-Israel, S., Chesson, A., Quan, S.F., 2007. The AASM Manual for Scoring of sleep And Associated Events: Rules, Terminology and Technical Specifications. American Academy of Sleep Medicine.

Jellinger, K., 1988. The pedunculopontine nucleus in Parkinson's disease, progressive supranuclear palsy and Alzheimer's disease. J. Neurol. Neurosurg. Psychiatr. 51, 540–543.

Kezunovic, N., Hyde, J., Simon, C., Urbano, F.J., Williams, D.K., et al., 2012. Gamma band activity in the developing parafascicular nucleus. J. Neurophysiol. 107, 772–784.

Lee, M.S., Rinne, J.O., Marsden, C.D., 2000. The pedunculopontine nucleus: its role in the genesis of movement disorders. Yonsei Med. J. 41, 167–184.

Lehmann, D., Faber, P.L., Tei, S., Pascual-Marqui, R.D., Milz, P., et al., 2012. Reduced functional connectivity between cortical sources in five meditation traditions detected with lagged coherence using EEG tomography. Neuroimage 60, 1574–1586.

Lim, A.S., Lozano, A.M., Moro, E., Hamani, C., Hutchison, W.D., et al., 2007. Characterization of REM-sleep associated ponto-geniculo-occipital waves in the human pons. Sleep 30, 823–827.

Lim, A.S., Moro, E., Lozano, A.M., Hamani, C., Dostrovsky, J.O., et al., 2009. Selective enhancement of rapid eye movement sleep by deep brain stimulation of the human pons. Ann. Neurol. 66, 110–114.

Manaye, K.F., Zweig, R., Wu, D., Hersh, L.B., De Lacalle, S., et al., 1999. Quantification of cholinergic and select non-cholinergic mesopontine neuronal populations in the human brain. Neuroscience 89, 759–770.

Marrosu, F., Portas, C., Mascia, M.S., Casu, M.A., Fa, M., et al., 1995. Microdialysis measurement of cortical and hippocampal acetylcholine release during sleep-wake cycle in freely moving cats. Brain Res. 671, 329–332.

Mazzone, P., Paoloni, M., Mangone, M., Santilli, V., Insola, A., et al., 2014. Unilateral deep brain stimulation of the pedunculopontine tegmental nucleus in idiopathic Parkinson's disease: effects on gait initiation and performance. Gait Posture 40, 357–362.

Mazzone, P., Sposato, S., Insola, A., Scarnati, E., 2011. The deep brain stimulation of the pedunculopontine tegmental nucleus: towards a new stereotactic neurosurgery. J. Neural. Transm. 118, 1431–1451.

Mazzone, P., Sposato, S., Insola, A., Scarnati, E., 2013. The clinical effects of deep brain stimulation of the pedunculopontine tegmental nucleus in movement disorders may not be related to the anatomical target, leads location, and setup of electrical stimulation. Neurosurgery 73, 894–906.

Mazzone, P., Vilela, F.O., Viselli, F., Insola, A., Sposato, S., et al., 2016. Our first decade of experience in deep brain stimulation of the brainstem: elucidating the mechanism of action of stimulation of the ventrolateral pontine tegmentum. J. Neural. Transm. 123, 751–767.

Mazzone, P., Vitale, F., Capozzo, A., Viselli, F., Scarnati, E., 2018. Deep brain stimulation of the pedunculopontine tegmental nucleus improves static balance in Parkinson's disease. In: Krames, E., Hunter Peckham, P., Rezai, A. (Eds.), Neuromodulation. Academic Press Elsevier, New York, pp. 967–976.

Mena-Segovia, J., Bolam, J.P., 2017. Rethinking the pedunculopontine nucleus: from cellular organization to function. Neuron 94, 7–18.

Mena-Segovia, J., Sims, H.M., Magill, P.J., Bolam, J.P., 2008. Cholinergic brainstem neurons modulate cortical gamma activity during slow oscillations. J. Physiol. 586, 2947–2960.

Mesulam, M.M., Geula, C., Bothwell, M.A., Hersh, L.B., 1989. Human reticular formation: cholinergic neurons of the pedunculopontine and laterodorsal tegmental nuclei and some cytochemical comparisons to forebrain cholinergic neurons. J. Comp. Neurol. 283, 611–633.

Moro, E., Hamani, C., Poon, Y.Y., Al-Khairallah, T., Dostrovsky, J.O., et al., 2010. Unilateral pedunculopontine stimulation improves falls in Parkinson's disease. Brain 133, 215–224.

Obeso, J.A., Rodriguez, M.C., Delong, M.R., 1997. Basal ganglia pathophysiology. A critical review. Adv. Neurol. 74, 3–18.

Pahapill, P.A., Lozano, A.M., 2000. The pedunculopontine nucleus and Parkinson's disease. Brain 123 (Pt 9), 1767–1783.

Paxinos, G., Huang, X.F., 1995. Atlas of the Human Brainstem. Academic Press, San Diego.

Perera, T., Tan, J.L., Cole, M.H., Yohanandan, S.A.C., Silberstein, P., et al., 2018. Balance control systems in Parkinson's disease and the impact of pedunculopontine area stimulation. Brain 141, 3009–3022.

Romigi, A., Placidi, F., Peppe, A., Pierantozzi, M., Izzi, F., et al., 2008. Pedunculopontine nucleus stimulation influences REM sleep in Parkinson's disease. Eur. J. Neurol. 15, e64–e65.

Sebille, S.B., Belaid, H., Philippe, A.C., Andre, A., Lau, B., et al., 2017. Anatomical evidence for functional diversity in the mesencephalic locomotor region of primates. Neuroimage 147, 66–78.

Simon, C., Kezunovic, N., Williams, D.K., Urbano, F.J., Garcia-Rill, E., 2011. Cholinergic and glutamatergic agonists induce gamma frequency activity in dorsal subcoeruleus nucleus neurons. Am. J. Physiol. Cell Physiol. 301, C327–C335.

Steriade, M., Datta, S., Pare, D., Oakson, G., Curro Dossi, R.C., 1990a. Neuronal activities in brainstem cholinergic nuclei related to tonic activation processes in thalamocortical systems. J. Neurosci. 10, 2541–2559.

Steriade, M., Dossi, R.C., Pare, D., Oakson, G., 1991. Fast oscillations (20–40 Hz) in thalamocortical systems and their potentiation by mesopontine cholinergic nuclei in the cat. Proc. Natl. Acad. Sci. U. S. A. 88, 4396–4400.

Steriade, M., Pare, D., Datta, S., Oakson, G., Curro, D.R., 1990b. Different cellular types in mesopontine cholinergic nuclei related to ponto-geniculo-occipital waves. J. Neurosci. 10, 2560–2579.

Thevathasan, W., Coyne, T.J., Hyam, J.A., Kerr, G., Jenkinson, N., et al., 2011. Pedunculopontine nucleus stimulation improves gait freezing in Parkinson disease. Neurosurgery 69, 1248–1253.

Urbano, F.J., Kezunovic, N., Hyde, J., Simon, C., Beck, P., et al., 2012. Gamma band activity in the reticular activating system. Front. Neurol. 3, 6.

Valencia, M., Chavez, M., Artieda, J., Bolam, J.P., Mena-Segovia, J., 2014. Abnormal functional connectivity between motor cortex and pedunculopontine nucleus following chronic dopamine depletion. J. Neurophysiol. 111, 434–440.

Vazquez, J., Baghdoyan, H.A., 2001. Basal forebrain acetylcholine release during REM sleep is significantly greater than during waking. Am. J. Physiol. Regul. Integr. Comp. Physiol. 280, R598–R601.

Vitale, F., Capozzo, A., Mazzone, P., Scarnati, E., 2018. Neurophysiology of the pedunculopontine tegmental nucleus. Neurobiol. Dis. https://doi.org/10.1016./Jnbd. 2018. 03.004. Epub ahead of print.

Wang, H.L., Morales, M., 2009. Pedunculopontine and laterodorsal tegmental nuclei contain distinct populations of cholinergic, glutamatergic and GABAergic neurons in the rat. Eur. J. Neurosci. 29, 340–358.

Wichmann, T., Delong, M.R., 2006. Basal ganglia discharge abnormalities in Parkinson's disease. J. Neural. Transm. Suppl. 21–25.

Zweig, R.M., Jankel, W.R., Hedreen, J.C., Mayeux, R., Price, D.L., 1989. The pedunculopontine nucleus in Parkinson's disease. Ann. Neurol. 26, 41–46.

Chapter 10

Arousal in REM sleep behavior disorder and narcolepsy

Muna Irfan*, Carlos H. Schenck†, Edgar Garcia-Rill‡
*Minnesota Regional Sleep Disorders Center, Department of Neurology, Hennepin County Medical Center, Veterans Affair Medical Center, Minneapolis, University of Minnesota Medical School, Minneapolis, MN, United States, †Minnesota Regional Sleep Disorders Center, Department of Psychiatry, Hennepin County Medical Center, University of Minnesota Medical School, Minneapolis, MN, United States, ‡Center for Translational Neuroscience, Department of Neurobiology and Developmental Sciences, University of Arkansas for Medical Sciences, Little Rock, AR, United States

REM sleep behavior disorder (RBD)

Symptoms, etiology, clinical manifestations, consequences

RBD is a parasomnia discovered in 1986 that features dream-enacting behaviors emerging with the loss of the generalized skeletal muscle paralysis of REM sleep, namely "REM-atonia" (Schenck and Mahowald, 2002). A person with RBD moves with eyes closed while attending to the inner dream environment and is unaware of the actual bedside surroundings, which poses a risk for injury (Schenck et al., 1989), especially since dream-enactment is often aggressive and can result in serious and life-threatening injuries, with forensic implications (Schenck et al., 2009). Video polysomnography (vPSG) is required to establish the diagnosis, with the diagnostic hallmark finding being the loss of REM-atony (AASM, 2014), as shown in Fig. 10.1. Clonazepam or melatonin therapy at bedtime is usually effective (Schenck and Mahowald, 2002).

The typical RBD clinical profile involves predominantly middle-aged and older men (Schenck and Mahowald, 2002; Fernández-Arcos et al., 2016), which most likely reflects a clinical referral bias on the account of more aggressive and injurious RBD behaviors in men compared with women. This is supported by the first epidemiological study of vPSG-confirmed RBD in the general population that found an equal gender ratio, along with a 1.06% prevalence (Haba-Rubio et al., 2018). Furthermore, a different clinical RBD profile has been found in patients under 50 years old, with greater gender parity, less severe RBD, greater association with narcolepsy-cataplexy, and greater association with antidepressant use and psychiatric disorders (Ju, 2018). RBD can also affect children and

Arousal in Neurological and Psychiatric Diseases. https://doi.org/10.1016/B978-0-12-817992-5.00010-6

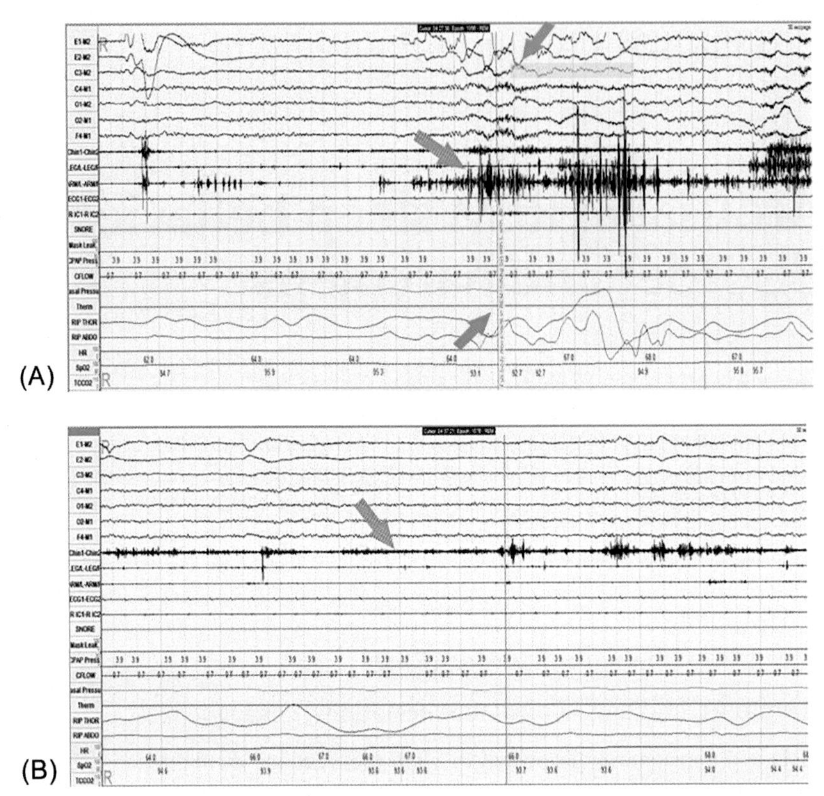

FIG. 10.1 Changes in muscle tone in RBD. (A) illustrates polysomnographic recording of increased phasic EMG tone in arm lead during REM sleep denoted by orange arrow. Green arrow denotes annotation by sleep technician when patient had body movement with arm flailing and hitting bed rail, enacting a dream. Blue arrow is pointing at rapid eye movements. (B) illustrates increased tonic EMG activity in chin lead in REM sleep of the same patient, denoted by orange arrow.

adolescents, albeit infrequently, and is usually associated with the parasomnia overlap disorder (AASM, 2014) (RBD + Non-REM parasomnia [sleepwalking and sleep terrors]), narcolepsy, antidepressant use (as therapy for depression or cataplexy), brain stem tumors, and various neuropsychiatric conditions (Shukla et al., 2018). Patients diagnosed with RBD can be idiopathic or symptomatic of a broad spectrum of neurological disorders (Schenck and Mahowald, 2002) that can affect the brain stem neuronal centers and circuitry subserving REM sleep atony, to be described in the succeeding text. Most antidepressants (especially serotonin reuptake inhibitors, venlafaxine, and tricyclic antidepressants) and other medications can trigger or aggravate RBD, most likely by disrupting the neurochemistry subserving REM sleep atony. Motor activity with RBD can range from subtle, jerky limb movements to flailing, kicking, punching, thrashing, and jumping off the bed. Episodes are typically brief with abrupt transition to alertness upon awakening (typical of most awakenings from REM sleep) at which time patients often recall vivid and elaborate details of their dreams that often involve confrontation and attack by unfamiliar people and animals. The frequency of RBD episodes can vary from several per night to once a month. Often the bed partner is more aware of the RBD episodes than the affected person who is only aware of the dream action and not the dream-enacting behaviors (Fernández-Arcos et al., 2016).

Experimental animal models of RBD have provided insights into the key brain stem nuclei and pathways subserving REM sleep atony that are disrupted by the spectrum of neurological disorders (and medications) causing RBD. The seminal work by Jouvet and colleagues in Lyon, who experimentally induced RBD in cats who had received peri-locus coeruleus lesions (Schenck and Mahowald, 2002), has been extended by other groups and including the current Lyon group led by Luppi, whose research has confirmed the critical roles played by the glutamate neurons of the sublateral dorsal (subcoeruleus) tegmental nucleus in the pons (Valencia Garcia et al., 2017) and by the ventromedial inhibitory nucleus in the medulla (Valencia Garcia et al., 2018). Genetic inactivation of these brain stem neuronal groups in rats generates behavioral release during REM sleep that replicates human RBD.

Longitudinal studies of patients ≥50 years old with idiopathic RBD (iRBD) have shown that >80% will develop alpha-synucleinopathy neurodegeneration (the Parkinson disease, dementia with Lewy bodies, and multiple system atrophy), with the mean interval from RBD onset to overt neurodegeneration being 12–14 years (Schenck et al., 2013; Iranzo et al., 2013). Thus, iRBD is now considered to be a prodromal feature and early biomarker of synucleinopathy neurodegeneration (Mahowald and Schenck, 2018), which has accelerated the search for promising disease-modifying/neuroprotective therapies to be tested in double-blind trials. An International RBD Study Group, founded in 2009, with yearly research symposia and nine peer-reviewed journal articles to date, is devoting a large share of its efforts to identifying the highest-risk iRBD patients for imminent conversion to overt synucleinopathy, to be enrolled in future studies testing disease-modifying therapies (Oertel et al., 2018).

Narcolepsy

Symptoms, etiology, clinical manifestations, consequences

Narcolepsy is a primary disorder of excessive daytime sleepiness that was independently described by Westphal in Germany (Berlin) and Gélineau in France (Paris) in the late 19th century (Schenck et al., 2007). The International Classification of Sleep Disorders, third edition, recognizes two subtypes: Narcolepsy type 1 (NT1) is caused by the loss of hypothalamic neurons that produce the neuropeptides orexin-A and orexin-B (also called hypocretin-1 and hypocretin-2). Narcolepsy type 2 (NT2) shares most of the same symptoms, except for the absence of cataplexy, without any identified cause to date (AASM, 2014).

Rare cases of narcolepsy have been described in patients with associated abnormalities of the diencephalic, hypothalamic, or pontine regions of the brain, and these cases are referred to as "secondary narcolepsy."

There is a clear genetic component to NT1, with more than 98% of patients carrying the human leukocyte antigen (HLA)-DQB1*06:02, a finding that makes this the strongest identified association between HLA and any disease (Scammel, 2015). DQB1*06:02 is also detected in about half of patients with NT2 but only in up to 30% of the general population in the United States, Europe, and Japan. Overall, this gene is predicted to increase the risk of narcolepsy by a factor of about 200. In persons who are positive for DQB1*06:02 and in whom it is difficult to make a diagnosis, measurement of orexin-A levels in cerebrospinal fluid can be diagnostic. NT1 has also been linked to polymorphisms in other genes. DQB1*06:02 is almost always accompanied by the linked DQA1*01:02 gene, and their gene products form a heterodimer that presents antigens to T-cell receptors on CD4 T cells. Variations in other HLA-DQ alleles, HLA-DP, and HLA class I also contribute to genetic susceptibility, as do polymorphisms in other genes that affect immune function, such as *TCRA*, *TCRB*, *P2RY11*, *EIF3G*, *ZNF365*, *IL10RB-IFNAR1*, *CTSH*, and *TNFSF4* (Scammel, 2015). Given these reported findings, it is not surprising that an autoimmune process resulting in the loss of dorsolateral hypothalamic hypocretin (orexin)-producing neurons, triggered by environmental factors, is the main pathogenic hypothesis. This is further supported by clinical evidence for environmental triggering factors such as streptococcal infections close to disease onset and by the increased incidence of narcolepsy following the Pandemrix® vaccination and H1N1 influenza infection (Pizza et al., 2015; Nohynek et al., 2012; Han et al., 2011; Partinen et al., 2012).

The usual age of narcolepsy onset is from 10 to 20 years of age, although the range extends from early childhood to senescence. There is usually the sudden onset of persistent daytime sleepiness, although it can also develop gradually. For many the sleepiness is severe, resulting in major dysfunction at school or work and during periods of inactivity, such as when reading or watching television. Narcolepsy is a chronic illness that rarely, if ever, remits completely.

Narcolepsy is usually sporadic, and the risk that an affected parent will have an affected child is only 1%. Even if one monozygotic twin has narcolepsy, there is only about a 30% chance that narcolepsy will develop in the other twin, which is analogous to schizophrenia, where there is a 50% chance that the identical twin of a schizophrenic proband will also develop schizophrenia. Clearly, environmental triggers play an interactive role with predisposing genetic factors.

NT1 has been depicted as a distinct neurological disorder characterized by disruption of the normal sleep-wakefulness rhythm and the loss of boundaries between sleep and wake, with frequent state transitions and intrusions of REM sleep (or REM sleep elements) into the other ongoing states of being (Plazzi, 2018). NT1 is a lifelong disorder, mainly arising during childhood and early adulthood, with a common diagnostic delay of many years after the onset of symptoms.

Narcolepsy is a prototypic example of dissociated sleep/wake phenomena in which components of one state (REM sleep) appear together with components of another state (wakefulness) (Mahowald et al., 2011). Narcolepsy features disordered regulation of REM sleep that is the basis for the classic "narcoleptic tetrad" often found in NT1, consisting of REM-onset sleep attacks, cataplexy, sleep paralysis, and hypnagogic/hypnopompic hallucinations. Double vision and blurred vision may accompany the onset of the sleep attack. During periods of excessive sleepiness a brief (10–30 min) nap is frequently very refreshing, if only for a short period of time. The restorative nature of a brief nap may be more characteristic of narcolepsy than other disorders of excessive daytime sleepiness. The symptom of sleepiness is accompanied by impaired performance due to impairment of sustained attention.

Cataplexy, considered to be a pathognomonic sign of NT1, is the sudden intrusion of REM sleep paralysis ("REM sleep atony") into wakefulness that manifests as partial or complete paralysis of voluntary muscles. Cataplectic episodes are triggered by strong emotions, such as laughing at a joke or being delighted or surprised. In the case reported by Gélineau, cataplexy was described as follows: "when laughing out loud or when anticipating a good business deal in his profession, he would feel weakness in his legs, which would buckle under him. Later, when playing cards, if he was dealt a good hand, he would freeze, unable to move his arms (Schenck et al., 2007)." Cataplexy can also be triggered by strong negative emotions, such as irritation or anger. The cataplectic paralysis usually evolves over many seconds, first affecting the face and neck and then progressing to weakness in the trunk and limbs. Respiratory muscles are spared. With partial cataplexy, slurred speech and a sagging face are common, and with complete episodes the person may slump to the ground with preserved consciousness during an immobile state that can last up to several minutes. Children with cataplexy can have long-lasting periods with reduced muscle tone and with a wobbly gait and perioral movements such as grimacing and tongue protrusion.

Sleep paralysis (the sleep analog to cataplexy in wakefulness) occurs during the transition from wake to sleep or from sleep (especially dreaming [REM] sleep) to wake. Sleep paralysis is considered to be the inappropriate intrusion of REM sleep atony into wake-sleep transitional states and can last up to several minutes and be frightening, at least during the first few episodes until the person becomes familiar with this new abnormal recurrent phenomenon. Sleep paralysis is often associated with dreamlike and often disturbing visual, auditory, and tactile hallucinations that occur at the onset of sleep (hypnagogic hallucinations) or upon awakening (hypnopompic hallucinations). A threatening figure in the bedroom, a burglar, or an attack by an animal is common scenarios. As with cataplexy, sleep paralysis and hypnagogic hallucinations rarely last more than several minutes. These hallucinations are considered to be the intrusion of REM sleep dream states into wake-sleep transitional states. Narcoleptics also often have fragmented sleep at night, which can be as much a prominent feature as daytime sleepiness.

Daytime automatic behavior occurs in most patients with narcolepsy and represents the simultaneous or rapidly oscillating occurrence of wake and sleep during which the individual appears to be awake but without full awareness. Wakefulness is interrupted by "microsleep episodes," resulting in unusual behavior, often prolonged and complex—such as driving long distances without awareness, making inappropriate statements, writing bizarre sentences on paper or on a computer, or even shoplifting. These spells of automatic behavior may occasionally be misdiagnosed as partial complex seizures or psychogenic fugue states. These symptoms result in significant socioeconomic consequences, particularly as regards performance in the workplace, in the classroom, or behind the wheel of an automobile. Falling asleep while operating machinery or while driving a motor vehicle is a major concern, with potentially catastrophic consequences. A major psychosocial problem involves family relationships. The relentless daytime sleepiness often results in irritability and may be misinterpreted by the family. The hallucinations and automatic behavior may lead to an erroneous primary psychiatric diagnosis. Narcoleptic sleepiness in children often manifests with attentional deficits and/or hyperactivity (as a countermeasure for the sleepiness) and may result in the erroneous diagnosis of attention deficit hyperactivity disorder.

Comorbid conditions

Narcolepsy-RBD

RBD is reported to occur in NT1 with a frequency ranging between 7% and 63% in different cohorts (Schenck and Mahowald, 1992; Mayer and Meier-Ewert, 1993; Nightingale et al., 2005; Knudsen et al., 2010). RBD in NT1 is characterized by the lack of sex predominance, less complex and more elementary movements during REM sleep, less aggressive behavior in REM sleep, earlier age of onset, and hypocretin deficiency (AASM, 2014; Plazzi, 2018). RBD may also be precipitated or worsened by the pharmacological treatment of cataplexy with antidepressants. In pediatric patients, RBD may be an initial manifestation of NT1 (AASM, 2014). Although data on phenoconversion of narcolepsy-RBD into a synucleinopathy neurodegenerative disorder are lacking, cross-sectional studies are reassuring, since there has been no reported increased risk to develop neurodegenerative diseases in elderly narcolepsy-RBD patients.

The association of RBD and narcolepsy has been considered as one of the multifaceted aspects of REM sleep motor dyscontrol of narcolepsy and was reported since the earliest RBD discovery and descriptions (Plazzi, 2018). Overall, narcolepsy is the second most frequent condition associated with RBD, after neurodegenerative diseases.

REM sleep behavior disorder in narcolepsy: The childhood phenotype

Only recently RBD has been systematically assessed in children with NT1 (Antelmi et al., 2017). In many children the RBD-related behaviors range from the "acting out" of a dream to almost continuous/subcontinuous "pantomime-like" activities. These latter episodes usually consist of calm, repetitive, and slow gesturing, resembling purposeful behaviors or reminiscent of lively interactions with the environment and/or with persons. Also in children as with adults, RBD episodes are not restricted to REM sleep in the latter part of the night, but occur during every REM sleep period throughout the night (and usually more than once per night). Overall, in narcoleptic children, RBD seems to have a common pattern to adult RBD. However, children with narcolepsy-RBD, despite a comparable sleep structure to those narcolepsy patients without RBD, complain of a greater amount of excessive daytime sleepiness and impaired nocturnal sleep, indicating that RBD in childhood NT1 is associated with greater narcolepsy disease burden. These narcoleptic children with RBD also had significantly higher rates of cataplexy during the daytime. Finally, RBD can also be the initial symptom heralding the development of fully declared NT1 in children (Nevsimalova et al., 2007).

REM sleep behavior disorder in narcolepsy: Pathophysiology

Hypocretin (orexin) neurons are excitatory and active during wakefulness with strong projections to the brain stem structures implicated in REM sleep motor modulation. A decreased hypocretinergic tone due to the loss of hypocretin-producing neurons in NT1 may cause atony during wakefulness, leading to cataplexy and to the loss of muscle atony and RBD in REM sleep. Moreover a dysfunction of the amygdala has also been suggested in NT1. Several functional studies, indeed, are convergent in identifying an amygdala-hypothalamic dysfunction during wakefulness and during cataplexy in NT1 patients (Meletti et al., 2015), which could help explain the mechanisms of emotional triggers of cataplexy. Hence, we may suppose a wide and complex network dysfunction responsible for RBD and cataplexy, unique to narcolepsy. In secondary and iRBD the persistence of REM sleep with muscle tone and/or RBD may be the consequence of direct lesions in the subcoeruleus region, which has not been reported in NT1. Although hypocretin network dysfunction is crucial for NT1, it remains unclear how in these patients hypocretin deficiency might cause REM sleep without atony, cataplexy, and RBD.

RBD in NT1 seems to have a different pathophysiology from that of iRBD, and it does not seem to represent a marker of impending synucleinopathies. The high frequency of RBD in NT1, indeed, is a plausible result of the decreased hypocretinergic activity input to brain stem structures that may contribute to dissociated sleep/wake states and motor disinhibition during REM sleep.

Narcolepsy and other hypoaroused sleep-related states

In a case control study involving 65 NT1 patients and 65 controls (Palaia et al., 2011), significant differences were found between the two groups in the following hyperaroused states: sleep-related eating disorder (AASM, 2014), 32%

versus 2%; nocturnal smoking (Provini et al., 2008), 21% versus 0%; and restless legs syndrome (AASM, 2014), 18% versus 5%. Therefore NT1 is associated with a spectrum of motor-behavioral dyscontrol during wakefulness, REM and non-REM sleep, and wake-sleep transitional states.

The following are excerpts from a recorded and transcribed interview (with the patient's written permission) of a patient with NT2 who presented to one of the authors (CHS) on account of sleep-related eating disorder, and narcolepsy was also documented during her clinical and polysomnographic evaluation (Schenck, 2005; pp. 314–319):

> *"I've always gotten up at night, ever since I can remember, since at least 5 years old. I've been able to fix a meal when I would get up from bed at night...I never thought it was a problem until I started doing it unconsciously, within the last 2 years... had a friend visit me and she found a bowl of rice crispies in the oven. I have also fallen asleep in the middle of eating and spilled milk and wheat checks all over myself."*
>
> *"One night, I took lettuce out of the kitchen and went into the bathroom, and I remember saying to myself, 'I have to find a safe place for this lettuce.' I couldn't find any place clean. I realized I was in the bathroom, but I had no rationale for why I was there. So I laid this big pile of lettuce in the bathroom—I have no idea why—then I went back to the kitchen and proceeded to eat my cereal. A little bit further back in time, I woke up in the middle of the night, and there was a burning smell, and my Mom said there had been a pan on the stove, the oatmeal was gone from the pan, the gas on the stove was on high, and the pan was burning up. And I had been sleeping on the couch, probably 15 ft from the kitchen and never smelled anything nor remembered doing it, nothing. Three weeks ago when I woke up, the pizza pan was on the stove—red hot—and so I moved the pizza pan off, and for some reason, I melted butter in the frying pan and poured Special K in there. I don't know if I thought I was making stir-fry, but the weird thing is that then I ate the stuff. When I woke up, I had the worst stomachache: Special K fried in butter."*
>
> *"When I noticed that I was gaining weight, I thought, 'gee, this is getting serious.' And it's not just a bowl of cereal; I mean I put tons of sugar on top of it."*

This patient was one of five family members (both parents and three siblings) documented to have NT2 (Schenck and Mahowald, 1997), and besides narcolepsy, she had two other autoimmune diseases consisting of autoimmune hepatitis and celiac disease.

RBD and narcolepsy

Potential pathophysiological mechanisms

From the foregoing discussion, we can glean some indications on the potential pathophysiological mechanisms involved in RBD and narcolepsy from the fairly effective treatments for each condition. In the case of RBD, many patients respond to benzodiazepines, suggesting that GABAergic synapses and/or neurons are at least primarily involved. This tends to suggest that GABAergic tone needs to increase. On the other hand, in the case of narcolepsy the frontline treatment option nowadays is the stimulant modafinil that increases electric coupling between GABAergic neurons that are coupled through gap junctions. By increasing electric coupling, GABA release decreases, suggesting that GABAergic tone in narcolepsy needs to decrease. However, the location of these neurons differs, and the manner in which they are regulated by other cells also differs. That is, the network in which these cells operate needs to be considered. The reticular activating system (RAS) is made up of the pedunculopontine nucleus (PPN), locus coeruleus (LC), and raphe nucleus, but the PPN is the most active during waking and REM sleep (Garcia-Rill, 2015). The LC and raphe nucleus both fire during waking and somewhat during slow-wave sleep, but not during REM sleep. As discussed in the preceding text, patients with RBD appear to manifest decreased LC output [as evidenced also by postmortem decrease in LC neurons in RBD (Schenck et al., 1996)], which normally inhibits the PPN. This suggests that the PPN is abnormally active in RBD, at least during REM sleep.

In Chapter 2, we highlighted two major discoveries made on the control of waking and sleep that have helped understand the physiological substrates of the lack of atony in RBD and the hypoarousal of narcolepsy. This research was focused on the partly cholinergic PPN, the portion of the RAS that is active during waking and REM sleep but less active during slow-wave sleep, and at its REM sleep-related target, the subcoeruleus nucleus dorsalis (SubCD) (Garcia-Rill, 2015). As such the PPN modulates the manifestation of waking through ascending projections to the intralaminar thalamus and the manifestation of REM sleep through descending projections to the SubCD (Garcia-Rill et al., 2013). These regions possess a proportion of cells that are electrically coupled through gap junctions, thus promoting coherence within each nucleus, and every cell in these nuclei shows gamma band activity that can export gamma frequency activity to its targets (Garcia-Rill et al., 2014).

Electrical coupling

The discovery of electric coupling in the RAS was one of the recent major discoveries in the control of sleep and waking, with some cells in the PPN and in the SubC demonstrated to be electrically coupled (Heister et al., 2007; Garcia-Rill et al., 2008). The presence of dye coupling and spikelets was reported in some PPN and SubC neurons and in the parafascicular nucleus (Pf), an ascending target of the PPN involved in thalamocortical oscillations (Urbano et al., 2007). The presence of electric coupling was demonstrated by using patch-clamp recordings of pairs of neurons and blocking action potentials with tetrodotoxin (TTX) and fast synaptic transmission with excitatory amino acid and GABA receptor blockers. In these conditions, intracellular pulses delivered to one cell induced a response in the other cell and vice versa, suggesting the presence of gap junctions in 7%–10% of these neurons (Heister et al., 2007; Garcia-Rill et al., 2008). Neuronal gap junction protein connexin 36 (Cx36) was also found to be present in these regions and decreased in level along with the developmental decrease in REM sleep (Heister et al., 2007; Garcia-Rill et al., 2008; Urbano et al., 2007).

Modafinil is approved for treating excessive sleepiness in narcolepsy, sleepiness in obstructive sleep apnea, and shift work sleep disorder. Modafinil was found to increase electric coupling between cortical interneurons, thalamic reticular neurons, and inferior olive neurons (Evans and Boitano, 2001). Following pharmacological blockade of connexin permeability, modafinil restored electrotonic coupling. The effects of modafinil were counteracted by the gap junction blocker mefloquine. Thus modafinil may be acting in a wide variety of cerebral areas by increasing electrotonic coupling in such a way that the high input resistance typical of GABAergic neurons is reduced. Modafinil also decreased the resistance of PPN and SubCD neurons. The effects of modafinil were evident in the absence of changes in resting membrane potential or of changes in the amplitude of induced EPSCs and were blocked by low concentrations of the gap junction blocker mefloquine and also in the absence of changes in resting membrane potential or in the amplitude of induced excitatory postsynaptic currents (Heister et al., 2007; Garcia-Rill et al., 2008; Urbano et al., 2007). This suggests that these compounds do not act indirectly by affecting voltage-sensitive channels such as potassium channels, but rather modulate electric coupling via gap junctions.

Gap junctions can be blocked through membrane fluidization such as that induced by the fast-acting anesthetics halothane and propofol (He and Burt, 2000; Boger et al., 1998). Oleamide promotes sleep and blocks gap junctions (Murillo-Rodriguez et al., 2003). Anandamide enhances adenosine levels to induce sleep (Gigout et al., 2006) and blocks gap junctions (He and Burt, 2000). One possibility is that a mechanism by which these agents may induce sleep and/or anesthesia is through blockade of electric coupling in the RAS. Carbenoxolone, a putative gap junction blocker, decreases the synchronicity of gamma oscillations (Gareri et al., 2004) and seizure activity (Rozental et al., 2000). 18-Glycyrrhetinic acid is a glycyrrhetinic acid derivative that blocks gap junctions (Srinivas et al., 2001). Quinine and a related compound, mefloquine, also block gap junctions (Garcia-Rill et al., 2015). These considerations suggest that gap junction blockers induce sleep and gap junction promoters act as stimulants.

Intrinsic gamma oscillations

The discovery of gamma band activity in the RAS was another major recent discovery in the control of waking and sleep. The PPN is composed of different populations of cholinergic, glutamatergic, and GABAergic neurons

(Wang and Morales, 2009). The PPN contains three basic cell types based on in vitro intrinsic membrane properties (Leonard and Llinas, 1990; Kamondi et al., 1992; Takakusaki and Kitai, 1997). Extracellular recordings of PPN neurons in vivo identified six categories of thalamic projecting PPN cells distinguished by their firing properties relative to ponto-geniculo-occipital (PGO) wave generation (Steriade et al., 1990). Some of these neurons had low rates of spontaneous firing (<10 Hz), but most had high rates of tonic firing in the beta/gamma range (20–80 Hz). PPN neurons increase firing during REM sleep ("REM-on") or both waking and REM sleep ("wake/REM-on") but decrease during slow-wave sleep (SWS) (Steriade, 1999; Steriade et al., 1990; Sakai et al., 1990; Datta and Siwek, 2002; Boucetta et al., 2014), suggestive of increased excitation only during activated states, and such firing is relayed to its targets, including the SubCD. Stimulation of the PPN potentiated the appearance of fast (20–40 Hz) oscillations in the cortical EEG, outlasting stimulation by 10–20 s (Steriade et al., 1991). Injections of glutamate into the PPN increased waking and REM sleep (Datta et al., 2001a), while injections of the glutamatergic receptor agonist N-methyl-D-aspartic acid (NMDA) increased only waking (Datta et al., 2001b), and injections of the glutamatergic receptor agonist kainic acid increased only REM sleep (Datta, 2002).

Oscillations mediated by voltage-dependent, high threshold N- and P/Q-type calcium channels are present in every PPN neuron, regardless of cell or transmitter type. These channels are distributed along the dendrites of PPN cells (Hyde et al., 2013). It has been shown that PPN neurons exhibit beta/gamma frequencies in vivo during active waking and REM sleep, but not during slow-wave sleep (Steriade et al., 1990; Sakai et al., 1990; Datta and Siwek, 2002; Boucetta et al., 2014). Similarly the presence of gamma band activity has been confirmed in the cortical EEG of the cat in vivo when the animal is active (Steriade et al., 1990, 1991) and in the region of the PPN in humans during stepping, but not at rest (Fraix et al., 2013). A recent study showed that PPN neurons fired at low frequencies ~10 Hz at rest, but the same neurons increased firing to gamma band frequencies when the animal woke up or when the animal began walking on a treadmill (Goetz et al., 2016). That is, the same PPN cells were involved in both arousal and motor control.

Thus, there is ample evidence for gamma band activity during active waking and movement in the PPN in vitro, in vivo, and across species, including man. However, gamma band activity during waking has different mechanisms than gamma band activity during REM sleep. Intracellularly, protein kinase C (PKC), which modulates kainic acid receptors (recall that kainic acid injected in the PPN-induced REM sleep), enhances N-type channel activity and has no effect on P/Q-type channel function (Stea et al., 1995), but Ca^{2+}/calmodulin-dependent protein kinase II (CaMKII), which modulates NMDA receptors (NMDA-induced waking when injected into the PPN), was shown to modulate P/Q-type channel function (Jiang et al., 2008). That is, the two calcium channel subtypes in the PPN are modulated by different intracellular pathways, N-type by the cAMP/PK pathway and P/Q-type via the CaMKII pathway. Moreover, there are three cell types in the PPN, those bearing only N-type calcium channels, those with both N- and P/Q-type, and those with only P/Q-type calcium channels (Luster et al., 2015, 2016). The implications from all of these results are that (a) there is a "waking" pathway mediated by CaMKII and P/Q-type channels and a "REM sleep" pathway mediated by cAMP/PK and N-type channels and (b) different PPN cells fire during waking (those with N+P/Q and only P/Q-type) versus REM sleep (those with N+P/Q and only N-type) (Luster et al., 2015, 2016) (see Chapter 2). Abnormalities in these pathways must be present in narcolepsy. In addition, PPN projections to glutamatergic and GABAergic neurons in the SubCD must normally impart gamma band activity onto these cells, which manifest intrinsic gamma oscillations that are subserved by sodium-dependent leak currents. Fig. 10.2 is a diagram of the ascending and descending projections from the PPN, whose cells generate gamma oscillations upon receiving sensory information. Pf neurons also manifest P/Q-type channels and relay gamma activity to the upper layers of the cortex. PPN neurons project to the SubCD that contains glutamatergic and GABAergic neurons (some of which are electrically coupled) but whose gamma activity is generated by sodium-dependent subthreshold oscillations, which is relayed to the hippocampus.

In regard to the cortex the difference between gamma band activity during waking versus REM sleep appears to be a lack of coherence (Castro et al., 2013). That is, brain stem driving of gamma band activity during waking carries with it coherence across distant cortical regions, while driving of gamma band activity during REM sleep does not include coherence across distant regions (Castro et al., 2013; Cavelli et al., 2015). Also, carbachol-induced REM sleep

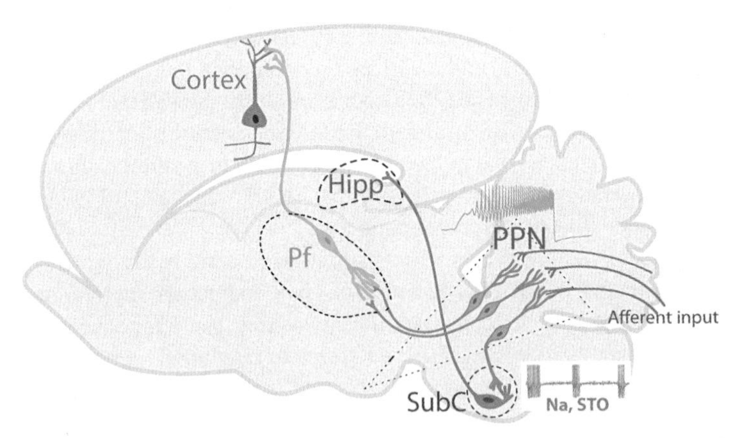

FIG. 10.2 Waking and REM sleep pathways and channels. Sagittal view of the rat brain showing the location of the PPN with its ramp-induced gamma oscillations (*blue recording and cells*) that receive sensory afferent input (*orange arrows*). The PPN projects downstream to the SubC that manifests sodium-dependent subthreshold oscillations (*purple recording and cells*), which in turn send projections to the hippocampus (*Hipp*). The PPN also projects to the intralaminar parafascicular nucleus (*Pf in green*), which in turn projects to the upper layers of the cortex.

with cataplexy is characterized by decreased gamma band coherence in the cortex (Torterolo et al., 2015). These results suggest that (a) brain stem centers drive gamma band activity that is manifested in the cortical EEG; (b) during waking, brain stem-thalamic projections include coherence across regions; and (c) during REM sleep, which is controlled by the SubCD region (as described in the preceding text, lesion of this region eliminates REM sleep, while injection of carbachol induces REM sleep signs), drives cortical EEG rhythms without coherence. Thus, PPN gamma activity is mediated by high threshold P/Q- and N-type calcium channels, while SubCD gamma is mediated by sodium-dependent subthreshold oscillations. Both mechanisms are present in cortical cells to promote gamma band firing, although, unlike PPN and SubCD, not every cortical cell has these intrinsic membrane properties. Moreover the coherent gamma band activity during waking is different from the incoherent gamma band activity during REM sleep, as each is modulated by different intracellular pathways (CaMKII vs. cAMP/PK, respectively). Narcolepsy should reflect abnormalities in these pathways.

In the case of comorbid conditions, both LC-PPN interactions and PPN-SubCD cross talk must be involved. One common factor is the presence of GABAergic neurons in the PPN and SubCD that may be dysregulated to manifest both types of symptoms. This seems to be a potentially productive target for future research.

Acknowledgment

Supported by NIH award P30 GM110702 from the IDeA program at NIGMS to EGR.

References

American Academy of Sleep Medicine (Eds.), 2014. International Classification of Sleep Disorders. third ed. American Academy of Sleep Medicine, Darien, IL.

Antelmi, E., Pizza, F., Vandi, S., Neccia, G., Ferri, R., et al., 2017. The spectrum of REM sleep-related episodes in children with type 1 narcolepsy. Brain 140, 1669–1679.

Boger, D.L., Henriksen, S.J., Cravatt, B.F., 1998. Oleamide: an endogenous sleep-inducing lipid and prototypical member of a new class of biological signaling molecules. Curr. Pharm. Des. 4, 303–314.

Boucetta, S., Cisse, Y., Mainville, L., Morales, M., Jones, B.E., 2014. Discharge profiles across the sleep-waking cycle of identified cholinergic, gabaergic, and glutamatergic neurons in the pontomesencephalic tegmentum of the rat. J. Neurosci. 34, 4708–4727.

Castro, S., Falconi, A., Chase, M., Torterolo, P., 2013. Coherent neocortical 40-Hz oscillations are not present during REM sleep. Eur. J. Neurosci. 37, 1330–1339.

Cavelli, M., Castro, S., Schwartzkopf, N., Chase, M., Falconi, A., et al., 2015. Coherent cortical oscillations decrease during REM sleep in the rat. Behav. Brain Res. 281, 318–325.

Datta, S., 2002. Evidence that REM sleep is controlled by the activation of brain stem pedunculopontine tegmental kainate receptor. J. Neurophysiol. 87, 1790–1798.

Datta, S., Patterson, E.H., Spoley, E.E., 2001a. Excitation of the pedunculopontine tegmental NMDA receptors induces wakefulness and cortical activation in the rat. J. Neurosci. Res. 66, 109–116.

Datta, S., Spoley, E.E., Patterson, E.H., 2001b. Microinjection of glutamate into the pedunculopontine tegmentum induces REM sleep and wakefulness in the rat. Amer. J. Physiol. Reg. Integ. Comp. Physiol. 280, R752–R759.

Datta, S., Siwek, D.F., 2002. Single cell activity patterns of pedunculopontine tegmentum neurons across the sleep-wake cycle in the freely moving rats. J. Neurosci. Res. 70, 79–82.

Evans, W.H., Boitano, S., 2001. Connexin mimetic peptides: specific inhibitors of gap-junctional intercellular communication. Bioch. Soc. Trans. 29, 606–612.

Fernández-Arcos, A., Iranzo, A., Serradell, M., Gaig, C., Santamaria, J., 2016. The clinical phenotype of idiopathic rapid eye movement sleep behavior disorder at presentation: a study in 203 consecutive patients. Sleep 39, 121–132.

Fraix, V., Bastin, J., David, O., Goetz, L., Ferraye, M., et al., 2013. Pedunculopontine nucleus area oscillations during stance, stepping and freezing in Parkinson's disease. PLoS One 8, e83919.

Garcia-Rill, E., 2015. Waking and the Reticular Activating System. Academic Press, New York, p. 330.

Garcia-Rill, E., Charlesworth, A., Heister, D., Ye, M., Hayar, A., 2008. The developmental decrease in REM sleep: the role of transmitters and electrical coupling. Sleep 31, 673–690.

Garcia-Rill, E., Kezunovic, N., Hyde, J., Beck, P., Urbano, F.J., 2013. Coherence and frequency in the reticular activating system (RAS). Sleep Med. Rev. 17, 227–238.

Garcia-Rill, E., Kezunovic, N., D'Onofrio, S., Luster, B., Hyde, J., Bisagno, V., et al., 2014. Gamma band activity in the RAS-intracellular mechanisms. Exptl. Brain Res. 232, 1509–1522.

Garcia-Rill, E., Luster, B., D'Onofrio, S., Mahaffey, S., Bisagno, V., et al., 2015. Implications of gamma band activity in the pedunculopontine nucleus. J. Neural Transm. 123, 655–665.

Gareri, P., Condorelli, D., Belluardo, N., Russo, E., Loiacono, A., et al., 2004. Anticonvulsant effects of carbenoxolone in genetically epilepsy prone rats (GEPRs). Neuropharmacology 47, 1205–1216.

Gigout, S., Louvel, J., Kawasaki, H., D'Antuono, A., Armand, V., et al., 2006. Effects of gap junction blockers on human neocortical synchronization. Neurobiol. Dis. 22, 496–508.

Goetz, L., Piallat, B., Bhattacharjee, M., Mathieu, H., David, O., et al., 2016. The primate pedunculopontine nucleus region: towards a dual role in locomotion and waking state. J. Neural Transm. 123, 667–678.

Haba-Rubio, J., Frauscher, B., Marques-Vidal, P., et al., 2018. Prevalence and determinants of REM sleep behavior disorder in the general population. Sleep 41 (2), zsx197.

Han, F., Lin, L., Warby, S.C., Faraco, J., Dong, S.X., et al., 2011. Narcolepsy onset is seasonal and increased following the 2009 H1N1 pandemic in China. Ann. Neurol. 703, 410–417.

He, D.S., Burt, J.M., 2000. Mechanism and selectivity of the effects of halothane on gap junction channel function. Circ. Res. 86, 1–10.

Heister, D.S., Hayar, A., Charlesworth, A., Yates, C., Zhou, Y., et al., 2007. Evidence for electrical coupling in the Sub Coeruleus (Sub C) nucleus. J. Neurophysiol. 97, 3142–3147.

Hyde, J.R., Kezunovic, N., Urbano, F.J., Garcia-Rill, E., 2013. Spatiotemporal properties of high speed calcium oscillations in the pedunculopontine nucleus. J. Appl. Physiol. 115, 1402–1414.

Iranzo, A., Tolosa, E., Gelpi, E., 2013. Neurodegenerative disease status and post-mortem pathology in idiopathic rapid-eye-movement sleep behaviour disorder: an observational cohort study. Lancet Neurol. 12, 443–453.

Jiang, X., Lautermilch, N.J., Watari, H., Westenbroek, R.E., Scheuer, T., et al., 2008. Modulation of $Ca_v2.1$ channels by Ca^+/calmodulin-dependent kinase II bound to the C-terminal domain. Proc. Natl. Acad. Sci. USA 105, 341–346.

Ju, Y.-E., 2018. RBD in adults under 50 years old. In: Schenck, C.H., Högl, B., Videnovic, A. (Eds.), Rapid-Eye-Movement Sleep Behavior Disorder. Springer, Heidelberg, Berlin, New York, pp. 201–214.

Kamondi, A., Williams, J., Hutcheon, B., Reiner, P., 1992. Membrane properties of mesopontine cholinergic neurons studied with the whole-cell patch-clamp technique: implications for behavioral state control. J. Neurophysiol. 68, 1359–1372.

Knudsen, S., Gammeltoft, S., Jennum, P.J., 2010. Rapid eye movement sleep behaviour disorder in patients with narcolepsy is associated with hypocretin-1 deficiency. Brain 133, 568–579.

Leonard, C.S., Llinas, R.R., 1990. Electrophysiology of mammalian pedunculopontine and laterodorsal tegmental neurons in vitro: implications for the behavior of REM sleep. In: Steriade, M., Biesold, D. (Eds.), Brain Cholinergic Systems. Oxford Science, Oxford, pp. 205–223.

Luster, B., D'Onofrio, S., Urbano, F.J., Garcia-Rill, E., 2015. High-threshold Ca^{2+} channels behind gamma band activity in the pedunculopontine nucleus (PPN). Physiol. Rep. 3, e12431.

Luster, B., Urbano, F.J., Garcia-Rill, E., 2016. Intracellular mechanisms modulating gamma band activity in the pedunculopontine nucleus (PPN). Physiol. Rep. 4, e12787.

Mahowald, M.W., Cramer Bornemann, M.A., Schenck, C.H., 2011. State dissociation, human behavior, and consciousness. Curr. Topics Med. Chem. 11, 2392–2402.

Mahowald, M.W., Schenck, C.H., 2018. The "when" and "where" of α-synucleinopathies: insights from REM sleep behavior disorder [Editorial]. Neurology 91, 435–436.

Mayer, G., Meier-Ewert, K., 1993. Motor dyscontrol in sleep of narcoleptic patients (a lifelong development?). J. Sleep Res. 2, 143–148.

Meletti, S., Vaudano, A.E., Pizza, F., Ruggieri, A., Vandi, S., et al., 2015. The brain correlates of laugh and cataplexy in childhood narcolepsy. J. Neurosci. 35, 11583–11594.

Murillo-Rodriguez, E., Blanco-Centurion, C., Sanchez, C., Piomelli, D., Shiromani, P.J., 2003. Anandamide enhances extracellular levels of adenosine and induces sleep: an in vivo microdialysis study. Sleep 26, 943–947.

Nevsimalova, S., Prihodova, I., Kemlink, D., Lin, L., Mignot, E., 2007. REM behavior disorder (RBD) can be one of the first symptoms of childhood narcolepsy. Sleep Med. 8, 784–786.

Nightingale, S., Orgill, J.C., Ebrahim, I.O., de Lacy, S.F., Agrawal, S., et al., 2005. The association between narcolepsy and REM behavior disorder (RBD). Sleep Med. 6, 253–258.

Nohynek, H., Jokinen, J., Partinen, M., Vaarala, O., Kirjavainen, T., et al., 2012. AS03 adjuvanted AH1N1 vaccine associated with an abrupt increase in the incidence of childhood narcolepsy in Finland. PLoS One 7, e33536.

Oertel, W., Mayer, G., Salminen, A.V., Schenck, C.H., 2018. The foundation of the international RBD study group. In: Schenck, C.H., Högl, B., Videnovic, A. (Eds.), Rapid-Eye-Movement Sleep Behavior Disorder. Springer Verlag, Cham, Switzerland, pp. 19–32.

Palaia, V., Poli, F., Pizza, F., Antelmi, E., Franceschini, C., et al., 2011. Narcolepsy with cataplexy associated with nocturnal compulsive behaviours: a case-control study. Sleep 34, 1365–1371.

Partinen, M., Saarenpa-Heikkila, O., Ilveskoski, I., Hublin, C., Linna, M., et al., 2012. Increased incidence and clinical picture of childhood narcolepsy following the 2009 H1N1 pandemic vaccination campaign in Finland. PLoS One 7, e33723.

Pizza, F., Vandi, S., Iloti, M., Franceschini, C., Liguori, R., et al., 2015. Nocturnal sleep dynamics identify narcolepsy type 1. Sleep 38, 1277–1284.

Plazzi, G., 2018. REM sleep behavior disorder in narcolepsy. In: Schenck, C.H., Högl, B., Videnovic, A. (Eds.), Rapid-Eye-Movement Sleep Behavior Disorder. Springer Verlag, Cham, Switzerland, pp. 135–152.

Provini, F., Vetrugno, R., Montagna, P., 2008. Sleep-related smoking syndrome. Sleep Med. 9, 903–905.

Rozental, R., Srinivas, M., Spray, D.C., 2000. How to close a gap junction channel. In: Bruzzone, R., Giaume, C. (Eds.), Methods in Molecular Biology. Connexin Methods and Protocols, vol. 154. Humana Press, NJ, pp. 447–477.

Sakai, K., El Mansari, M., Jouvet, M., 1990. Inhibition by carbachol microinjections of presumptive cholinergic PGO-on neurons in freely moving cats. Brain Res. 527, 213–223.
Scammel, T.E., 2015. Narcolepsy. N. Engl. J. Med. 373, 2654–2662.
Schenck, C.H., 2005. Paradox lost: midnight. In: The Battleground of Sleep and Dreams. Extreme-Nights. LLC, Minneapolis, MN. (available from the author schen010@umn.edu).
Schenck, C.H., Bassetti, C.L., Arnulf, I., Mignot, E., 2007. English translations of the first clinical reports on narcolepsy and cataplexy by Westphal and Gélineau in the late 19th century, with commentary. J. Clin. Sleep Med. 3, 301–311.
Schenck, C.H., Boeve, B.F., Mahowald, M.W., 2013. Delayed emergence of a parkinsonian disorder or dementia in 81% of older men initially diagnosed with idiopathic rapid eye movement sleep behavior disorder: a 16-year update on a previously reported series. Sleep Med. 14, 744–748.
Schenck, C., Garcia-Rill, E., Skinner, R.D., Anderson, M., Mahowald, M.W., 1996. A case of REM sleep behavior disorder with autopsy-confirmed Alzheimer's disease: post mortem brainstem histochemical analyses. Biol. Psychiatr. 40, 422–425.
Schenck, C.H., Lee, S.A., Bornemann, M.A., Mahowald, M.W., 2009. Potentially lethal behaviors associated with rapid eye movement sleep behavior disorder: review of the literature and forensic implications. J. Forensic Sci. 54, 1475–1484.
Schenck, C.H., Mahowald, M.W., 1992. Motor dyscontrol in narcolepsy: rapid-eye-movement (REM) sleep without atonia and REM sleep behaviour disorder. Ann. Neurol. 32, 3–10.
Schenck, C.H., Mahowald, M.W., 1997. Familial narcolepsy affecting all three siblings and both biological parents. Sleep Res. 26, 496.
Schenck, C.H., Mahowald, M.W., 2002. REM sleep behavior disorder: clinical, developmental, and neuroscience perspectives 16 years after its formal identification in *SLEEP*. Sleep 25, 120–138.
Schenck, C.H., Milner, D.M., Hurwitz, T.D., Bundlie, S.R., Mahowald, M.W., 1989. A polysomnographic and clinical report on sleep-related injury in 100 adult patients. Am. J. Psychiatr. 146, 1166–1173.
Shukla, G., Kotagal, S., Schenck, C.H., 2018. RBD in childhood and adolescence. In: Schenck, C.H., Högl, B., Videnovic, A. (Eds.), Rapid-Eye-Movement Sleep Behavior Disorder. Springer Verlag, Cham, Switzerland, pp. 187–200.
Srinivas, M., Hopperstad, M.G., Spray, D.C., 2001. Quinine blocks specific gap junction channel subtypes. Proc. Natl. Acad. Sci. 98, 10942–10947.
Stea, A., Soomg, T.W., Snutch, T.P., 1995. Determinants of PKC-dependent modulation of a family of neuronal Ca^{2+} channels. Neuron 15, 929–940.
Steriade, M., 1999. Cellular substrates of oscillations in corticothalamic systems during states of vigilance. In: Lydic, R., Baghdoyan, H.A. (Eds.), Handbook of Behavioral State Control. Cellular and Molecular Mechanisms. CRC Press, New York, pp. 327–347.
Steriade, M., Paré, D., Datta, S., Oakson, G., Curro Dossi, R., 1990. Different cellular types in mesopontine cholinergic nuclei related to ponto-geniculo-occipital waves. J. Neurosci. 10, 2560–2579.
Steriade, M., Curro Dossi, R., Paré, D., Oakson, G., 1991. Fast oscillations (20–40 Hz) in thalamocortical systems and their potentiation by mesopontine cholinergic nuclei in the cat. Proc. Natl. Acad. Sci. U. S. A. 88, 4396–4400.
Takakusaki, K., Kitai, S.T., 1997. Ionic mechanisms involved in the spontaneous firing of tegmental pedunculopontine nucleus neurons of the rat. Neuroscience 78, 771–794.
Torterolo, P., Castro-Zaballa, S., Cavelli, M., Chase, M., Falconi, A., 2015. Neocortical 40 Hz oscillations during carbachol-induced rapid eye movement sleep and cataplexy. Eur. J. Neurosci. 281, 318–325.
Urbano, F.J., Leznik, E., Llinas, R., 2007. Modafinil enhances thalamocortical activity by increasing neuronal electrotonic coupling. Proc. Natl. Acad. Sci. 104, 12554–12559.

Valencia Garcia, S., Libourel, P.-A., Lazarus, M., Grassi, D., Luppi, P.H., et al., 2017. Genetic inactivation of glutamate sublaterodorsal nucleus recapitulates REM sleep behaviour disorder. Brain 140, 414–420.

Valencia Garcia, S., Brischoux, F., Clément, O., Libourel, P.A., Arthauld, S., et al., 2018. Ventromedial medulla inhibitory neuron inactivation induces REM sleep without atonia and REM sleep behavior disorder. Nat. Commun. 9, 504.

Wang, H.L., Morales, M., 2009. Pedunculopontine and laterodorsal tegmental nuclei contain distinct populations of cholinergic, glutamatergic and GABAergic neurons in the rat. Eur. J Neurosci. 29, 340–358.

Chapter 11

Arousal and movement disorders

Edgar Garcia-Rill
Center for Translational Neuroscience, Department of Neurobiology and Developmental Sciences, University of Arkansas for Medical Sciences, Little Rock, AR, United States

Introduction

A convincing model about how the brain functions proposes that the brain manifests intrinsic activity that must jive or smoothly interact with external sensory afferent information (Sperry, 1969). This eliminates proposing a dual mind versus brain dichotomy, introducing the idea that thought is an emergent property of a single process, the integration between internal and external signals in the brain (see Chapter 14 for a more detailed discussion). Assuming that this integration is widespread in the brain, one of the first levels at which intrinsic and external signals interact is the reticular activating system (RAS). Sensory information impinges on the RAS to modulate not only waking and sleep but also fight-or-flight responses, the startle response, and motor set, the ability to prepare for movement. The RAS is the first stage at which our very survival is secured by integrating threats in the environment with motion in order to first detect and then to fight or flee. The RAS, therefore, must control not only our ability to perceive threats but also our ability to act on them quickly.

What is required to react at short latency to salient threats yet carry out tasks of daily living? (1) The most basic requirement is that we be awake, so that arousal is essential to survival. Beyond *arousal and maintaining waking*, a basic property of the RAS, we also need some degree of ongoing background activity onto which to superimpose novel stimuli, in order to retain the capacity to perceive them. (2) This ongoing background is known as *preconscious awareness*, the dynamic stream of information that supports a reliable state needed to integrate sensorimotor processing essential for interaction with the world. (3) In addition, this process must be ready to generate movement and have the capacity to modulate *posture and locomotion*. The RAS and in particular the PPN possess the properties necessary for fulfilling these functions.

Arousal in Neurological and Psychiatric Diseases. https://doi.org/10.1016/B978-0-12-817992-5.00011-8

Arousal and maintaining waking

Stimulation of the RAS resulted in waking (Moruzzi and Magoun, 1949), while lesions of the area eliminated tonic arousal (Watson et al., 1974). Thus sensory afferent information and intrinsic activity in the form of gamma oscillations are essential for perception, learning, and memory, as well as consciousness (see Chapter 2). Consciousness is associated with continuous gamma band activity, as opposed to an interrupted pattern of activity (Vanderwolf, 2000a, b). That is, the gamma band activity provided by the PPN in the form of bottom-up gamma is necessary for maintaining the waking state. Disturbance of this system thus can result in abnormal sensory and motor responses, both tonic and phasic, that can account for a number of symptoms expressed in schizophrenia, bipolar disorder, PTSD, the Parkinson's disease (PD), drug abuse, autism, and certain strokes (Garcia-Rill, 2015). The PPN, as the source of bottom-up gamma, is critically involved in arousal and maintaining waking, leading to disturbances in these processes if its activity is compromised. The previous chapters have described the specific disturbances that lead to the disorders listed. In all of these disorders, the essential mechanism for navigating in the world and being able to deal with its challenges is preconscious awareness.

Gamma activity may participate in a number of processes such as sensory perception including binding of information from separate cortical regions, problem solving, and even memory (Eckhorn et al., 1988; Gray and Singer, 1989; Llinás and Paré, 1991; Llinas et al., 2002; Philips and Takeda, 2009; Palva et al., 2009; Voss et al., 2009). Coherence activity at gamma frequencies has been found to originate at both cortical or thalamocortical levels (Llinas et al., 1991; Singer, 1983). While on the one hand, cortical gamma band activity may participate in these processes, the RAS on the other hand generates the background of gamma activity essential for supporting a constant state capable of reliably assessing the world around us. These mechanisms may underlie the process of preconscious awareness (Urbano et al., 2014; Garcia-Rill et al., 2015). Therefore sensory activation of the RAS provides the background level of activity necessary for perception and voluntary movement (Garcia-Rill et al., 2015). When that level is not met, both perception and motor control are impaired.

Preconscious awareness

Preconscious awareness refers to a process that originates subcortically; we hypothesized that it is a role of the PPN (Urbano et al., 2014) and is a mechanism that remains below the level of consciousness (Civin and Lombardi, 1990). Upon waking and throughout the day, sensory input impinges on PPN dendrites and increases the excitability of these cells (Hyde et al., 2013; Garcia-Rill, 2015). Because every PPN neuron manifests intrinsic gamma oscillations, their firing increases from their usual resting state ~10 Hz and reaches the beta/gamma range as information activates the neurons and their output triggers arousal and movement (Simon, et al., 2010; Kezunovic, et al., 2011, 2013).

Based on the presence of electric coupling, intrinsic oscillation properties, and ensemble circuitry that is capable of generating and maintaining gamma band activity, we proposed a novel role for the RAS: supporting a contextual background state necessary for reliably assessing and interacting with the world, preconscious awareness (Garcia-Rill, 2015; Garcia-Rill, et al., 2013, 2014; Urbano, et al., 2014). We hypothesized that the activation of the RAS during waking induces coherent activity (through electrically coupled cells) and high-frequency oscillations (through N- and P/Q-type Ca^{2+} channel and subthreshold oscillations) to sustain gamma activity (through the activation of intracellular pathways and gene transcription) and support a persistent, reliable state needed to integrate the sensorimotor process of preconscious awareness that is instrumental for interaction with our world.

This process was likened to a "stream of consciousness," like "a river flowing forever through a man's conscious waking hours" (James, 1890). This ongoing stream of information reaches the entire brain, yet we do not pay specific attention, letting much sensorimotor information go unnoticed. Once we pay specific attention to particular elements of that stream, that is, when we actively pay tribute to a piece of sensorimotor information, we then become fully "conscious" of it (Civin and Lombardi, 1990). The origin of this preconscious process is the RAS, a phylogenetically conserved area of the brain that generates endogenous or intrinsic activity and melds it with external information. This process modulates wake-sleep cycles, the startle response, and fight-or-flight responses including changes in muscle tone and locomotion necessary for survival. The background of activity provided by this mechanism serves as a platform from which to plan and elicit movements at the shortest possible latencies.

Preparation for movement

The RP is recorded at the vertex in humans and is a DC wave that occurs 600–800 ms in advance of a voluntary movement (Kornhuber and Deecke, 1965; Deecke et al., 1976). More recent studies proposed that the RP is related to intentional binding or the process of linking movement to a sensory cue (Jo et al., 2014). The RP is, therefore, a noninvasive physiological measure of volition and has been used to assess voluntary motor activation in neurological disorders. For example, the RP is reduced or absent in PD (Simpson and Khuraibet, 1987), indicative of the deficit to initiate timely voluntary movements in PD. Another example of RP dysfunction is its reduction in HD (Turner et al., 2015). These results imply that intentional binding is reduced in PD and HD.

Another noninvasive measure of RAS activation in neurological disorders is the P50 potential. The P50 potential is also recorded at the vertex, in the same location as the RP is generated (Garcia-Rill et al., 2016). The electric P50 potential and the magnetic M50 response are localized to the region of the vertex

(Garcia-Rill et al., 2008). The following characteristics of the P50 midlatency auditory evoked potential are detailed in Garcia-Rill et al. (2001). (1) The P50 potential is present during waking and REM sleep, the states during which the PPN is active. (2) In addition, the P50 potential is reduced or blocked by cholinergic antagonists, suggesting that it is generated by cholinergic projections. In animal studies, the feline and rodent equivalents of the P50 potential were blocked by lesion or inactivation of the PPN. (3) The P50 potential manifests rapid habituation, suggesting that it is generated by a reticular, multisynaptic pathway. The conclusion is that the P50 potential is generated by the PPN and is abnormal in PD and HD. How is this sensory response related to movement?

The P50 potential is elicited by an auditory click stimulus, and the amplitude of the response has been assumed to represent a measure of level of arousal (Garcia-Rill et al., 2001). Thus the P50 potential is reduced in disorders that exhibit decreases in level of arousal, such as narcolepsy (Boop et al., 1993), the Alzheimer's disease (Buchwald et al., 1989), and autism (Buchwald et al., 1992). However, when using a paired stimulus paradigm, the first stimulus elicits a full-blown response, but if a second stimulus is applied soon after the first, the second response is reduced in amplitude. That is, the response to the second stimulus is "gated" due to the descending modulation induced by the first response. Thus the paired stimulus paradigm measures both the initial level of arousal (response to first stimulus) and the degree of habituation (response to the second stimulus) induced by sensory gating. If there is a reduction in habituation or a sensory gating deficit, the implication is that the second response is increased or unfiltered. Therefore disorders in which habituation of the P50 potential is decreased are presumed to involve a deficit in sensory gating.

The two movement disorders mentioned in the preceding text, PD and HD, in which the RP is reduced or absent, also manifest decreased habituation of the P50 potential. We demonstrated that PD patients had decreased habituation of the P50 midlatency auditory evoked potential [57], and this decrease depended on stage, such that PD patients in stage 5 showed more pronounced decreases in habituation than those in stage 3 or earlier (Teo et al., 1997). We also showed that in PD patients who received bilateral pallidotomy for the relief of their motor symptoms, habituation of the P50 potential returned to normal levels (Teo et al., 1998). Fig. 11.1 shows the degree of habituation at various interstimulus intervals in PD patients across stages 3–5 of the disease.

Just as PD manifests sleep dysregulation with abnormal reflexes and abnormal arousal, HD patients show similar disturbed sleep and reflex function (for review, see Garcia-Rill et al., 2016). The fact that both diseases exhibit abnormalities in PPN function in a similar direction is emphasized in a similar deficit in P50 potential habituation to PD in HD patients (Uc et al., 2003). Fig. 11.2 is a graph of the percent habituation of the amplitude of the response to the second stimulus in relation to the response to the first stimulus. Note that the sensory gating deficit in HD patients was similar to the decrease in habituation exhibited by PD patients in late stages (4 or 5) of that disease. Thus, in these two

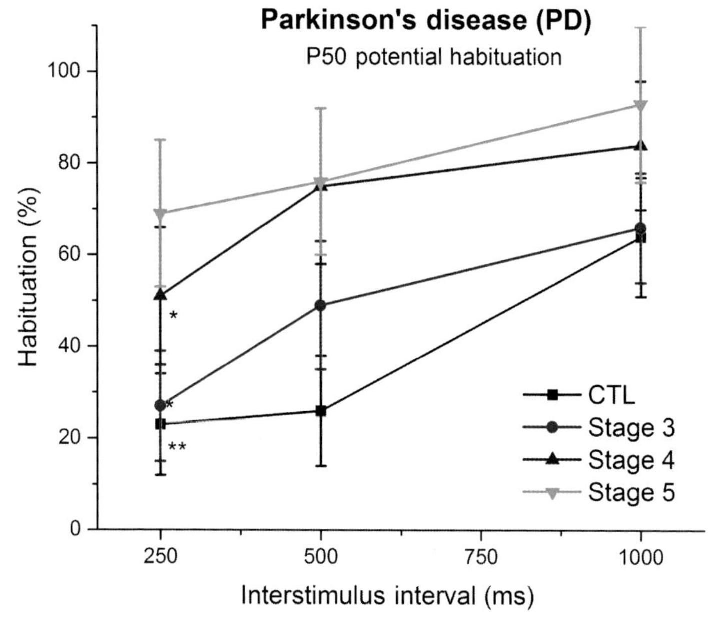

FIG. 11.1 P50 potential habituation in PD. Graph of the percent amplitude of the response to the second stimulus in relation to the amplitude of the response to the first stimulus of a pair. Three ISIs of 250, 500, and 1000 ms were tested in PD patients at stage 3 (red circles), stage 4 (blue triangles), and stage 5 (green inverted triangles) compared with age- and gender-matched controls (black squares). The percent habituation of the stage 5 patients was significantly higher compared with the control (**$P<0.01$) and stage 3 and 4 patients (*$P<0.05$) at the 250-ms ISI. *(Data from Teo, C., Rasco, A.I., Al-Mefty, K., Skinner, R.D., Garcia-Rill, E., 1997. Decreased habituation of midlatency auditory evoked responses in Parkinson's disease. Mov. Disord. 12, 655–664; Teo C., Rasco, A.I., Skinner, R.D., Garcia-Rill, E., 1998. Disinhibition of the sleep state-dependent P1 potential in Parkinson's disease-improvement after pallidotomy. Sleep Res. Online 1, 62–70.)*

movement disorders, the sleep-wake abnormalities, the suppression of the RP, and the sensory gating deficit of the P50 potential are all in the same direction and at similar levels. These results emphasize the utility of sleep studies and of P50 potential and RP recordings in determining the severity of the disease and potential response to treatment (if any or all measures are improved after therapeutic intervention).

Posture and locomotion

PPN activity, as described in the preceding text, is relayed to ascending targets to modulate arousal, but it is also relayed to descending targets to modulate posture and locomotion. Some of the regions affected by descending projections of the PPN include the subcoeruleus dorsalis (SubCD) nucleus; the pontine and medullary reticular formation; and, to a lesser extent, the spinal cord. PPN projections to the SubCD modulate its generation of ponto-geniculo-occipital

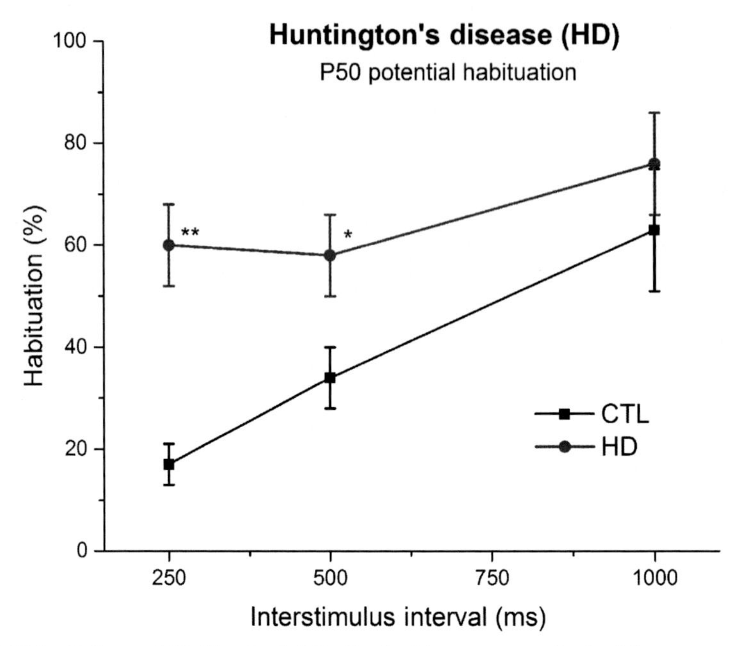

FIG. 11.2 P50 potential habituation in HD. Graph of the percent amplitude of the response to the second stimulus in relation to the amplitude of the response to the first stimulus of a pair. Three ISIs of 250, 500, and 1000ms were tested in HD patients (red circles) compared with age- and gender-matched controls (black squares). Habituation was decreased at the 250-ms ISI (**$P < 0.01$) and the 500-ms ISI (*$P < 0.05$) compared with control. *(Data from Uc, E.Y., Skinner, R.D., Rudnitzky, R.L., Garcia-Rill, E., 2003. The midlatency auditory evoked potential P50 is abnormal in Huntington's disease. J. Neurol. Sci. 212, 1–5.).*

(PGO) waves, atony of REM sleep, and SubCD outputs to the hippocampus, manifested especially in REM sleep. That is, this role of the PPN-SubCD projection is related to its activity during REM sleep, which triggers SubCD induction of PGO waves during REM sleep and its modulation of atony. We discussed the role of the PPN in such disorders as REM sleep behavior disorder (RBD) and narcolepsy in Chapter 10, and will not be addressed further.

PPN projections to pontine and medullary reticular formation, in particular the nucleus magnocellularis and the nucleus gigantocellularis, modulate reticulospinal projections that affect locomotion and postural control. Descending efferents from the PPN to the pontine and medioventral medulla are known to induce increases in stepping. In turn, outputs form the pontine, and medioventral medulla reticulospinal systems are known to activate spinal pattern generators to induce locomotion (Garcia-Rill and Skinner, 1991; Garcia-Rill et al., 2004; Reese et al., 1995). Because of the existence of spinal central pattern generators, direct projections to the spinal cord are not needed to induce stepping; these need to only trigger the pattern generators via reticulospinal projections.

The pontomedullary region can be activated using electric or chemical (in particular cholinergic agonists) stimuli to induce decreased muscle tone at some sites and stepping movements at different sites (Garcia-Rill et al., 2001; Reese et al., 1995). For example, PPN projections induce long-duration hyperpolarization in large reticulospinal neurons presumably involved in the atony of REM sleep and depolarization in medium-sized interneurons that in turn send reticulospinal projections to activate spinal pattern generators (Garcia-Rill et al., 2001; Mamiya et al., 2005). That is, the inhibition of large neurons that drive extensor muscle tone and standing are depressed at the same time as smaller neurons that drive locomotion are activated, creating a push-pull reciprocity between standing and postural drive and alternation and walking drive (Mamiya et al., 2005). The inhibition of PPN outputs of large reticulospinal neurons (Mamiya et al., 2005) may also be involved in modulating the auditory startle reflex in which the inhibition of standing is essential to preparing for movement by unlocking the knees and assuming a flexed posture ready to fight or flee. Fig. 11.3 shows the types of effects descending PPN projections exert on nucleus reticularis magnocellularis (NRM) (smaller reticulospinal cells are depolarized) and nucleus reticularis gigantocellularis (NRG) (giant cells are hyperpolarized) neurons in relation to driving locomotion.

Early studies employed decerebrate animals to identify locomotion-specific regions using long-duration pulses at 40- to 60-Hz frequency of a site originally named the mesencephalic locomotor region (MLR) (Shik et al., 1966; Garcia-Rill, 1986). Much of the misunderstanding on the role of this region stems from a lack of appreciation for the strict parameters that had to be employed to locate

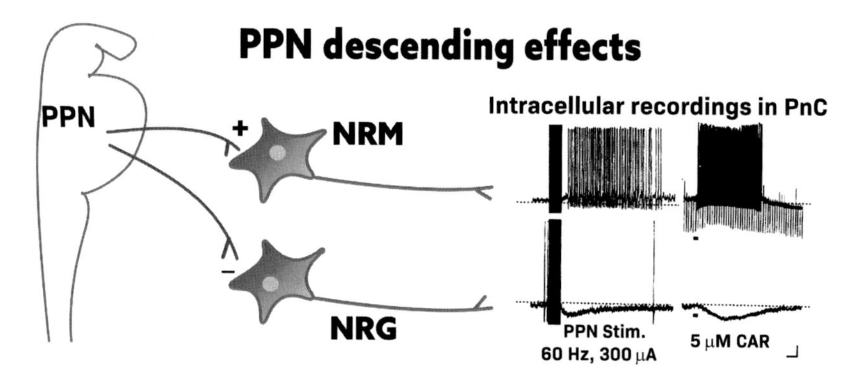

FIG. 11.3 Descending PPN output to reticulospinal pathways. Drawing of a sagittal view of the brain stem and descending pathways from the PPN, to pontine NRM, to NRG. Intracellular recordings in these neurons in the pontine nuclei (PnC) showed that both PPN electric stimulation (60-Hz pulses at 300 μA) and chemical stimulation of PnC using 5-μM carbachol induced long-lasting depolarizations in medium-sized magnocellular neurons. The same electric and chemical stimulation instead induced prolonged inhibition of large PnC neurons. *(Data from Mamiya, N., Bay, K., Skinner, R.D., Garcia-Rill, E., 2005. Induction of long-lasting depolarization in medioventral medulla (MED) neurons by cholinergic input from the pedunculopontine nucleus (PPN). J. App. Physiol. 99, 1127–1137.)*

locomotion-specific sites. For example, identification of "controlled locomotion" (presumably indicative of a site that could specifically control stepping) required (a) alternation between antagonists in the same limb; (b) alternation between agonists in opposite limbs; and (c) a proximodistal delay in contraction of muscles in the hip, knee, and ankle joints during walking. Stepping also needed to be activated in the absence of higher brain centers and only under the control of a specific region. Locomotion activated in intact animals (nondecerebrate) introduces potential confounding factors, such that responses to pain, escape responses, and processes involved in behavioral acts such as rearing and turning could originate in other brain regions and confound the localization of a locomotion-specific region.

Moreover, some authors have made assertions on the location of the so-called MLR without labeling the stimulated region or even carrying out locomotion studies these demanding criteria for controlled stepping (Noga et al., 2017; Takakusaki et al., 2003). We used uncompromising criteria in decerebrate animals to determine which stimulation sites in the mesopontine region could elicit "controlled locomotion." Such sites were located dorsal to the PPN, in the region of the ventral inferior colliculus near the cuneiform nucleus, and within the PPN, but locomotion on a treadmill could not be produced following stimulation of more medial sites such as the laterodorsal tegmental nucleus or more ventral sites in the area of the midbrain extrapyramidal region (MEA) or substantia nigra. We identified the PPN using immunocytochemical and histochemical labeling (Garcia-Rill et al., 1987; Garcia-Rill and Skinner, 1991). In addition, in decerebrate animals, we recorded PPN neurons in relation to locomotion (Garcia-Rill et al., 1983) and could chemically activate the same site to induce locomotion, which eliminated the idea that electric stimulation activated fibers of passage to elicit stepping (Garcia-Rill et al., 1985).

Our findings led to the subsequent use of PPN DBS for the treatment of gait and postural disorders in PD. Our discovery that every PPN cell manifests intrinsic gamma oscillations (see Chapter 2) helped explain why the optimal frequencies for inducing locomotion on a treadmill were 40–60 Hz and why this is also one of the optimal frequencies for some of the beneficial effects of PPN DBS (Garcia-Rill et al., 2015, 2018a, b).

Frequencies that move us

In Chapter 2, we saw how RAS activity undergoes a stepwise process across states of arousal, rather than a gradual operation. Evidence that waking is stepwise comes from clinical data showing that lesions can induce distinguishable stages, for example, from coma (complete absence of wakefulness), to a vegetative state (partial consciousness), to locked-in syndrome (awareness without movement), to full waking. In general, normally manifested electroencephalography (EEG) frequencies range from delta, to theta, to alpha, to beta/gamma, in which 10 Hz is the fulcrum between waking and sleep. These conditions represent stages of waking based on the degree of damage to gamma band generating processes (Steriade, 1999). Lower frequencies result in sleep; higher frequencies allow awareness, perception, and movement (Garcia-Rill, 2017). This 10-Hz frequency is considered an idling frequency of the PPN, while higher frequencies elicit more complex processes.

PPN neurons manifested low spontaneous firing rates <10Hz, but most showed high-frequency tonic firing in the beta/gamma range (20–80Hz) (Steriade et al., 1990). Steriade also observed the presence of gamma band activity in correlation with the cortical EEG in the cat in vivo when the animal was active (Steriade and Llinás, 1988; Steriade et al., 1990). Studies in humans observed gamma frequency activity in the region of the PPN during stepping but lower frequencies when they were at rest (Fraix et al., 2013). A study in primates reported that PPN neurons fired at low frequencies ~10Hz when the animal was at rest but fired at gamma frequencies when the animal either awakened or when it walked on a treadmill (Goetz et al., 2016). Therefore the same PPN neurons participated in both arousal and motor control. There is considerable evidence reporting the presence of gamma band activity during active waking and movement in the PPN in vitro, in vivo, and in various species, including human.

Deep brain stimulation

While DBS of the subthalamic region is well established for alleviating tremor and rigidity in PD, a number of centers began to use PPN DBS for alleviating the axial and gait symptoms of PD. The first groups to report the use of PPN DBS in PD patients showed that PD patients with severe axial symptoms and locomotion deficits, which were no longer responsive to L-dopa therapy, responded with some alleviation of symptoms (Mazzone, et al., 2005; Plaha and Gill, 2005). Another series of studies began using imaging and intracranial recordings to verify stimulation sites (Mazzone et al., 2008, 2009; Zrinzo et al., 2008). Recordings from DBS electrodes showed that neurons in the PPN neurons fired in relation to passive and voluntary movements (Weinberger et al., 2008) and during imagined locomotion (Karachi et al., 2010). Later studies showed that cells in this region fired in phase with alpha (~10Hz) oscillations following passive and imagined movements in quiescent patients (Tattersall et al., 2014).

One group found that bilateral stimulation of the PPN at 15 and 25Hz decreased falls and freezing of gait (Ferraye et al., 2010). Similar findings were described by others showing that there was improvement in motor scores and falls using 50 and 70Hz (Moro et al., 2010). Others found a modest improvement in motor scores but significant improvements in sleep scores when using 10 and 25Hz (Stefani et al., 2007, 2013). On the other hand, some found no motor improvement but marked improvements in sleep scores and cognitive function (Alessandro et al., 2010). Stimulation at 20–35Hz was reported to decrease reaction time and improve falls along with freezing of gait (Thevathasan et al., 2010), and the same group carried out a double-blind analysis to establish that bilateral stimulation was more effective than unilateral stimulation (Thevathasan et al., 2012). A review of the PPN DBS data requested by the Movement Disorder Society described the surgical and clinical aspects of this field (Hamani et al., 2016a, b). The latest MRI-based study of the optimal area for PPN DBS has shown it to be centered on the PPN (Goetz et al., 2018). These studies also showed that stimulation frequencies in the 10–30 Hz were optimal for alleviating freezing of gait.

Basically, the results to date are encouraging but suffer from the typical low numbers of patients, the variability in axial and gait disturbances manifested by patients with a heterogeneous pathology, and the lack of certainty in localization of stimulation sites. Nevertheless, PPN DBS shows promise and needs to be pursued since few therapies appear to affect sleep parameters and cognitive measures in addition to gait and postural stability. In the interest of optimizing the methodology for the benefit of patients, we propose a more thorough assessment of the parameters studied in PD patients and their responses to a variety of frequencies of stimulation. Fig. 11.4 shows examples of EEG recordings at various sleep-wake stages, in general, from deep sleep (delta), to light sleep (theta), to sleep-wake transition (alpha), to beta and gamma (full waking), with the

The 10 Hz frequency fulcrum

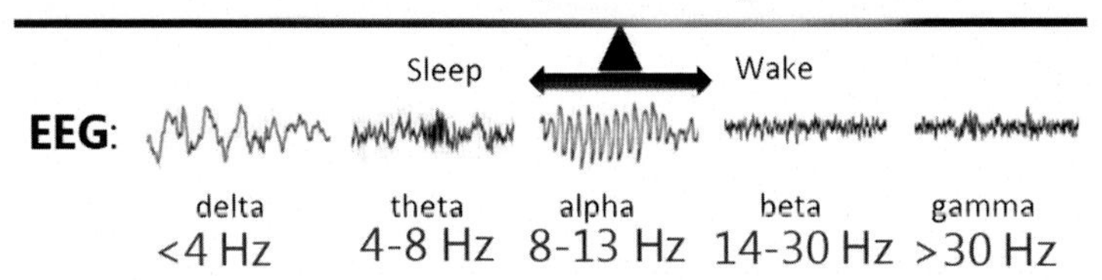

Potential PPN stimulation frequencies

PPN	cell activity	10 Hz	15-30 Hz > 30 Hz
	role	**REST**	**ACTIVE**
RAS	**role**	**TONIC**	**PHASIC**
	potential functions	stand fog	walk cognitive

FIG. 11.4 The PPN and EEG frequencies. The 10-Hz frequency fulcrum is shown at the top, with examples of delta, theta, alpha, beta, and gamma frequency EEG recordings. Recordings in PPN showed that cells at rest fired ~10 Hz but increased firing to beta/gamma upon waking or moving. The role of the RAS in tonic and phasic functions calls for a more detailed workup of each patient's symptoms and responses to different frequencies when using DBS for the treatment of either postural or locomotor symptoms.

10-Hz fulcrum marking the border between sleep and waking. Recordings in the PPN at rest indicate that cells fire at ~10 Hz, while they fire at beta/gamma frequencies when the animal is active (aroused or walking). In addition, the RAS manifests both tonic and phasic driving of postural and locomotor systems.

An important piece of information is that symptoms and functions respond best to DBS at frequencies in the 10-Hz range compared with the 40-Hz range and may be blocked by stimulation at high frequencies (~100 Hz) that may depolarize block the PPN. A standing individual fights gravity with knees locked and tonic extensor muscle tone. Too much or too little postural tone may lead to instability and sway. On the other hand, in order to walk, extensor tone must be inhibited so that flexion of the legs can be initiated. This is one consequence of the startle response. Once locomotion has begun, the feedback generated by leg swing facilitates locomotion, so that stepping requires only a stop signal to cease walking. Given these facts, PPN DBS at various frequencies may elicit different effects on descending reticulospinal systems. Testing such ranges of stimuli is important for optimizing the therapy for the appropriate deficit.

In addition, it needs to be remembered that stimulation of the PPN induces activity at a host of ascending targets. It is likely that stimulation at gamma band could promote waking activity and may also improve cognitive function. Given the multiple roles of gamma activity in the brain, PPN DBS at gamma frequencies could be potentiating bottom-up gamma and drive a number of processes. This information is important because PPN DBS can be seen as a gamma generator and in that way help regulate overall brain gamma activity. We recently proposed that PPN DBS could be used not only to normalize REM sleep and waking in PD but also to normalize gamma activity in diseases in which gamma band is impaired, such as in schizophrenia, bipolar disorder, neglect, coma, and even epilepsy (Garcia-Rill et al., 2018b). Wide raging clinical studies are called for in order to determine the optimal frequencies of stimulation for the appropriate process to be activated or deactivated.

References

Alessandro, S., Ceravolo, R., Brusa, L., Pierantozzi, M., Costa, A., et al., 2010. Non-motor functions in parkinsonian patients implanted in the pedunculopontine nucleus: focus on sleep and cognitive problems. J. Neurol. Sci. 289, 44–48.

Boop, F.A., Garcia-Rill, E., Dykman, R., Skinner, R.D., 1993. The P1: insights into attention and arousal. J. Pediat. Neurosurg. 20, 57–62.

Buchwald, J.S., Erwin, R., Read, S., Van Lancker, J., Cummings, J.L., 1989. Midlatency auditory evoked responses: differential abnormality of P1 in Alzheimer's disease. Electroenceph. Clin. Neurophyiol. 74, 378–384.

Buchwald, J.S., Erwin, R., Van Lancker, D., Guthrie, D., Scwafel, J., et al., 1992. Midlatency auditory evoked responses: P1 abnormalities in adult autistic subjects. Electroencephalogr. Clin. Neurophysiol. 84, 164–171.

Civin, M., Lombardi, K.L., 1990. The preconscious and potential space. Psychoanal. Rev. 77, 573–585.

Deecke, L., Grozinger, B., Kornhuber, H.H., 1976. Voluntary finger movement in man: cerebral potentials and theory. Biol. Cybern. 23, 99.

Eckhorn, R., Bauer, R., Jordan, W., Brosch, M., Kruse, W., et al., 1988. Coherent oscillations: a mechanism of feature linking in the visual system? Multiple electrode and correlation analyses in the cat. Biol. Cybern. 60, 121–130.

Ferraye, M.U., Debu, B., Fraix, V., Goetz, L., Ardouin, C., et al., 2010. Effects of pedunculopontine nucleus area stimulation on gait disorders in Parkinson's disease. Brain 133, 205–214.

Fraix, V., Bastin, J., David, O., Goetz, L., Ferraye, M., et al., 2013. Pedunculopontine nucleus area oscillations during stance, stepping and freezing in Parkinson's disease. PLoS One 8, e83919.

Garcia-Rill, E., 1986. The basal ganglia and the locomotor regions. Brain Res. Rev. 11, 47–63.

Garcia-Rill, E., 2015. Waking and the Reticular Activating System. Academic Press, New York, p. 330.

Garcia-Rill, E., 2017. Bottom-up gamma and stages of waking. Med. Hypotheses 104, 58–62.

Garcia-Rill, E., Homma, Y., Skinner, R.D., 2004. Arousal mechanisms related to posture and movement. I. Descending modulation. In: Mori, S., Stuart, D.G., Wiesendanger, M. (Eds.), Brain Mechanisms for the Integration of Posture and Movement. Prog. Brain Res.143, pp. 283–290.

Garcia-Rill, E., Houser, C.R., Skinner, R.D., Smith, W., Woodward, D.J., 1987. Locomotion-inducing sites in the vicinity of the pedunculopontine nucleus. Brain Res. Bull. 18, 731–738.

Garcia-Rill, E., Kezunovic, N., Hyde, J., Beck, P., Urbano, F.J., 2013. Coherence and frequency in the reticular activating system (RAS). Sleep Med. Rev. 17, 227–238.

Garcia-Rill, E., Kezunovic, N., D'Onofrio, S., Luster, B., Hyde, J., et al., 2014. Gamma band activity in the RAS-intracellular mechanisms. Exp. Brain Res. 232, 1509–1522.

Garcia-Rill, E., Luster, B., D'Onofrio, S., Mahaffey, S., Bisagno, V., et al., 2015. Implications of gamma band activity in the pedunculopontine nucleus. J. Neural Transm. 123, 655–665.

Garcia-Rill, E., Mahaffey, S., Hyde, J., Urbano, F.J., 2018b. Bottom-up gamma maintenance in various disorders. Neurobiol. Dis. (in press).

Garcia-Rill, E., Moran, K., Garcia, J., Findley, W.M., Walton, K., et al., 2008. Magnetic sources of the M50 response are localized to frontal cortex. Clin. Neurophysiol. 119, 388–398.

Garcia-Rill, E., Saper, C., Rye, D., Kofler, M., Nonnekes, J., et al., 2018a. Focus on the pedunculopontine nucleus. Clin. Neurophysiol. (in press).

Garcia-Rill, E., Skinner, R.D., 1991. Modulation of rhythmic functions by the brainstem. In: Shimamura, M., Grillner, S., Edgerton, V.R. (Eds.), Neurobiology of Human Locomotion. Japan Scientific Societies Press, Tokyo, pp. 137–158.

Garcia-Rill, E., Skinner, R.D., Fitzgerald, J.A., 1983. Activity in the mesencephalic locomotor region during locomotion. Exp. Neurol. 82, 606–622.

Garcia-Rill, E., Skinner, R.D., Fitzgerald, J.A., 1985. Chemical activation of the mesencephalic locomotor region. Brain Res. 330, 43–54.

Garcia-Rill, E., Skinner, R.D., Miyazato, H., Homma, Y., 2001. Pedunculopontine stimulation induces prolonged activation of pontine reticular neurons. Neuroscience 104, 455–465.

Garcia-Rill, E., Virmani, T., Hyde, J.R., D'Onofrio, S., Mahaffey, S., 2016. Arousal and the control of perception and movement. Curr. Trends Neurol. 10, 53–64.

Goetz, L., Piallat, B., Bhattacharjee, M., Mathieu, H., David, O., et al., 2016. The primate pedunculopontine nucleus region: towards a dual role in locomotion and waking state. J. Neural Transm. 123, 667–678.

Goetz, L., Bhattacharjee, M., Ferraye, M.U., Fraiz, V., Maineri, C., et al., 2018. Deep brain stimulation of the pedunculopontine nucleus area in Parkinson's disease: MRI-based anatomical correlations and optimal target. Neurosurgery 25, (e-pub ahead of print).

Gray, C.M., Singer, W., 1989. Stimulus-specific neuronal oscillations in orientation columns of cat visual cortex. Proc. Natl. Acad. Sci. U. S. A. 86, 1698–1702.

Hamani, C., Aziz, T., Bloem, B.R., Brown, P., Chabardes, S., et al., 2016a. Pedunculopontine nucleus region deep brain stimulation in Parkinson disease: surgical anatomy and terminology. Stereotact. Funct. Neurosurg. 94, 298–306.

Hamani, C., Lozano, A.M., Mazzone, P.A., Moro, E., Hutchison, W., et al., 2016b. Pedunculopontine nucleus region deep brain stimulation in Parkinson disease: surgical techniques, side effects, and postoperative imaging. Stereotact. Funct. Neurosurg. 94, 307–319.

Hyde, J.R., Kezunovic, N., Urbano, F.J., Garcia-Rill, E., 2013. Spatiotemporal properties of high speed calcium oscillations in the pedunculopontine nucleus. J. Appl. Physiol. 115, 1402–1414.

James, W., 1890. In: Holt, H. (Ed.), The Principles of Psychology. 1. Henry Holt and Company Press, pp. 225.

Jo, H.G., Wittmann, M., Hinterberger, T., Schmidt, S., 2014. The readiness potential reflects intentional binding. Front. Hum. Neurosci. 8, (421(1-9).

Karachi, C., Grabli, D., Bernard, F.A., Tandee, D., Wattiez, N., et al., 2010. Cholinergic mesencephalic neurons are involved in gait and postural disorders in Parkinson's disease. J. Clin. Invest. 120, 2745–2754.

Kezunovic, N., Urbano, F.J., Simon, C., Hyde, J., Smith, K., et al., 2011. Mechanism behind gamma band activity in the pedunculopontine nucleus (PPN). Eur. J. Neurosci. 34, 404–415.

Kezunovic, N., Hyde, J., Goitia, B., Bisagno, V., Urbano, F.J., et al., 2013. Muscarinic modulation of high frequency activity in the pedunculopontine nucleus (PPN). Front. Neurol. Sleep Chronobiol. 4, 176(1–13.

Kornhuber, H.H., Deecke, L., 1965. Changes in the brain potential in voluntary movements and passive movements in man: readiness potential and reafferent potentials. Pflugers Arch. Gesamte Physiol. Menschen Tiere 284, 1–17.

Llinas, R.R., Grace, A.A., Yarom, Y., 1991. In vitro neurons in mammalian cortical layer 4 exhibit intrinsic oscillatory activity in the 10- to 50-Hz frequency range. Proc. Natl. Acad. Sci. U. S. A. 88 (199), 897–901.

Llinas, R.R., Leznik, E., Urbano, F.J., 2002. Temporal binding via cortical coincidence detection of specific and nonspecific thalamocortical inputs: a voltage-dependent dye-imaging study in mouse brain slices. Proc. Natl. Acad. Sci. U. S. A. 99, 449454.

Llinás, R.R., Paré, D., 1991. Of dreaming and wakefulness. Neuroscience 44, 521–535.

Mamiya, N., Bay, K., Skinner, R.D., Garcia-Rill, E., 2005. Induction of long-lasting depolarization in medioventral medulla (MED) neurons by cholinergic input from the pedunculopontine nucleus (PPN). J. App. Physiol. 99, 1127–1137.

Mazzone, P., Insola, A., Sposato, S., Scarnati, E., 2009. The deep brain stimulation of the pedunculopontine tegmental nucleus. Neuromodulation 12, 191–204.

Mazzone, P., Lozano, A., Sposato, S., Scarnati, E., Stefani, A., 2005. Brain stimulation and movement disorders: where we going? In: Proceedings of 14th Meeting of the World Society of Stereotactic and Functional Neurosurgery (WSSFN). Monduzzi, Bologna.

Mazzone, P., Sposato, S., Insola, A., Dilazzaro, V., Scarnati, E., 2008. Stereotactic surgery of nucleus tegmenti pedunculopontine [corrected]. Brit. J. Neurosurg. 22 ((2008) Suppl. 1), S33–S40.

Moro, E., Hamani, C., Poon, Y.Y., Al-Khairallah, T., Dostrovsky, J.O., et al., 2010. Unilateral pedunculopontine stimulation improves falls in Parkinson's disease. Brain 133, 215–224.

Moruzzi, G., Magoun, H.W., 1949. Brain stem reticular formation and activation of the EEG. Electroencephalogr. Clin. Neurophysiol. 1, 455–473.

Noga, B., Sanchez, F., Villamil, L., O'Toole, C., Kasicki, C., et al., 2017. LFP oscillations in the mesencephalic locomotor region during voluntary locomotion. Front. Neural Circuits 11, 34.

Palva, S., Monto, S., Palva, J.M., 2009. Graph properties of synchronized cortical networks during visual working memory maintenance. Neuroimage 49, 3257–3568.

Philips, S., Takeda, Y., 2009. Greater frontal-parietal synchrony at low gamma-band frequencies for inefficient than efficient visual search in human EEG. Int. J. Psychophysiol. 73, 350.

Plaha, P., Gill, S.S., 2005. Bilateral deep brain stimulation of the pedunculopontine nucleus for Parkinson's disease. Neuroreport 16, 1883–1887.

Reese, N.B., Garcia-Rill, E., Skinner, R.D., 1995. The pedunculopontine nucleus-auditory input, arousal and pathophysiology. Prog. Neurobiol. 47, 105–133.

Shik, M.L., Severin, F.V., Orlovskii, G.N., 1966. Control of walking and running by means of electric stimulation of the midbrain. Biofizika 11, 659–666.

Simpson, J.A., Khuraibet, A.J., 1987. Readiness potential of cortical area 6 preceding self paced movement in Parkinson's disease. J. Neurol. Neurosurg. Psychiatry 50, 1184–1191.

Singer, W., 1983. Synchronization of cortical activity and its putative role in information processing and learning. Annu. Rev. Physiol. 55, 349–374.

Simon, C., Kezunovic, N., Ye, M., Hyde, J., Hayar, A., et al., 2010. Gamma band unit and population responses in the pedunculopontine nucleus. J. Neurophysiol. 104, 463–474.

Sperry, R.W., 1969. A modified concept of consciousness. Psychol. Rev. 76, 532–536.

Stefani, A., Lozano, A.M., Peppe, A., Stanzione, P., Galati, S., et al., 2007. Bilateral deep brain stimulation of the pedunculopontine and subthalamic nuclei in severe Parkinson's disease. Brain 130, 1596–1607.

Stefani, A., Peppe, A., Galati, S., Stampanoni, A., Basso, M., et al., 2013. The serendipity case of the pedunculopontine nucleus low-frequency brain stimulation: chasing a gait response, finding sleep, and cognitive improvement. Front. Neurol. 4, 68.

Steriade, M., Paré, D., Datta, S., Oakson, G., Curro Dossi, R., 1990. Different cellular types in mesopontine cholinergic nuclei related to ponto-geniculo-occipital waves. J.Neurosci. 10, 2560–2579.

Steriade, M., 1999. Cellular substrates of oscillations in corticothalamic systems during states of vigilance. In: Lydic, R., Baghdoyan, H.A. (Eds.), Handbook of Behavioral State Control. Cellular and molecular mechanisms. CRC Press, New York, pp. 327–347.

Steriade, M., Llinás, R.R., 1988. The functional states of the thalamus and the associated neuronal interplay. Physiol. Rev. 68, 649–742.

Takakusaki, K., Habaguchi, T., Ohtinata-Sugimoto, J., Saitoh, K., Sakamoto, T., 2003. Basal ganglia efferents to the brainstem centers controlling postural muscle tone and locomotion: a new concept for understanding motor disorders in basal ganglia dysfunction. Neuroscience 119, 293–308.

Tattersall, T.L., Stratton, P.G., Coyne, T.J., Cook, R., Silberstein, P., et al., 2014. Imagined gait modulates neuronal network dynamics in the human pedunculopontine nucleus. Nat. Neurosci. 17, 449–454.

Teo, C., Rasco, A.I., Al-Mefty, K., Skinner, R.D., Garcia-Rill, E., 1997. Decreased habituation of midlatency auditory evoked responses in Parkinson's disease. Mov. Disord. 12, 655–664.

Teo, C., Rasco, A.I., Skinner, R.D., Garcia-Rill, E., 1998. Disinhibition of the sleep state-dependent P1 potential in Parkinson's disease-improvement after pallidotomy. Sleep Res. Online 1, 62–70.

Thevathasan, W., Silburn, P.A., Brooker, H., Coyne, T.J., Kahn, S., et al., 2010. The impact of low-frequency stimulation of the pedunculopontine nucleus region on reaction time in Parkinsonism. J. Neurol. Neurosurg. Psychiatry 81, 1099–1104.

Thevathasan, W., Cole, M.H., Grapel, C.L., Hyam, J.A., Jenkinson, N., et al., 2012. A spatiotemporal analysis of gait freezing and the impact of pedunculopontine nucleus stimulation. Brain 135, 1446–1454.

Turner, L.M., Croft, R.J., Churchyard, A., Looi, J.C., Apthorp, D., et al., 2015. Abnormal electrophysiological motor responses in Huntington's disease: evidence of premanifest compensation. PLoS One 10, e0138563.

Uc, E.Y., Skinner, R.D., Rudnitzky, R.L., Garcia-Rill, E., 2003. The midlatency auditory evoked potential P50 is abnormal in Huntington's disease. J. Neurol. Sci. 212, 1–5.

Urbano, F.J., D'Onofrio, S.M., Luster, B.R., Hyde, J.R., Bosagno, V., et al., 2014. Pedunculopontine nucleus gamma band activity- preconscious awareness, waking, and REM sleep. Front. Sleep Chronobiol. 5, 210.

Vanderwolf, C.H., 2000a. Are neocortical gamma waves related to consciousness? Brain Res. 855, 217–224.

Vanderwolf, C.H., 2000b. What is the significance of gamma wave activity in the pyriform cortex? Brain Res. 877, 125–133.

Voss, U., Holzmann, R., Tuin, I., Hobson, J.A., 2009. Lucid dreaming: a state of consciousness with features of both waking and non-lucid dreaming. Sleep 32, 1191–2000.

Watson, R.T., Heilman, K.M., Miller, B.D., 1974. Neglect after mesencephalic reticular formation lesions. Neurolpgy 24, 294–298.

Weinberger, M., Hamani, C., Hutchison, W.D., Moro, E., Lozano, A.M., et al., 2008. Pedunculopontine nucleus microelectrode recordings in movement disorder patients. Exp. Brain Res. 188, 165–174.

Zrinzo, L., Zrinzo, L.V., Tisch, S., Limousin, P.D., Yousry, T.A., et al., 2008. Stereotactic localization of the human pedunculopontine nucleus: atlas-based coordinates and validation of a magnetic resonance imaging protocol for direct localization. Brain 131, 1588–1598.

Further reading

Garcia-Rill, E., 1991. The pedunculopontine nucleus. Prog. Neurobiol. 36, 363–389.

Garcia-Rill, E., Skinner, R.D., 1988. Modulation of rhythmic function in the posterior midbrain. Neuroscience 17, 639–654.

Garcia-Rill, E., Skinner, R.D., 2001. The sleep state-dependent P50 midlatency auditory evoked potential. In: Lee-Chiong, T.L., Carskadon, M.A., Sateia, M.J. (Eds.), Sleep Medicine. Hanley & Belfus, Philadelphia, pp. 697–704.

Chapter 12

Epilepsy and arousal

Sang-Hun Lee, Ethan M. Clement, Young-Jin Kang
Department of Neurology, University of Arkansas for Medical Sciences, Little Rock, AR, United States

Introduction

Epilepsy, as defined by the International League Against Epilepsy (ILAE), is "enduring tendency to have recurrent seizures" (Fisher et al., 2014). Based on a 2015 estimate, 1.2% of the US population and 6.3 out of 1000 children suffer from active epilepsy (Zack and Kobau, 2017). A significant number of patients with epilepsy live with a reduced quality of life due not only to debilitating seizures but also to comorbidities such as increased anxiety, depression, and memory deficits (Keezer et al., 2016). In agreement with evidence showing that brain regions that are affected by certain types of epilepsy (e.g., absence epilepsy) also mediate consciousness, patients experiencing a seizure event will often have transient decreased or complete loss of consciousness, whereas patients with other types of epilepsy appear to spare dysregulation of consciousness. Accordingly, in 2017, the ILAE reclassified seizure types with an important move from simple partial versus complex partial seizures to a classification that dictated focal aware versus focal impaired awareness seizures, stating that this new classification included the "practical importance of awareness" (Fisher et al., 2017).

In this chapter, we will discuss absence epilepsy and mesial temporal lobe epilepsy (TLE) regarding dysregulation of consciousness. See Chapter 1 for a more complete discussion of arousal and consciousness. Seizures in absence epilepsy are accompanied by largely distinct behaviors, EEG patterns, and dysregulation of consciousness compared with those in temporal lobe epilepsy (Blair, 2012; Halász, 2015; Tatum, 2012; Unterberger et al., 2018). Seizures in both types of epilepsy, on the other hand, are commonly characterized by hypersynchronous abnormal network activity that is similar to that observed during non-REM sleep in key brain regions for maintaining consciousness (Blumenfeld, 2012). Thus, we will consider how the abnormal network activity may arise from neuronal circuits and contribute to impaired consciousness in epilepsy. According to current literature, consciousness consists of three key aspects: arousal (level of wakefulness), awareness (the

Arousal in Neurological and Psychiatric Diseases. https://doi.org/10.1016/B978-0-12-817992-5.00012-X

content of wakefulness), and attention (ability to respond and perform tasks) to the external world (Blumenfeld, 2012). We will use those definitions of arousal, awareness, and attention throughout this chapter when necessary.

Types of altered consciousness in epilepsy

As informed earlier, practical importance of consciousness in seizure classification is well received (Fisher et al., 2017). However, current literature lacks unifying definitions regarding types of altered consciousness in epilepsy. Lüders and his colleagues proposed a classification in which the loss of consciousness in epilepsy is categorized into five types according to severity and presentation: aura, dyscognitive seizure, ictal delirium, dialeptic seizure, and coma during epileptic seizures (Lüders et al., 2014). These alterations of consciousness can be as mild such as an aura or as severe as epileptic coma, each presenting with specific clinical manifestations and different underlying mechanisms. Thus, we briefly describe the proposed five classifications in conjunction with seizure types.

Auras during a seizure event are sensory alterations while maintaining level and content of consciousness. This is the lowest level of alteration in consciousness during a seizure. Auras typically occur before a seizure and can include visual, auditory, gustatory, or déjà vu sensations (Dugan et al., 2014). Sixty-five percent of generalized epilepsy patients surveyed self-reported aura symptoms directly before a seizure (Dugan et al., 2014). Dyscognitive seizures interfere with awareness as they disrupt high-order processes while sparing the arousal system. This classification is meant to describe an acute and specific loss of a cognitive ability such as language, voluntary movement, or memory formation (Lüders et al., 2014). Ictal delirium during a seizure interferes with the perception of reality while still maintaining wakefulness and posture. Patients who experience ictal delirium have hallucinations indistinguishable from reality but can still interact with reality. Lüders and his colleagues suggest that greater cortical involvement during ictal delirium, compared with that during auras may be associated with the misperception (Lüders et al., 2014). Dialepsis is similar to ictal delirium as it alters the content of consciousness; however, instead of active hallucinations, patients "zone out" as they become completely unresponsive to their environment while still maintaining arousal and posture. Lüders et al. proposed this term as it not only encompasses "absences," seen as ictal 3-Hz spike-and-wave discharges (SWDs) on EEG, but also includes similar presentations seen in focal seizures that do not have the same ictal pattern (Lüders et al., 2014). Impaired consciousness in absence seizures will be discussed in detail in the section of "absence seizures" of this chapter. Lastly, epileptic coma disrupts the arousal system, muscle tone, and awareness. Epileptic coma is most commonly seen in generalized epilepsy, where the frontoparietal association cortices, thalamus, and brain stem are involved in ictal activity (Blumenfeld, 2012). It is important to note that each of the aforementioned five classifications of consciousness impairments can be found in multiple seizure types. On the other hand, certain seizure types (e.g., secondarily generalized tonic-clonic seizures observed in mesial TLE) are accompanied by multiple types of consciousness impairments as a seizure progresses, which will be discussed in the section of "Mesial TLE" of this chapter.

Finally, based on the classifications proposed by Lüders et al. (2014), fully developed classifications will require much research to delineate specific clinical manifestations regarding each consciousness impairment and to identify distinct mechanisms underlying each class of altered consciousness. Given there are no appropriate treatments for altered consciousness in epilepsy, such efforts are urgently needed to identify the classes of impaired consciousness and develop evidence-based targeted treatment of altered consciousness in epilepsy.

Absence epilepsy

The term "absence" of absence seizures is largely used for an absence of consciousness (Unterberger et al., 2018). Thus, the name itself stresses the importance of consciousness impairments. Childhood absence seizures are accompanied by a stereotypical EEG pattern, 3- to 4-Hz SWDs that only last approximately 10 s and are typically seen in children from ages 6 to 7 (Crunelli and Leresche, 2002). Children with absence epilepsy generally manifest behavioral arrest during an absence seizure, along with eyelid movement and some hand or face repetitive movement (Tenney and Glauser, 2013). These seizures can occur from tens to hundreds of times a day, often without the patient being able to sustain awareness throughout the events or having any recollection of their occurrence (Crunelli and Leresche, 2002).

Although absence seizures are categorized as "generalized" seizures, evidence from human and animal studies supports the notion that selective focal bilateral cortical and subcortical networks are involved, leading to specific consciousness impairments (Blumenfeld, 2012; Guo et al., 2016; Unterberger et al., 2018). The severity of consciousness impairments during absence seizures are associated with amplitude and distribution of SWDs that arise from hypersynchronous thalamocortical circuits (Browne et al., 1974; Guo et al., 2016; Sadleir et al., 2008). The etiology of absence seizures is primarily genetic, which is associated with mutations in ion channels or neurotransmitter receptors (Crunelli and Leresche, 2002). We will first discuss how the thalamocortical circuit is functionally organized and how consciousness impairments during absence seizures may arise resulting from compromised thalamocortical network activity.

Thalamocortical circuits and thalamocortical oscillations

The prevailing view of circuit mechanisms underlying cortical 3–4 Hz SWDs is that the thalamocortical circuit is critically involved in SWDs observed during absence seizures (Chang and Lowenstein, 2003). This circuit also contributes to the generation of sleep spindles (12–16Hz oscillations) observed during non-REM sleep (Beenhakker and Huguenard, 2009). Importantly, the presence of these rhythms is closely correlated with the level of arousal (McCormick and Bal, 1997). Three major cell types within the thalamocortical circuits (i.e., cortical pyramidal cells, thalamocortical relay cells, and reticular thalamic (RT) nucleus neurons) form their functional local and long-range connectivity. The reciprocal connectivity between the thalamus and the cortex is formed by glutamatergic principal neurons in the thalamus (i.e., thalamocortical relay neurons) and in the cortex (i.e., pyramidal cells). RT neurons are GABAergic and form GABAergic synapses onto thalamocortical relay neurons. Thalamocortical relay neurons, in turn, form glutamatergic synapses onto RT neurons on the way to the cortex by branching off their axons, resulting in an inhibitory-excitatory loop between RT neurons and thalamocortical relay neurons (Fig. 12.1A). Reciprocally, excitatory inputs from cortical pyramidal cells form excitatory synapses onto RT neurons on the way to thalamocortical relay neurons. This reciprocal long-range thalamocortical and intrathalamic connectivity between thalamocortical relay neurons and RT neurons are the key anatomical bases to the 3- to 4-Hz SWDs seen in absence epilepsy and sleep spindles (Beenhakker and Huguenard, 2009).

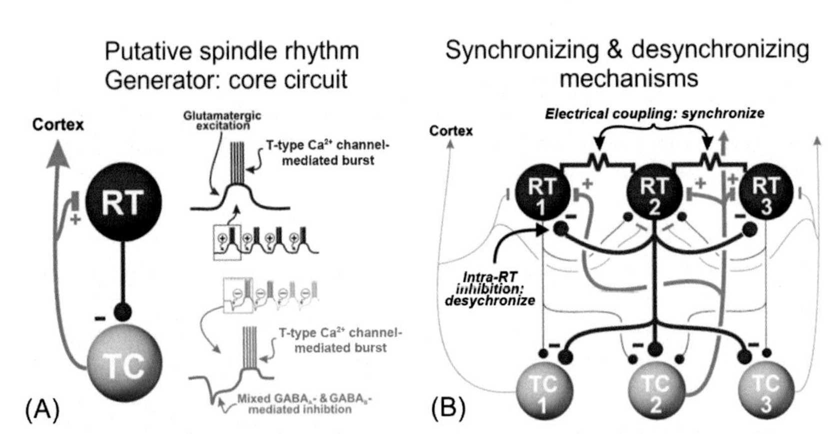

FIG. 12.1 Thalamic rhythm generator. (A) A pair of synaptically coupled thalamocortical relay neurons and RT neuron are shown. Note that T-type Ca^{2+} channel-mediated bursting in both cell types is required to maintain spindles, along with RT neuron-mediated inhibition of the thalamocortical relay neuron. The corticothalamic inputs from pyramidal cells, which are not shown in this figure, form excitatory synapses on both the RT neuron and the thalamocortical relay neuron, and these excitatory inputs are thought to excite the spindle rhythm generator. (B) Key regulatory mechanisms underlying synchronizing and desynchronizing RT neurons. Electric coupling among RT neurons are thought to synchronize their activity, whereas $GABA_A$-mediated inhibition among the RT neurons is thought to desynchronize their activity. Plus and minus signs refer to glutamatergic excitation and GABAergic inhibition, respectively. *(Parts (A) and (B) are from Beenhakker, M.P., Huguenard, J.R., 2009. Neurons that fire together also conspire together: is normal sleep circuitry hijacked to generate epilepsy? Neuron 62, 612–632.)*

Intrinsic and synaptic properties of thalamic neurons along with reciprocal thalamocortical connectivity critically contribute to thalamocortical oscillations (McCormick and Bal, 1997). Thalamocortical relay neurons and RT neurons manifest two distinct types of brain state-dependent firing modes: burst and tonic (Llinas and Steriade, 2006). They both express low-voltage-activated T-type Ca^{2+} channels. During non-REM sleep, activation of T-type Ca^{2+} channels lead to low-threshold calcium spike, triggering high-frequency action potential discharges (burst mode) superimposed on the low-threshold calcium spike. Thus, thalamocortical relay neurons activate the cortex in a rhythmic way during non-REM sleep, leading to synchronous EEG signals (e.g., sleep spindles). In contrast, thalamocortical relay neurons fire tonically during wakefulness and REM sleep due to reduced activation of T-type Ca^{2+} channels, allowing for the transfer of information to the cortex. The firing mode of thalamocortical relay neurons is regulated primarily by RT neuron-mediated inhibition. For example, T-type Ca^{2+} channels in thalamocortical relay neurons remain largely inactivated at a resting membrane potential (Sherman, 2001). When burst firing of RT neurons evokes activation of $GABA_B$ receptors on thalamocortical relay neurons, $GABA_B$ receptor-mediated inhibition persists for approximately several hundreds of milliseconds, which is long enough to remove inactivation of T-type Ca^{2+} channels, resulting in postinhibitory rebound burst firing upon release of thalamocortical relay neurons from inhibition (Sherman, 2001). Since thalamocortical relay neurons excite RT neurons, the burst from thalamocortical relay neurons leads to subsequent burst discharges of RT neurons, resulting in the next cycle of thalamocortical oscillations (Fig. 12.1A).

Thalamocortical oscillations are regulated by intrinsic properties of the thalamocortical circuit and external inputs from the brain stem. RT neurons form both chemical and electric synapses onto RT neurons. Electric synapses via gap junctions are known to play a prooscillatory role in thalamocortical oscillations, whereas $GABA_A$ receptor-associated chemical synapses are known to play an antioscillatory role in thalamocortical network oscillations (Fig. 12.1B). Importantly, this thalamocortical circuit activity is also regulated by ascending cholinergic, noradrenergic, and serotonergic inputs from various brain stem structures including the pedunculopontine nuclei, laterodorsal tegmental nuclei, the locus coeruleus, and the dorsal raphe nuclei, respectively, which are parts of a key system for arousal, known as the reticular activating system (RAS) (McCormick and Bal, 1997). In addition, several distinct types of neuropeptides (e.g., cholecystokinin and vasoactive intestinal polypeptide) and their receptors are highly expressed in the thalamus (Cox et al., 1997; Lee and Cox, 2003, 2006, 2008; Sun et al., 2003). These peptidergic systems regulate thalamocortical oscillations (Lee and Cox, 2003, 2006, 2008), which may be associated with the level of arousal. Thus, these results suggest that brain state-dependent thalamocortical network oscillations are controlled by inputs from neuromodulatory systems in the brain stem and intrinsic peptidergic systems.

Thalamocortical functions in consciousness: Content and context

What is the role of the thalamocortical circuit in consciousness during wakefulness? An answer to this question may arise from a proposal by Rodolfo Llinas and his colleagues indicating that the thalamocortical circuit subserves temporal binding of contents and context, which flow to the thalamus, during wakefulness (Llinás et al., 1998; see Fig. 12.2, revised by Edgar Garcia-Rill, illustrating Llinas' model in conjunction with the role of the RAS in consciousness, Garcia-Rill, 2015). According to the model, the "specific" (also called "first-order" by some authors) thalamic nuclei (e.g., lateral geniculate nucleus) receives sensory content (e.g., visual information), which is sent to cortical layer IV, whereas

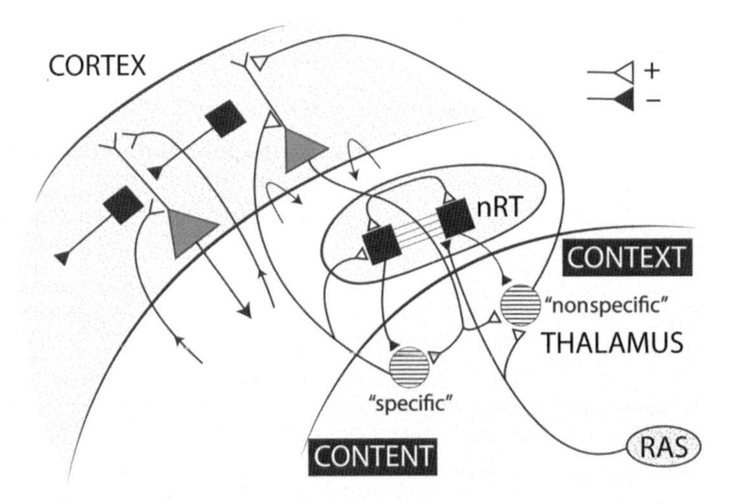

FIG. 12.2 Content and context in the thalamocortical circuits and the RAS. The specific thalamus and nonspecific thalamus relay content of sensory experience and context of sensory experience to layer IV and layer I in the cortex, respectively. These coherent excitatory inputs coincide to produce action potentials in pyramidal cells at the cortical level, sending the outputs back to the thalamus. Thalamocortical reverberation at gamma frequencies is maintained by this reciprocal connectivity. The RAS facilitates the thalamocortical reverberation by continuously activating the nonspecific thalamus while inhibiting RT neurons. Note that electrically coupled RT neurons (shown as black squares with horizontal lines) produce coherent inhibition of thalamocortical relay neurons, resulting in synchronous slow thalamocortical oscillations if the thalamus is released from continuous activation from the RAS. The lateral inhibition (shown as two black squares with laterally projecting axons) is necessary for sensory discrimination. Such inhibition is absent when selective activation of cortical columns is lacking. nRT, reticular thalamic nucleus.*(The figure is from Garcia-Rill, E., 2015. Governing principles of brain activity. In: Garcia-Rill, E. (Ed.), Waking and the Reticular Activiating System in Health and Disease. Elsevier, London.)*

the "nonspecific" (also called "higher-order") thalamic nuclei (e.g., the intralaminar nuclei) receive the context of sensory experience, which is transferred to layer I of the cortex. Temporal coherence between these two distinct pathways would be achieved by coincident depolarization of thalamocortical relay neurons via both pathways (Garcia-Rill, 2015; Llinás et al., 1998). Thus, the coherent inputs from both pathways will summate along the proximodistal dendritic axis of cortical pyramidal cells, promoting cortical gamma oscillations. The cortical outputs from pyramidal cells in layer V and layer VI flow back to specific and nonspecific thalamic nuclei, respectively, establishing resonant loops. Therefore, the specific system and the nonspecific system would provide content and context of cognitive experience, respectively.

According to Llinas and his colleagues' model of consciousness, the RAS is an indispensable system for consciousness as also previously described by Edgar Garcia-Rill (2015). When the RAS is active, it inhibits RT neurons, thus blocking slow thalamocortical oscillations (e.g., sleep spindles), while it excites thalamocortical relay neurons in the intralaminar nuclei, thus promoting tonic activity and the transfer of the context of conscious experience to the cortex. Some authors pointed out the importance of coherent oscillations (e.g., gamma oscillations) for such a binding during wakefulness (Garcia-Rill, 2015; Llinás et al., 1998; see also Chapter 1 of this book for the critical roles of gamma oscillations in consciousness). Therefore, the RAS and its long-range targets, the nonspecific thalamic nuclei, might be required to integrate contents into context in a normal cognitive experience through directly regulating the thalamocortical circuit.

Hypersynchronous thalamocortical circuit in absence epilepsy

There is evidence indicating that the same thalamocortical circuit generates two distinct types of network oscillations in the same patient with absence epilepsy: pathological 3-Hz SWDs and normal 11- to 14-Hz sleep spindles (Beenhakker and Huguenard, 2009). Distinctive types of GABAergic inhibition of thalamocortical relay neurons can be produced depending on how the RT neurons fire (Kim et al., 1997). Kim et al. showed that brief action potential discharges in the GABAergic cells produced short-lasting (~100 ms) $GABA_A$ receptor-mediated inhibitory postsynaptic potentials (IPSPs) in thalamocortical relay neurons, which are conducive to sleep spindle frequency of ~10 Hz, primarily due to the short duration of the IPSPs. Similarly, strong sustained action potential discharges in the RT neurons produces long-lasting (~ 300 ms) $GABA_B$ receptor-mediated IPSPs in thalamocortical relay neurons, which are conducive to 3- to 4-Hz thalamocortical oscillations. These findings suggest that enhanced burst firing of the GABAergic cells might play a key role in transitions from sleep spindles to SWDs (Blumenfeld, 2003). In agreement with these results from human and animal studies, extracellular multiple-unit recordings from the ventrobasal nucleus in isolated mouse thalamic slices revealed that electric stimuli in corticothalamic axons produced spindle-like rhythms (5–8 Hz). Also, bath application of the $GABA_A$ receptor antagonist, bicuculline, caused a shift from the spindle-like rhythms to absence-like 3- to 4-Hz rhythms (Lee and Cox, 2006).

Which condition(s) can trigger the transformation of normal thalamocortical oscillations to pathological oscillations? Blumenfeld and McCormick (2000) found that single stimuli in corticothalamic axons generated 6- to 10-Hz thalamocortical oscillations, which are similar to sleep spindles. In contrast, a high frequency (200 Hz) of multiple stimuli in corticothalamic axons in response to each thalamic burst shifted the spindle-like rhythms to absence-like 3- to 4-Hz rhythms. They also found that a large increase in the firing of the GABAergic thalamic cells were associated with a switch in the duration of IPSPs of thalamocortical relay neurons from brief (~100–150 ms) to slow (~300 ms). Given that sleep spindles and SWDs are commonly observed during non-REM sleep (Halász, 2015), their findings raise the possibility that increased cortical firing might transform sleep spindles to SWDs.

In agreement with the critical role of the thalamocortical circuit in absence epilepsy, mutations in neurotransmitter receptors and ion channels in thalamic neurons have all been related to childhood absence epilepsy. Published studies suggest that the T-type Ca^{2+} channels expressed in thalamic neurons may be the primary cause (Cain and Snutch, 2012; Kim et al., 2001; Zamponi et al., 2010). Accordingly, a blocker of T-type Ca^{2+} channels, ethosuximide, has been commonly used to selectively prevent absence seizures (Glauser et al., 2010). The importance of altered $GABA_A$ and $GABA_B$ receptor function in the thalamocortical circuit for absence seizures has been also reported (Caddick and Hosford, 1996; Wallace et al., 2001). Thus, published results support the notion that dysregulation of the thalamocortical circuit can lead to abnormal 3- to 4-Hz SWDs observed during absence seizures (Beenhakker and Huguenard, 2009).

Impaired awareness in absence epilepsy

Consciousness during absence seizures is not completely lost unlike in coma, a medical condition defined by the complete absence of arousal. It is important to note that the degree of impairment differs depending on tasks in given patients. For example, complex tasks (e.g., tasks that require complex brain processes including a verbal response) are more affected during absence seizures compared with simple repetitive tasks (Bai et al., 2010). The impairments start and stop with SWD paroxysms, and the degree of altered consciousness depends on both amplitude and spread of the SWD as described earlier in this chapter. Thus, it supports the notion that SWDs directly contribute to the impairments, which will be discussed later in detail.

Simultaneous functional magnetic resonance imaging (fMRI) and EEG recordings from children with absence epilepsy were used to determine anatomical and physiological bases of the impairments. Those studies revealed that absence seizures selectively involve specific brain regions while sparing other brain regions. Recent studies showed distinct changes in cortical and subcortical activity on fMRI during absence seizures. In general, fMRI studies revealed increased activity in the thalamus along with reduced activity in the association cortices including the medial frontal, medial parietal, anterior and posterior cingulate, and lateral parietal cortices, while some association cortices showed mixed activity (Blumenfeld, 2012; Cavanna and Monaco, 2009; Unterberger et al., 2018). In agreement with human studies, results from animal studies also suggest that absence seizures selectively involve specific brain regions (e.g., the whisker barrel cortex) (Meeren et al., 2002). Based on human and animal studies, some authors advanced the hypothesis that impaired consciousness in absence seizures is due to compromised normal information processing in specific brain areas including association cortices and association thalamic nuclei (e.g., intralaminar nuclei) (Cavanna and Monaco, 2009).

SWD paroxysms in the thalamocortical circuit contribute to compromised thalamocortical functions in absence epilepsy. A prominent underlying mechanism is simply absence of gamma oscillations during SWD paroxysms, which can lead to altered consciousness, in part, via the two following mechanisms. First, SWDs are expected to disrupt temporal summation of specific and nonspecific afferents from the thalamus, as previously predicted (Kostopoulos, 2001). Increased cortical excitability during SWD paroxysms may make proper summation of specific and nonspecific afferent inputs impossible since both specific and nonspecific inputs to cortical pyramidal cells are expected to be suprathreshold, largely due to burst firing of thalamocortical relay neurons. Such an inappropriate summation may disrupt proper binding of the content of conscious experience to its context, contributing to compromised consciousness in absence epilepsy. Second, it is thought that low-frequency (e.g., theta frequency) thalamocortical stimulations produce neuronal synchronization in larger cortical areas compared with those evoked by gamma-frequency thalamocortical stimulations, which produce columnar activation in part by lateral inhibition (Llinás et al., 2005). These results are in agreement with the notion that gamma oscillations give rise to discrete clusters of cortical activity being indispensable to the complex brain activity during wakefulness. Similar to when one is asleep, the lack of selective activation of cortical columns, caused by SWDs, may be a key mechanism underlying impaired consciousness. Together, these results suggest that thalamocortical hypersynchronous neuronal activity at 3- to 4-Hz SWDs, caused by a switch of firing mode of thalamocortical relay neurons from tonic to burst firing, leads to impaired consciousness in absence epilepsy.

Mesial temporal lobe epilepsy

TLE is the most prevalent form of epilepsy in adults (Chang and Lowenstein, 2003). Mesial TLE represents approximately 80% of TLE (Blair, 2012). Mesial TLE is primarily caused by head trauma, stroke, brain tumor, or central nervous system infection (Blair, 2012). Febrile seizures are also common in most patients with mesial TLE (French et al., 1993). Among anatomical changes in the mesial temporal lobe, hippocampal sclerosis is the most common lesion detected by MRI (Williamson et al., 1993). Selective neuronal loss in the dentate hilus and pyramidal cell layers in the hippocampus are typically found in hippocampal sclerosis along with gliosis (Alexander et al., 2016). Moreover, direct recordings with depth electrodes from the mesial temporal lobe in patients revealed that seizures started from mesial temporal lobe including the hippocampus, the amygdala, and parahippocampal gyrus (Chang and Lowenstein, 2003). Furthermore, patients with a seizure focus or focal lesions within the mesial temporal lobe, surgical resection of those affected brain regions generally eliminated the seizures that were refractory to antiepileptic drugs (Wiebe et al., 2001).

At the onset of focal seizures in mesial TLE, an aura with visual, auditory, gustatory, or déjà vu sensations commonly occurs (Dugan et al., 2014; Tatum, 2012). As the seizures evolve into an altered level of consciousness, patients may manifest a blank stare, oral and manual automatisms (e.g., lip smacking and repetitive limb movements), and/or speech arrest (Blair, 2012). Unilateral anterior temporal theta activity (5–9 Hz) is the cardinal ictal semiology in medial TLE, which generally starts with a blank stare (Selwa, 2010). Secondary generalization of seizures can occur in mesial TLE. Such a seizure type is referred as to focal to bilateral tonic-clonic seizures (also called secondarily generalized tonic-clonic seizures) (Fisher et al., 2017). Some patients with mesial TLE show secondarily generalized tonic-clonic seizures (GTCS) (Blumenfeld, 2012). Behaviorally, these seizures have a tonic phase characterized by brief muscle contraction where the patient will generally lose consciousness. This is followed by a clonic phase where the patient will rhythmically jerk. Along with the ictal phase, patients will have profound conscious impairment during the postictal phase. The short loss of consciousness experienced by these patients has been likened to a coma-like state (Blumenfeld, 2012). The observation of GTCS inducing the loss of consciousness has drawn the attention of researchers who wish to delineate the reason that an epileptic event causes a loss of consciousness.

In the next section, we will first discuss structural and physiological properties of the hippocampus, along with their changes in mesial TLE, leading to hyperexcitability and spontaneous seizures. How seizures originating from the temporal lobe can lead to dysregulation of consciousness will then be discussed especially in relation to long-range connectivity of the hippocampus with the thalamus, the brain stem, and other regions.

Hippocampal circuits, neuronal excitability, and network oscillations

The hippocampus plays key roles in learning, memory, and spatial processing that are supported by coordinated neuronal network activity, including theta (4–12 Hz) and gamma (30–100 Hz) (Colgin, 2016). The hippocampus is also critically involved in hyperexcitability and spontaneous seizures in medial TLE (Chang and Lowenstein, 2003). The general structural properties of the hippocampus become evident in transverse hippocampal slices. As shown in Fig. 12.3A, the dentate gyrus layers and the cornu ammonis layers resemble two interlocking horseshoes. The cornu ammonis (CA) is divided into subregions (e.g., CA3 and CA1). The principal cells of the dentate gyrus and CA are granule cells and pyramidal cells, respectively. These excitatory cells are key types of neurons that form trisynaptic circuits. Excitatory inputs from the entorhinal cortex come to the hippocampus through the perforant pathway. The granule cells in the dentate gyrus receive the cortical inputs and project their axons (i.e., mossy fibers) to the CA3. From there, axons of pyramidal cells in CA3, called Schaffer collaterals, project to the CA1. Then, axons of pyramidal cells in the CA1 project back to the long-range targets of the CA1, including the entorhinal cortex. Since these circuits are all excitatory, runaway excitation may readily arise within the temporal lobe if local GABAergic interneurons in the hippocampus are absent or inactive.

There is significant evidence indicating that hippocampal interneurons regulate neuronal excitability and subserve key circuit operations including generation of hippocampal network oscillations (e.g., gamma oscillations), feedback inhibition, and feedforward inhibition (Pelkey et al., 2017). Hippocampal interneurons are diverse as indicated by the fact that there are at least 21 distinct subtypes of GABAergic interneurons in the CA1 alone based on their neurochemical, morphological, and electrophysiological properties, along with other properties (Klausberger and Somogyi, 2008). Parvalbumin-expressing basket cells (PVBCs) are a major type of GABAergic interneurons in the hippocampus (Bezaire and Soltesz, 2013). PVBCs are ubiquitously found in the hippocampus (Fig. 12.3) and other brain regions including the neocortex and the amygdala (Lewis et al., 2012; Veres et al., 2017). Since PVBCs form perisomatic synapses onto the somata and proximal dendrites of principal cells, they are able to powerfully control the outputs of principal cells via two key forms of inhibition: feedforward and feedback inhibition (Pelkey et al., 2017). For example, CA1 PVBCs are excited by CA3 pyramidal cells, thus providing feedforward inhibition to CA1 pyramidal cells. The same PVBCs are also excited by local CA1 pyramidal cells, thus providing feedback inhibition to the pyramidal cells. In this way, PVBCs contribute to the balance of excitation and inhibition, thus preventing uncontrollable excitation in all critical nodes of trisynaptic circuits: the dentate gyrus, CA3, and CA1. Accordingly, animal studies showed that optogenetic activation of PV+ interneurons in CA1 reduced seizures in a well-established animal model of TLE, the intrahippocampal kainate mouse model (Krook-Magnuson et al., 2013).

Gamma oscillations are 30- to 100-Hz network oscillations that commonly occur in numerous brain regions including the hippocampus (Colgin, 2016). They support memory, attention, sensory perception, and arousal (Buzsáki and Wang, 2012; Colgin, 2016; Lee et al., 2018). How such network oscillations

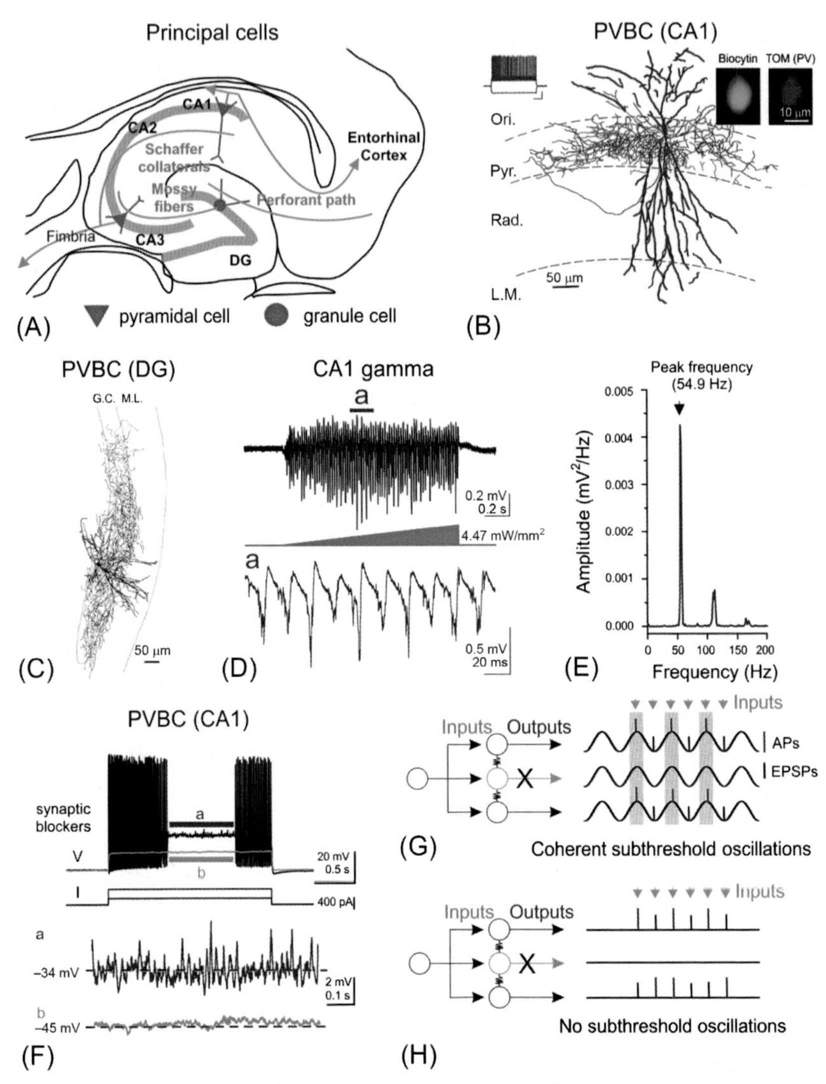

FIG. 12.3 Hippocampal circuits and gamma oscillations. (A) Trisynaptic circuits. The somata of a dentate gyrus granule cell and of pyramidal cells in CA3 and CA1 are shown as an orange circle and triangles, respectively. (B and C) Parvalbumin-expressing basket cells (PVBCs) in the CA1 and the dentate gyrus. Left inset shows the fast-spiking pattern and the lack of a depolarizing sag in response to current steps of +300 pA and −300 pA from −60 mV. Right inset shows expression of a red fluorescent protein tdTomato (TOM) in the PVBC as TOM is expressed under parvalbumin promoter control. Axonal arbors of PVBCs in the CA1 and the dentate gyrus are shown in blue and red, respectively. Note that axonal arbors are preferentially located within pyramidal or granule cell layers. Scales in B indicate 20 mV and 0.2 s. *Ori.*, stratum oriens; Pyr., stratum pyramidale; Rad., stratum radiatum; L. M., stratum lacunosum moleculare; G. C., granule cell layer; and M. L., molecular layer. (D and E) A representative CA1 gamma burst recorded with a glass pipette filled with

See the legend on page 207.

arise from individual neurons and/or through their interactions is incompletely understood. Recent evidence suggests that hippocampal CA1 and CA3 are able to generate these oscillations, at least in part through GABAergic interneurons (Butler et al., 2016; Buzsáki and Wang, 2012; Colgin, 2016). In agreement with those published studies, our optogenetic studies revealed the presence of CA1 gamma oscillations (Fig. 12.3D, E). We used a mouse line in which channelrhodopsin2 (ChR2) is selectively expressed in CaMKIIalpha-expressing cells (e.g., CA1 pyramidal cells) as previously described (Butler et al., 2016). Horizontal hippocampal slices from 4- to 8-week-old mice were used. 1.4-s-long ramps of blue light were delivered to the CA1 to produce CA1 gamma. Our local field potential recordings revealed that such light ramps evoked stable CA1 gamma for the duration of ramp stimulation in all tested hippocampal slices, which is in agreement with prior studies (Butler et al., 2016). In addition, our recordings revealed that bath application of excitatory synaptic blockers (APV and NBQX) or inhibitory synaptic blockers (SR95531 and CGP55845) strongly inhibited the CA1 gamma, supporting the idea that our CA1 gamma was generated via synaptic interactions between GABAergic interneurons and pyramidal cells (i.e., pyramidal-interneuron network gamma, also known as PING model). Thus, our results support the critical roles of GABAergic interneurons in generation of CA1 gamma.

Based on our and others' published studies (Kang et al., 2018; Kezunovic et al., 2011; Pike et al., 2000), we proposed an intrinsic subthreshold oscillation-based gamma model (Lee et al., 2018). Intrinsic subthreshold

---

FIG. 12.3, CONT'D ACSF from CA1 pyramidal cell layer in response to an 1.4-s-long 470 nm blue light ramp (from near zero to 4.47 mW/mm^2). A representative section of the gamma recording, indicated by a red line above the recording, is shown on a faster time base (bottom). A prominent gamma band peak (54.9 Hz) is shown in the power spectrum. (F) Intrinsic gamma subthreshold oscillations following typical bursts of fast action potential discharges in a representative PVBC, in response to a current step (+400 pA, 2 s) from −60 mV. When the membrane potential remained above −40 mV, the PVBC showed intrinsic gamma subthreshold oscillations (a), but not when the membrane potential was held below −40 mV (b). (G and H) Two distinct types of cell groups with (G) or without (H) intrinsic coherent subthreshold oscillations. Three output cells in both groups commonly receive excitatory inputs. They are electrically coupled to other cells by gap junctions, raising the possibility that coherent subthreshold oscillations can arise in all output cells shown in the G cell group. Most cells (top and bottom cells) can produce coherent action potential discharges dictated by the preferred frequency of intrinsic oscillations if excitatory inputs arrive around the peaks of intrinsic oscillations. However, in the H cell group, it is unlikely that the output cells will produce coherent action potentials due to the lack of intrinsic oscillations even when they show similar firing frequency to that in the G cell group. If excitatory transmission fails in a specific circuit (marked by X), that circuit will still keep coherent intrinsic oscillations at preferred frequencies due to electric coupling. When the synaptic failure recovers, the circuit reassumes coherent firing at those frequencies.*(Parts (B) and (F) are from Kang, Y.-J., Lewis, H.E.S., Young, M.W., Govindaiah, G., Greenfield, L.J., Garcia-Rill, E., Lee, S.-H., 2018. Cell type-specific intrinsic perithreshold oscillations in hippocampal GABAergic interneurons. Neuroscience 376, 80–93. Part (C) is from Szabo, G.G., Du, X., Oijala, M., Varga, C., Parent, J.M., Soltesz, I., 2017. Extended interneuronal network of the dentate gyrus. Cell Rep. 20, 1262–1268. Parts (G) and (H) are from Lee, S.-H., Urbano, F.J., Garcia-Rill, E., 2018. The critical role of intrinsic membrane oscillations. Neurosignals 26, 66–76.)*

membrane oscillations and intrinsic subthreshold resonance of neurons in other regions throughout the CNS have been revealed, since first discovered, in the inferior olive by Llinás and Yarom (1986). The resonance properties of individual cells cause them not only to preferentially respond to external stimuli at specific frequencies, thus resulting in spiking resonance (Hutcheon and Yarom, 2000; Stark et al., 2013), but also to produce self-sustaining intrinsic subthreshold oscillations at the single-cell level without synaptic interactions (Hutcheon and Yarom, 2000). Our recent studies showed that specific subtypes of hippocampal interneurons (e.g., PVBCs) produced distinct frequency bands of intrinsic subthreshold oscillations at theta, beta, or gamma frequencies (Govindaiah et al., 2018; Kang et al., 2018). We also showed that within-subtype differences in intrinsic perithreshold oscillations arose in specific interneuron subtypes including PVBCs, which showed intrinsic theta or gamma oscillations (Fig. 12.3F). These findings are in general agreement with earlier studies (Pike et al., 2000). Given that intrinsic subthreshold oscillations are thought to influence the timing of action potential discharges (Llinás, 1988), it is expected that those intrinsic oscillations facilitate coherent action potential discharges as shown in Fig. 12.3G. Full development of such gamma models will have a broad impact beyond hippocampal physiology and require further experimental and computational studies. Such efforts will provide novel insight into the mechanisms underlying memory, attention, sensory perception, arousal, and brain disorders, which are associated with compromised gamma oscillations.

Reorganization of neuronal circuits in mesial temporal lobe epilepsy

Hyperexcitability and spontaneous seizures in mesial TLE stem from structural and functional changes in the temporal lobe (Alexander et al., 2016; Dengler and Coulter, 2016). Hippocampal sclerosis is commonly observed in patients with mesial TLE and animal models of TLE (e.g., pilocarpine or kainate model of TLE). Recent human and animal studies on hippocampal sclerosis observed in epileptic tissues have revealed functionally important changes in molecules (e.g., ion channels, neurotransmitter receptors, and endogenous modulatory molecules), intrinsic and synaptic properties of individual cells, local and long-range connectivity, and circuit operations in TLE. A number of those changes contribute to spontaneous seizures, whereas others prevent hippocampal circuits from seizures.

In human TLE, hippocampal sclerosis is evident in the dentate, CA3, and CA1 with selective cell loss and gliosis as shown in Fig. 12.4B. Pyramidal cells in the hippocampus proper are in large part lost in CA3 and CA1, along with interneuron loss (Alexander et al., 2016; Andrioli et al., 2007). In the dentate hilus, there is also significant loss of excitatory mossy cells and interneurons. Due to the loss of mossy cells, which form excitatory synapses on basket cells in the dentate gyrus (see Fig. 12.3C for an example of PVBC in the dentate gyrus), surviving basket cells in TLE remain hypoactive, resulting in hyperexcitability

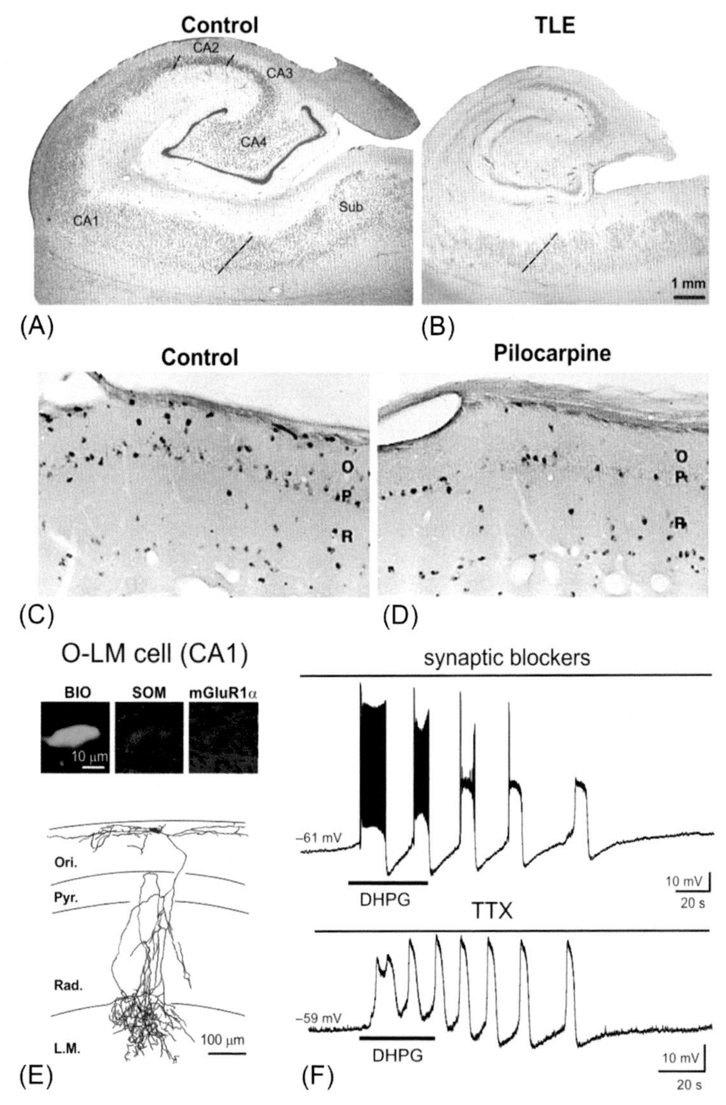

FIG. 12.4 Hippocampal sclerosis and slow oscillations in hippocampal interneurons mediated by group I metabotropic glutamate receptors. (A and B) Nissl stain of normal human hippocampus (A) and damaged hippocampus due to hippocampal sclerosis (B). Cell loss is evident in the hilus and pyramidal cell layers in CA3 and CA1. (C and D) GAD67 mRNA stain of the CA1 in control (control) and pilocarpine-treated (pilocarpine) rats (D). Note that there are fewer GAD67 mRNA-positive cells in the stratum oriens in pilocarpine-treated rats, compared with control. O, stratum oriens; P, stratum pyramidale; and R, stratum radiatum. (E and F) Reconstruction of a biocytin-filled CA1 O-LM interneuron. Axons and somatodendritic compartment are shown in black and red, respectively. Note extensive axonal branching in the stratum lacunosum moleculare and soma with dendrites projecting horizontally in the stratum oriens. The top panel shows

See the legend on page 210.

in the dentate gyrus (also known as "the dormant basket cell hypothesis"), along with interneuron loss. On the other hand, surviving mossy cells in TLE are thought to manifest increased excitability (also known as "irritable mossy cell hypothesis; Howard et al., 2007; Ratzliff et al., 2002).

An additional major change in the dentate gyrus is the recurrent sprouting of granule cells' axon fibers, called mossy fibers (Buckmaster, 2012). The sprouted fibers form excitatory synapses primarily onto other granule cells, likely contributing to hyperexcitability and spontaneous seizures (Scharfman et al., 2003). As in the DG, axon fibers of pyramidal cells in both CA3 and CA1 are known to sprout in TLE and contribute to increased connectivity between local pyramidal cells (Long et al., 2014; Siddiqui and Joseph, 2005).

Hippocampal interneurons constitute approximately 11% of neurons in the hippocampus (Bezaire and Soltesz, 2013) and critically regulate excitability of hippocampal principal neurons. These interneurons are also known to be vulnerable to excitotoxicity in TLE (Fig. 12.4D; see the Table 1 of a review by Alexander et al., 2016, presenting interneuron loss in detail in TLE). For example, major types of hippocampal interneurons (e.g., somatostatin-expressing interneurons and cholecystokinin-expressing interneurons) are lost in TLE (Buckmaster and Jongen-Rêlo, 1999; Cossart et al., 2001; Sun et al., 2014; Wyeth et al., 2010). As some authors pointed out, "cells are often identified by staining for cellular markers, and the loss of marker-stained cells could be subsequent to changed expression of that marker rather than cell death" (Alexander et al., 2016); published studies have limitations regarding the identification of interneuron subtypes in TLE. Thus, further studies are required to fully address whether cells expressing specific cellular markers are lost, whether expression of specific cellular markers is simply changed, and what are the functional consequences of interneuron loss in TLE. Nevertheless, selective activation of hippocampal interneurons, or transplantation of GABAergic interneuron precursors into the hippocampus, has been shown to reduce spontaneous seizures and/or comorbidities in animal models

FIG. 12.4, CONT'D expression of somatostatin (SOM) and metabotropic glutamate receptor type 1α (mGluR1α) in a biocytin-filled neuron. Voltage trace from an O-LM interneuron shows that selective group I mGluR agonist, (*S*)-3,5-dihydroxyphenylglycine (DHPG, 10 μM) produced large-amplitude slow oscillatory responses (<0.1 Hz), along with significant firing during the initial depolarization in the presence of synaptic blockers (F). The DHPG-induced slow oscillations are insensitive to the sodium channel blocker tetrodotoxin (TTX, 1 μM). These results suggest that activation of group I mGluRs in O-LM interneurons generates intrinsic, sodium channel-independent oscillations.*(Parts (A) and (B) are from Andrioli, A., Alonso-Nanclares, L., Arellano, J.I., DeFelipe, J., 2007. Quantitative analysis of parvalbumin-immunoreactive cells in the human epileptic hippocampus. Neuroscience 149, 131–143. Parts (C) and (D) are from Cossart, R., Dinocourt, C., Hirsch, J.C., Merchan-Perez, A., De Felipe, J., Ben-Ari, Y., Esclapez, M., Bernard, C., 2001. Dendritic but not somatic GABAergic inhibition is decreased in experimental epilepsy. Nat. Neurosci. 4, 52–62. Parts (E) and (F) are from Govindaiah, G., Kang, Y.-J., Lewis, H.E.S., Chung, L., Clement, E.M., Greenfield, L.J., Garcia-Rill, E., Lee, S.-H., 2018. Group I metabotropic glutamate receptors generate two types of intrinsic membrane oscillations in hippocampal oriens/alveus interneurons. Neuropharmacology 139, 150–162.)*

of TLE (Hunt et al., 2013; Krook-Magnuson et al., 2013). Thus, interneuron-associated disinhibition likely contributes to spontaneous seizures in TLE and is expected to be a viable target of the treatment of TLE.

Our recent studies revealed a potential link from expression of group I metabotropic glutamate receptors (i.e., $mGluR_1$ and $mGluR_5$) in somatostatin-expressing oriens-lacunosum moleculare (O-LM) interneurons to their vulnerability to excitotoxicity. Importantly, O-LM interneurons are critically involved in the generation of ictal discharges (Ziburkus et al., 2006), and group I metabotropic glutamate receptors contribute to vulnerability of hippocampal interneurons to excitotoxicity in seizures (Sanon et al., 2010). We showed not only that selective activation of group 1 metabotropic glutamate receptors induced intrinsic slow (< 0.1 Hz) oscillations in O-LM interneurons (Fig. 12.4E and F) but also that nonselective cation-conducting transient receptor potential channels, L-type Ca^{2+} channels, and ryanodine receptors contribute to those slow oscillations (Govindaiah et al., 2018). Given that somatostatin-expressing interneurons are particularly vulnerable in epilepsy (Best et al., 1993; Buckmaster and Jongen-Rêlo, 1999; Hofmann et al., 2016; Houser and Esclapez, 1996; Morin et al., 1998), our results suggest that intracellular calcium overload in O-LM interneurons, in part due to the involvement of nonselective cation-conducting transient receptor potential channels, L-type Ca^{2+} channels, and ryanodine receptors, triggers excitotoxicity in O-LM cells, as previously suggested by some authors (Choi, 1994). Increased Ca^{2+} levels in O-LM cells contribute to interneuron loss in TLE likely by necrotic cell death, apoptotic process, and other mechanisms that will not be discussed further since those topics are beyond the current scope (but reviewed in detail in Zhivotovsky and Orrenius, 2011).

Mechanisms underlying impaired consciousness

Dysregulation of consciousness in TLE has been extensively examined particularly by Blumenfeld's group. The results from their human and animal research, which used an interdisciplinary approach, combining brain imaging, EEG, behavioral tests, juxtacellular recordings from neurochemically identified cells, enzyme-based amperometry, and other methods, have revealed some of the mechanisms behind TLE and other types of epilepsy. Especially, an evidence-based "network inhibition hypothesis" was proposed by his group as potential explanation for impaired consciousness in TLE (Englot et al., 2010; see Fig. 12.5). According to their hypothesis, consciousness is constantly maintained through functional reciprocal interactions between the cortical and subcortical systems including the upper brain stem and diencephalic activating systems (Fig. 12.5A). Fast polyspike discharges arise from the affected temporal lobe with increased neuronal excitability (Fig. 12.5B). Abnormal seizure activity can spread from the temporal lobe to its long-range midline subcortical targets (Fig. 12.5C). The spread of epileptiform discharges causes activation of GABAergic neurons in subcortical targets, which leads to the inhibition

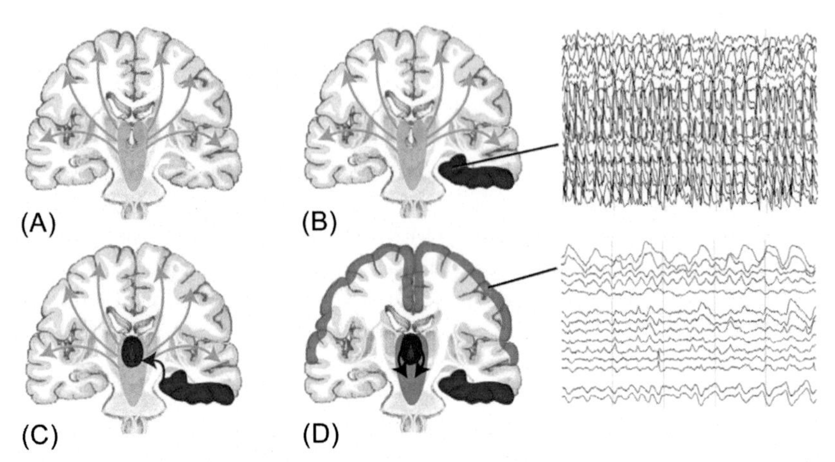

FIG. 12.5 Network inhibition hypothesis. (A) Normally, the level and content of consciousness in a patient are maintained by a large network stemming from the brain stem such as the RAS to the cortex through the thalamus. (B) If seizures are limited to the temporal lobe without secondary generalization, consciousness remains relatively intact. Depth electrode recordings show fast polyspike waves in the temporal lobe. (C) Evidence supports the hypothesis that seizures can be propagated to midline subcortical regions. (D) Activation of midline subcortical regions (red) leads to the inhibition of subcortical activating systems (blue), resulting in a decrease in activity in bilateral frontoparietal association cortices (blue). Loss of consciousness in TLE is mechanistically associated with the presence of slow-wave activity in the frontoparietal association cortex (right), which is similar to that observed during deep sleep. *(Parts A–D are reprinted from The Lancet Neurology, 11, Hal Blumenfeld, Impaired consciousness in epilepsy, 814–826, Copyright (2012), With permission from Elsevier.)*

of subcortical arousal systems (Fig. 12.5D). To this end, depressed subcortical arousal systems critically contribute to suppressed neuronal activity in bilateral frontoparietal association cortex, resulting in the loss of consciousness.

Over the last decade, key evidence supporting the network inhibition hypothesis has emerged. First, high-field blood-oxygen-level-dependent fMRI studies showed increased cerebral blood flow, during limbic seizures, to the anterior hypothalamus and lateral septum (Motelow et al., 2015). These brain areas are long-range targets of the temporal lobe and are rich in GABAergic cells that project to subcortical arousal system including intralaminar thalamic nuclei and the brain stem pedunculopontine tegmental nucleus (Motelow et al., 2015). Second, lateral septal stimulation in rodents has been shown to cause a decrease in choline levels (used as an alternative measurement of extracellular acetylcholine levels) and induce slow-wave activity in the frontal cortex (Li et al., 2015). Third, a similar decrease in choline levels and slow-wave activity in the frontal cortex were observed during temporal lobe seizures (Motelow et al., 2015). These changes were accompanied by a decrease in firing of identified cholinergic cells in both the pedunculopontine tegmental nucleus and basal forebrain during temporal lobe seizures (Motelow et al., 2015). That supports the notion that inhibitory inputs to the brain stem and thalamus may originate from the lateral septum and hypothalamus during temporal lobe seizures and reduced activity in the subcortical arousal system may lead to the loss of consciousness.

Given that the loss of consciousness in TLE is accompanied by increased slow-wave activity in the thalamocortical circuit during limbic seizures, which is similar to that observed during deep sleep, slow-wave brain activity likely contributes to a loss of consciousness. Recent studies showed how such abnormal thalamocortical network activity might arise in TLE. Specifically, animal studies showed that different thalamic nuclei manifested distinct changes in neuronal activity during focal limbic seizures (Feng et al., 2017). In agreement with known hippocampal-thalamic connectivity, the anterior nuclei of the thalamus showed increased firing during limbic seizures. In contrast, the central lateral thalamic nuclei, a part of the intralaminar thalamic nuclei (nonspecific thalamic nuclei), showed reduced overall firing rate. A relay nucleus of the thalamus, the ventral posteromedial nuclei (specific thalamic nuclei), showed increased spindle waves during limbic seizures. These distinct changes in neuronal activity in different thalamic nuclei may suggest their specific roles in limbic seizures. Since the central lateral thalamic nuclei are a part of the subcortical arousal system, its reduced neuronal activity may contribute to a decrease in cortical function. On the other hand, the anterior nuclei may contribute to seizure propagation and sleep spindles observed in relay nuclei during a limbic seizure, which may disrupt normal thalamocortical information flow. These complex changes in neuronal activity in the thalamus, along with changes in the pedunculopontine tegmental nucleus and basal forebrain, may combine to contribute to seizure severity and the loss of consciousness.

Some temporal lobe seizures are not accompanied by reduced overall level of consciousness. According to the 2017 ILAE seizure classification, those seizures are classified as focal aware seizures (previously classified as simple partial seizures). They are thought to be mainly limited to the temporal lobe and not associated with widespread dysregulation of the frontoparietal association cortices (Cavanna and Monaco, 2009) that is commonly evident in focal impaired awareness seizures, as shown in Fig. 12.5. Focal aware seizures are rather accompanied by more limited changes in the content of consciousness.

Clinical consequences

Impaired consciousness in epilepsy can cause serious clinical consequences from social stigma to serious injury, thus preventing patients from maintaining a normal quality of life. Patients may lose the ability to drive as some states have laws prohibiting driving for an extended time after a seizure (Krumholz, 2009). This, along with vocational exclusion, limits a patient's ability to function in society and maintain gainful employment, thereby affecting their socioeconomic status (Jennum et al., 2011). Furthermore, epilepsy patients may attempt to conceal their diagnoses from employers or friends for fear of social exclusion (Baker, 2002; Jacoby et al., 2005). Outside of the socioeconomic implications, physical injury due to the loss of consciousness during a seizure is common. Falls associated with the loss of consciousness may cause lacerations, concussions, or even fractures (Wirrell, 2006). A study recently released found that 14% of followed patients had experienced an injury related to their epilepsy within a 3-month period where 5.8% of injured patients had mild injuries and 6.8% required immediate treatment or hospitalization (Willems et al., 2018). However, the most freighting complication from seizures with the loss of consciousness is sudden unexpected death in epilepsy (SUDEP), which can occur in patients who lose consciousness during a generalized or secondary generalized seizure (Tomson et al., 2005). With the understanding that epilepsy greatly reduces the quality of life for patients and their families, it is critically important to not only manage epilepsy frequency but also implement treatments to address the loss of consciousness in patients who cannot achieve total seizure control, as it is the root of most social and physical disturbances for these patients.

Future directions

Currently, the first line of epileptic management is antiepileptic drugs, which properly manages two-thirds of patients (Kwan and Sander, 2004). However, one-third of patients do not respond to classical drug therapy (Kwan and Sander, 2004). For refractory epilepsy patients, surgical intervention may be indicated, and of those patients about half achieve complete seizure control (West et al., 2015). For refractory epilepsy patients who manifest impaired consciousness due to seizures, conscious control during seizures may be a treatment alternative. Recent animal and human studies demonstrated the feasibility and some benefits of deep brain stimulation (reviewed in detail in Gummadavelli et al., 2015). Although neurostimulation-based treatments for impaired consciousness in epilepsy have not been developed, the application of deep brain stimulation in other disorders of consciousness (e.g., the Parkinson's disease and minimally conscious state) could highlight its usefulness in regard to the maintenance of consciousness (Benabid et al., 1994; Chudy et al., 2018; Kleiner-Fisman et al., 2003; Lemaire et al., 2018). Furthermore, in rodent models of epilepsy, intralaminar thalamic or rostral pons deep brain stimulation was shown to increase behavioral arousal during the ictal or postictal period (Gummadavelli et al., 2015; Kundishora et al., 2017). Together, those results raise the possibility that neurostimulation is a treatment option that may address the loss of consciousness in epilepsy. Successful development of such treatments will require complete identification of, but not limited to, proper stimulation location, simulation pattern, stimulation duration, and stimulation timing for neurostimulation treatments based on evidence from animal and human research (for example, see Table 1 in Gummadavelli et al., 2015). If proper target location and/or cell types are identified, other therapeutic approaches could be also developed to selectively activate or inhibit those cell types for treatments for impaired consciousness in epilepsy.

In this chapter, we have stressed the cellular and circuit mechanisms underlying normal and pathological network oscillations, which support normal consciousness and impaired consciousness in epilepsy, respectively. It would be almost impossible to design proper treatments for impaired consciousness in epilepsy without mechanistic knowledge about the mechanisms underlying normal and pathological network activity. Thus, future studies should focus on (1) initiation and termination mechanisms of SWDs in absence epilepsy, (2) cell type-specific modulation of intrinsic gamma oscillations (e.g., a selective increase in intrinsic gamma oscillations in P/Q-only, "wake on" cells in the pedunculopontine nucleus, which are described in Chapter 2 of this book), and (3) cellular and circuit mechanisms underlying hyperexcitability and spontaneous seizures. With better mechanistic knowledge of cellular, circuit, and modulatory mechanisms underlying network activity, we will come closer to proper treatments for impaired consciousness in epilepsy.

Acknowledgments

Supported by the College of Medicine, University of Arkansas for Medical Sciences to S-HL, Bogard Neurology Research and Stroke Prevention Fund to S-HL, and Core Facilities of the Center for Translational Neuroscience, Award P30 GM110702 from the IDeA program at NIGMS to Dr. Edgar Garcia-Rill.

References

Alexander, A., Maroso, M., Soltesz, I., 2016. Organization and control of epileptic circuits in temporal lobe epilepsy. Prog. Brain Res. 226, 127–154.

Andrioli, A., Alonso-Nanclares, L., Arellano, J.I., DeFelipe, J., 2007. Quantitative analysis of parvalbumin-immunoreactive cells in the human epileptic hippocampus. Neuroscience 149, 131–143.

Bai, X., Vestal, M., Berman, R., Negishi, M., Spann, M., Vega, C., Desalvo, M., Novotny, E.J., Constable, R.T., Blumenfeld, H., 2010. Dynamic time course of typical childhood absence seizures: EEG, behavior, and functional magnetic resonance imaging. J. Neurosci. 30, 5884–5893.

Baker, G.A., 2002. The psychosocial burden of epilepsy. Epilepsia 43, 26–30.

Beenhakker, M.P., Huguenard, J.R., 2009. Neurons that fire together also conspire together: Is normal sleep circuitry hijacked to generate epilepsy? Neuron 62, 612–632.

Benabid, A.L., Pollak, P., Gross, C., Hoffmann, D., Benazzouz, A., Gao, D.M., Laurent, A., Gentil, M., Perret, J., 1994. Acute and long-term effects of subthalamic nucleus stimulation of Parkinson's disease. Stereotact. Funct. Neurosurg. 62, 76–84.

Best, N., Mitchell, J., Baimbridge, K.G., Wheal, H.V., 1993. Changes in parvalbumin-immunoreactive neurons in the rat hippocampus following a kainic acid lesion. Neurosci. Lett. 155, 1–6.

Bezaire, M.J., Soltesz, I., 2013. Quantitative assessment of CA1 local circuits: knowledge base for interneuron-pyramidal cell connectivity. Hippocampus 23, 751–785.

Blair, R., 2012. Temporal lobe epilepsy semiology. Epilepsy Res. Treat. 75, 1510.

Blumenfeld, H., 2003. From molecules to networks: cortical/subcortical interactions in the pathophysiology of idiopathic generalized epilepsy. Epilepsia 44, 7–15.

Blumenfeld, H., 2012. Impaired consciousness in epilepsy. Lancet Neurol. 11, 814–826.

Blumenfeld, H., McCormick, D.A., 2000. Corticothalamic inputs control the pattern of activity generated in thalamocortical networks. J. Neurosci. 20, 5153–5162.

Browne, T., Penry, J., Proter, R., Dreifuss, F., 1974. Responsiveness before, during, and after spike-wave paroxysms. Neurology 24, 659–665.

Buckmaster, P., 2012. Mossy fiber sprouting in the dentate gyrus. In: Noebels, J., Avoli, M., Rogawski, M. (Eds.), Jasper's Basic Mechanisms of the Epilepsies [Internet]. National Center for Biotechnology Information (US), Bethesda, MD, USA.

Buckmaster, P.S., Jongen-Rêlo, A.L., 1999. Highly specific neuron loss preserves lateral inhibitory circuits in the dentate gyrus of kainate-induced epileptic rats. J. Neurosci. 19, 9519–9529.

Butler, J.L., Mendonca, P.R.F., Robinson, H.P.C., Paulsen, O., 2016. Intrinsic cornu ammonis area 1 Theta-nested gamma oscillations induced by Optogenetic Theta frequency stimulation. J. Neurosci. 36, 4155–4169.

Buzsáki, G., Wang, X.-J., 2012. Mechanisms of gamma oscillations. Annu. Rev. Neurosci. 35, 203–225.

Caddick, S.J., Hosford, D.A., 1996. The role of GABABmechanisms in animal models of absence seizures. Mol. Neurobiol. 13, 23–32.

Cain, S., Snutch, T., 2012. Voltage-gated calcium channels in epilepsy. In: Noebels, J.L., Avoli, M., Rogawski, M.A., et al. (Eds.), Jasper's Basic Mechanisms of the Epilepsies [Internet]. National Center for Biotechnology Information, Bethesda, MD, USA.

Cavanna, A.E., Monaco, F., 2009. Brain mechanisms of altered conscious states during epileptic seizures. Nat. Rev. Neurol. 5, 267–276.

Chang, B.S., Lowenstein, D.H., 2003. Epilepsy. N. Engl. J. Med. 349, 1257–1266.

Choi, D., 1994. Calcium and excitotoxic neuronal injury. Ann. N. Y. Acad. Sci. 747, 162–171.

Chudy, D., Deletis, V., Almahariq, F., Marčinković, P., Škrlin, J., Paradžik, V., 2018. Deep brain stimulation for the early treatment of the minimally conscious state and vegetative state: experience in 14 patients. J. Neurosurg. 128, 1189–1198.

Colgin, L.L., 2016. Rhythms of the hippocampal network. Nat. Rev. Neurosci. 17, 239–249.

Cossart, R., Dinocourt, C., Hirsch, J.C., Merchan-Perez, A., De Felipe, J., Ben-Ari, Y., Esclapez, M., Bernard, C., 2001. Dendritic but not somatic GABAergic inhibition is decreased in experimental epilepsy. Nat. Neurosci. 4, 52–62.

Cox, C., Huguenard, J., Prince, D., 1997. Peptidergic modulation of intrathalamic circuit activity in vitro: actions of cholecystokinin. J. Neurosci. 17, 70–82.

Crunelli, V., Leresche, N., 2002. Childhood absence epilepsy: genes, channels, neurons and networks. Nat. Rev. Neurosci. 3, 371–382.

Dengler, C., Coulter, D., 2016. Normal and epilepsy-associated pathologic function of the dentate gyrus. Prog. Brain Res. 226, 155–178.

Dugan, P., Carlson, C., Bluvstein, J., Chong, D.J., Friedman, D., Kirsch, H.E., 2014. Auras in generalized epilepsy. Neurology 83, 1444–1449.

Englot, D.J., Yang, L., Hamid, H., Danielson, N., Bai, X., Marfeo, A., Yu, L., Gordon, A., Purcaro, M.J., Motelow, J.E., et al., 2010. Impaired consciousness in temporal lobe seizures: role of cortical slow activity. Brain 133, 3764–3777.

Feng, L., Motelow, J.E., Ma, C., Biche, W., McCafferty, C., Smith, N., Liu, M., Zhan, Q., Jia, R., Xiao, B., et al., 2017. Seizures and sleep in the thalamus: focal limbic seizures show divergent activity patterns in different thalamic nuclei. J. Neurosci. 37, 11441–11454.

Fisher, R.S., Acevedo, C., Arzimanoglou, A., Bogacz, A., Cross, J.H., Elger, C.E., Engel, J., Forsgren, L., French, J.A., Glynn, M., et al., 2014. ILAE official report: a practical clinical definition of epilepsy. Epilepsia 55, 475–482.

Fisher, R.S., Cross, J.H., D'Souza, C., French, J.A., Haut, S.R., Higurashi, N., Hirsch, E., Jansen, F.E., Lagae, L., Moshé, S.L., et al., 2017. Instruction manual for the ILAE 2017 operational classification of seizure types. Epilepsia 58, 531–542.

French, J.A., Williamson, P.D., Thadani, V.M., Darcey, T.M., Mattson, R.H., Spencer, S.S., Spencer, D.D., 1993. Characteristics of medial temporal lobe epilepsy. I. Results of history and physical examination. Ann. Neurol. 34, 774–780.

Garcia-Rill, E., 2015. Governing principles of brain activity. In: Garcia-Rill, E. (Ed.), Waking and the Reticular Activiating System in Health and Disease. Elsevier, London.

Glauser, T., Cnaan, A., Shinnar, S., Hirtz, D., Dlugos, D., Masur, D., Clark, P., Capparelli, E., Adamson, P., 2010. Ethosuximide, valproic acid, and lamotrigine in childhood absence epilepsy. N. Engl. J. Med. 362, 790–799.

Govindaiah, G., Kang, Y.-J., Lewis, H.E.S., Chung, L., Clement, E.M., Greenfield, L.J., Garcia-Rill, E., Lee, S.-H., 2018. Group I metabotropic glutamate receptors generate two types of intrinsic membrane oscillations in hippocampal oriens/alveus interneurons. Neuropharmacology 139, 150–162.

Gummadavelli, A., Kundishora, A.J., Willie, J.T., Andrews, J.P., Gerrard, J.L., Spencer, D.D., Blumenfeld, H., 2015. Neurostimulation to improve level of consciousness in patients with epilepsy. Neurosurg. Focus 38, E10.

Guo, J.N., Kim, R., Chen, Y., Negishi, M., Jhun, S., Weiss, S., Ryu, J.H., Bai, X., Xiao, W., Feeney, E., et al., 2016. Impaired consciousness in patients with absence seizures investigated by functional MRI, EEG, and behavioural measures: a cross-sectional study. Lancet Neurol. 15, 1336–1345.

Halász, P., 2015. Are absence epilepsy and nocturnal frontal lobe epilepsy system epilepsies of the sleep/wake system? Behav. Neurol. 231676.

Hofmann, G., Balgooyen, L., Mattis, J., Deisseroth, K., Buckmaster, P.S., 2016. Hilar somatostatin interneuron loss reduces dentate gyrus inhibition in a mouse model of temporal lobe epilepsy. Epilepsia 57, 977–983.

Houser, C.R., Esclapez, M., 1996. Vulnerability and plasticity of the GABA system in the pilocarpine model of spontaneous recurrent seizures. Epilepsy Res. 26, 207–218.

Howard, A.L., Neu, A., Morgan, R.J., Echegoyen, J.C., Soltesz, I., 2007. Opposing modifications in intrinsic currents and synaptic inputs in post-traumatic mossy cells: evidence for single-cell homeostasis in a hyperexcitable network. J. Neurophysiol. 97, 2394–2409.

Hunt, R.F., Girskis, K.M., Rubenstein, J.L., Alvarez-Buylla, A., Baraban, S.C., 2013. GABA progenitors grafted into the adult epileptic brain control seizures and abnormal behavior. Nat. Neurosci. 16, 692–697.

Hutcheon, B., Yarom, Y., 2000. Resonance, oscillation and the intrinsic frequency preferences of neurons. Trends Neurosci. 23, 216–222.

Jacoby, A., Gorry, J., Baker, G.A., 2005. Employers' attitudes to employment of people with epilepsy: still the same old story? Epilepsia 46, 1978–1987.

Jennum, P., Gyllenborg, J., Kjellberg, J., 2011. The social and economic consequences of epilepsy: a controlled national study. Epilepsia 52, 949–956.

Kang, Y.-J., Lewis, H.E.S., Young, M.W., Govindaiah, G., Greenfield, L.J., Garcia-Rill, E., Lee, S.-H., 2018. Cell type-specific intrinsic perithreshold oscillations in hippocampal GABAergic interneurons. Neuroscience 376, 80–93.

Keezer, M.R., Sisodiya, S.M., Sander, J.W., 2016. Comorbidities of epilepsy: current concepts and future perspectives. Lancet Neurol. 15, 106–115.

Kezunovic, N., Urbano, F.J., Simon, C., Hyde, J., Smith, K., Garcia-Rill, E., 2011. Mechanism behind gamma band activity in the pedunculopontine nucleus. Eur. J. Neurosci. 34, 404–415.

Kim, D., Song, I., Keum, S., Lee, T., Jeong, M.J., Kim, S.S., McEnery, M.W., Shin, H.S., 2001. Lack of the burst firing of thalamocortical relay neurons and resistance to absence seizures in mice lacking α1GT-type Ca^{2+} channels. Neuron 31, 35–45.

Kim, U., Sanchez-Vives, M.V., McCormick, D.A., 1997. Functional dynamics of GABAergic inhibition in the thalamus. Science 278, 130–134.

Klausberger, T., Somogyi, P., 2008. Neuronal diversity and temporal dynamics: the unity of hippocampal circuit operations. Science 321, 53–57.

Kleiner-Fisman, G., Fisman, D.N., Sime, E., Saint-Cyr, J.A., Lozano, A.M., Lang, A.E., 2003. Long-term follow up of bilateral deep brain stimulation of the subthalamic nucleus in patients with advanced Parkinson disease. J. Neurosurg. 99, 489–495.

Kostopoulos, G.K., 2001. Involvement of thalamocortical system in epileptic loss of consciousness. Epilepsia 42, 13–19.

Krook-Magnuson, E., Armstrong, C., Oijala, M., Soltesz, I., 2013. On-demand optogenetic control of spontaneous seizures in temporal lobe epilepsy. Nat. Commun. 4, 1376.

Krumholz, A., 2009. Driving issues in epilepsy: past, present, and future. Epilepsy Curr. 9, 31–35.

Kundishora, A.J., Gummadavelli, A., Ma, C., Liu, M., McCafferty, C., Schiff, N.D., Willie, J.T., Gross, R.E., Gerrard, J., Blumenfeld, H., 2017. Restoring conscious arousal during focal limbic seizures with deep brain stimulation. Cereb. Cortex 27, 1964–1975.

Kwan, P., Sander, J.W., 2004. The natural history of epilepsy: an epidemiological view. J. Neurol. Neurosurg. Psychiatry 75, 1376–1381.

Lee, S.-H., Cox, C.L., 2003. Vasoactive intestinal peptide selectively depolarizes thalamic relay neurons and attenuates intrathalamic rhythmic activity. J. Neurophysiol. 90, 1224–1234.

Lee, S.-H., Cox, C.L., 2006. Excitatory actions of vasoactive intestinal peptide on mouse thalamocortical neurons are mediated by VPAC2 receptors. J. Neurophysiol. 96, 858–871.

Lee, S.-H., Cox, C.L., 2008. Excitatory actions of peptide histidine isoleucine on thalamic relay neurons. Neuropharmacology 55, 1329–1339.

Lee, S.-H., Urbano, F.J., Garcia-Rill, E., 2018. The critical role of intrinsic membrane oscillations. Neurosignals 26, 66–76.

Lemaire, J., Sontheimer, A., Pereira, B., Coste, J., Rosenberg, S., Sarret, C., Coll, G., Gabrillargues, J., Jean, B., Gillart, T., et al., 2018. Deep brain stimulation in five patients with severe disorders of consciousness. Ann. Clin. Transl. Neurol. 10, 1013–1023.

Lewis, D.A., Curley, A.A., Glausier, J.R., Volk, D.W., 2012. Cortical parvalbumin interneurons and cognitive dysfunction in schizophrenia. Trends Neurosci. 35, 57–67.

Li, W., Motelow, J.E., Zhan, Q., Hu, Y.C., Kim, R., Chen, W.C., Blumenfeld, H., 2015. Cortical network switching: possible role of the lateral septum and cholinergic arousal. Brain Stimul. 8, 36–41.

Llinas, R.R., Steriade, M., 2006. Bursting of thalamic neurons and states of vigilance. J. Neurophysiol. 95, 3297–3308.

Llinás, R.R., 1988. The intrinsic electrophysiological properties of mammalian neurons: insights into central nervous system function. Science 242, 1654–1664.

Llinás, R., Yarom, Y., 1986. Oscillatory properties of Guinea-pig inferior olivary neurones and their pharmacological modulation: an in vitro study. J. Physiol. 376, 163–182.

Llinás, R., Ribary, U., Contreras, D., Pedroarena, C., 1998. The neuronal basis for consciousness. Philos. Trans. R. Soc. B Biol. Sci. 353, 1841–1849.

Llinás, R., Urbano, F.J., Leznik, E., Ramírez, R.R., Van Marle, H.J.F., 2005. Rhythmic and dysrhythmic thalamocortical dynamics: GABA systems and the edge effect. Trends Neurosci. 28, 325–333.

Long, L.L., Song, Y.M., Xu, L., Yi, F., Long, H.Y., Zhou, L., Qin, X.H., Feng, L., Xiao, B., 2014. Aberrant neuronal synaptic connectivity in CA1 area of the hippocampus from pilocarpine-induced epileptic rats observed by fluorogold. Int. J. Clin. Exp. Med. 7, 2687–2695.

Lüders, H.O., Amina, S., Bailey, C., Baumgartner, C., Benbadis, S., Bermeo, A., Carreño, M., Devereaux, M., Diehl, B., Eccher, M., et al., 2014. Proposal: different types of alteration and loss of consciousness in epilepsy. Epilepsia 55, 1140–1144.

McCormick, D.A., Bal, T., 1997. SLEEP AND AROUSAL: thalamocortical mechanisms. Annu. Rev. Neurosci. 20, 185–215.

Meeren, H.K., Pijn, J.P., Van Luijtelaar, E.L., Coenen, A.M., Lopes da Silva, F.H., 2002. Cortical focus drives widespread corticothalamic networks during spontaneous absence seizures in rats. J. Neurosci. 22, 1480–1495.

Morin, F., Beaulieu, C., Lacaille, J.C., 1998. Selective loss of GABA neurons in area CA1 of the rat hippocampus after intraventricular kainate. Epilepsy Res. 32, 363–369.

Motelow, J.E., Li, W., Zhan, Q., Mishra, A.M., Sachdev, R.N.S., Liu, G., Gummadavelli, A., Zayyad, Z., Lee, H.S., Chu, V., et al., 2015. Decreased subcortical cholinergic arousal in focal seizures. Neuron 85, 561–572.

Pelkey, K.A., Chittajallu, R., Craig, M.T., Tricoire, L., Wester, J.C., McBain, C.J., 2017. Hippocampal GABAergic inhibitory interneurons. Physiol. Rev. 97, 1619–1747.

Pike, F.G., Goddard, R.S., Suckling, J.M., Ganter, P., Kasthuri, N., Paulsen, O., 2000. Distinct frequency preferences of different types of rat hippocampal neurones in response to oscillatory input currents. J. Physiol. 529, 205–213.

Ratzliff, A.H., Santhakumar, V., Howard, A., Soltesz, I., 2002. Mossy cells in epilepsy: rigor mortis or vigor mortis? Trends Neurosci. 25, 140–144.

Sadleir, L.G., Scheffer, I.E., Smith, S., Carstensen, B., Carlin, J., Connolly, M.B., Farrell, K., 2008. Factors influencing clinical features of absence seizures. Epilepsia 49, 2100–2107.

Sanon, N.T., Pelletier, J.G., Carmant, L., Lacaille, J.C., 2010. Interneuron subtype specific activation of mGluR15 during epileptiform activity in hippocampus. Epilepsia 51, 1607–1618.

Scharfman, H.E., Sollas, A., Berger, R., Coodman, J., 2003. Electrophysiological evidence of monosynaptic excitatory transmission between granule cells after seizure-induced mossy fiber sprouting. J. Neurophysiol. 90, 2536–2547.

Selwa, L., 2010. Epileptiform activity, seizures, and epilepsy syndromes. In: Greenfield, L.J., Geyer, J., Carney, P. (Eds.), Reading EEGs: A Practical Approach. Lippincott Williams & Wikins, Philadelphia, PA, USA.

Sherman, S.M., 2001. Tonic and burst firing: dual modes of thalamocortical relay. Trends Neurosci. 24, 122–126.

Siddiqui, A.H., Joseph, S.A., 2005. CA3 axonal sprouting in kainate-induced chronic epilepsy. Brain Res. 1066, 129–146.

Stark, E., Eichler, R., Roux, L., Fujisawa, S., Rotstein, H.G., Buzsáki, G., 2013. Inhibition-induced theta resonance in cortical circuits. Neuron 80, 1263–1276.

Sun, C., Sun, J., Erisir, A., Kapur, J., 2014. Loss of cholecystokinin-containing terminals in temporal lobe epilepsy. Neurobiol. Dis. 62, 44–55.

Sun, Q., Baraban, S.C., Prince, D.A., Huguenard, J.R., 2003. Target-specific neuropeptide Y-ergic synaptic inhibition and its network consequences within the mammalian thalamus. J. Neurosci. 23, 39–49.

Tatum, W.O., 2012. Mesial temporal lobe epilepsy. J. Clin. Neurophysiol. 29, 356–365.

Tenney, J.R., Glauser, T.A., 2013. The current state of absence epilepsy: can we have your attention? Epilepsy Curr. 13, 135–140.

Tomson, T., Walczak, T., Sillanpaa, M., Sander, J.W.A.S., 2005. Sudden unexpected death in epilepsy: a review of incidence and risk factors. Epilepsia 46, 54–61.
Unterberger, I., Trinka, E., Kaplan, P.W., Walser, G., Luef, G., Bauer, G., 2018. Generalized nonmotor (absence) seizures—what do absence, generalized, and nonmotor mean? Epilepsia 59, 523–529.
Veres, J.M., Nagy, G.A., Hájos, N., 2017. Perisomatic GABAergic synapses of basket cells effectively control principal neuron activity in amygdala networks. elife 6, e20721.
Wallace, R.H., Marini, C., Petrou, S., Harkin, L.A., Bowser, D.N., Panchal, R.G., Williams, D.A., Sutherland, G.R., Mulley, J.C., Scheffer, I.E., et al., 2001. Mutant GABAA receptor γ2-subunit in childhood absence epilepsy and febrile seizures. Nat. Genet. 28, 49–52.
West, S., Sj, N., Cotton, J., Gandhi, S., Weston, J., Sudan, A., Ramirez, R., Newton, R., 2015. Surgery for epilepsy. Cochrane Database Syst. Rev. 7, CD010541.
Wiebe, S., Blume, W.T., Girvin, J.P., Eliasziw, M., 2001. A randomized, controlled trial of surgery for temporal-lobe epilepsy. N. Engl. J. Med. 345, 311–318.
Willems, L.M., Watermann, N., Richter, S., Kay, L., Hermsen, A.M., Knake, S., Rosenow, F., Strzelczyk, A., 2018. Incidence, risk factors and consequences of epilepsy-related injuries and accidents: a retrospective, single center study. Front. Neurol. 9, 414.
Williamson, P.D., French, J.A., Thadani, V.M., Kim, J.H., Novelly, R.A., Spencer, S.S., Spencer, D.D., Mattson, R.H., 1993. Characteristics of medial temporal lobe epilepsy. II. Interictal and ictal scalp electroencephalography, neuropsychological testing, neuroimaging, surgical results, and pathology. Ann. Neurol. 34, 781–787.
Wirrell, E.C., 2006. Epilepsy-related injuries. Epilepsia 47, 79–86.
Wyeth, M.S., Zhang, N., Mody, I., Houser, C.R., 2010. Selective reduction of cholecystokinin-positive basket cell innervation in a model of temporal lobe epilepsy. J. Neurosci. 30, 8993–9006.
Zack, M., Kobau, R., 2017. National and state estimates of the numbers of adults and children with active epilepsy—United States, 2015. MMWR—Morb. Mortal. Wkly Rep. 66, 821–825.
Zamponi, G.W., Lory, P., Perez-Reyes, E., 2010. Role of voltage-gated calcium channels in epilepsy. Pflugers Arch. Eur. J. Physiol. 460, 395–403.
Zhivotovsky, B., Orrenius, S., 2011. Calcium and cell death mechanisms: a perspective from the cell death community. Cell Calcium 50, 211–221.
Ziburkus, J., Cressman, J., Barreto, E., Schiff, S., 2006. Interneuron and pyramidal cell interplay during in vitro seizure-like events. J. Neurophysiol. 95, 3948–3954.

Further reading

Szabo, G.G., Du, X., Oijala, M., Varga, C., Parent, J.M., Soltesz, I., 2017. Extended interneuronal network of the dentate gyrus. Cell Rep. 20, 1262–1268.

Chapter 13

Neuroepigenetics of arousal and the formulation of the self

Edgar Garcia-Rill
Center for Translational Neuroscience, Department of Neurobiology and Developmental Sciences, University of Arkansas for Medical Sciences, Little Rock, AR, United States

Introduction

The third of three of the most significant discoveries in the reticular activating system (RAS) during the last 10 years is the presence of gene transcription when intrinsic gamma frequency oscillations are manifested (Urbano et al., 2018). The implications of this finding are monumental for future consideration of the role of arousal in brain function and in neurological and psychiatric disorders. Gamma oscillations play a critical role in higher functions such as perception, problem-solving, and memory (Eckhorn et al., 1986; Gray and Singer, 1989; Philips and Takeda, 2009; Palva et al., 2009; Voss et al., 2009), as well as subcortical thalamocortical (Llinas et al., 1991), cerebellar (Lang et al., 2006; Middleton et al., 2008), hippocampal (Chrobak and Buzsáki, 1988), and arousal-related reticular activating system (RAS) processes (Simon et al., 2010, 2011; Kezunovic et al., 2011, 2013; Urbano et al., 2012, 2014; Hyde et al., 2013; Garcia-Rill et al., 2013, 2014; D'Onofrio et al., 2015, 2016, 2017; Luster et al., 2015, 2016). As covered in Chapter 2, intrinsic gamma oscillations in the PPN are generated by voltage-dependent, high-threshold N- and P/Q-type Ca^{2+} channels. Epigenetic mechanisms (i.e., histone posttranslational modifications and DNA methylation) play a role in the regulation of gene expression in response to a wide range of environmental stimuli, such as learning, stress, or drugs of abuse (Cadet, 2016). Acetylation of histones is associated with the modulation of transcription in multiple ways, by relaxing chromatin structure that would increase the accessibility of transcription factors to their target genes and also by direct acetylation of transcription factors (Haberland et al., 2009). Histone acetylation is a dynamic process controlled by the antagonistic actions of two large families of enzymes: the histone acetyltransferases (HATs) and the histone deacetylases (HDACs). We provided novel results on the effects of the TSA that acts as a class I and II HDAC inhibitor. TSA blocked the manifestation

Arousal in Neurological and Psychiatric Diseases. https://doi.org/10.1016/B978-0-12-817992-5.00013-1

of Ca^{2+} channel-mediated intrinsic gamma oscillations in PPN neurons, both acutely and after preincubation exposure. TSA also significantly reduced Ca^{2+} currents in PPN neurons, suggesting a homeostatic effect of deacetylation on plasma membrane ion channels. These results open important new lines of research on the neuroepigenetics of an important survival mechanism, arousal (Urbano et al., 2018).

The epigenetic profile of numerous HDAC inhibitors is being explored for translational applications (West and Johnstone, 2014). HDACs and their inhibitors are promising candidates in treating cancer and several neurodegenerative disorders (Didonna and Opal, 2015). Some of these inhibitors have proved to have synergistic effects with other drugs (Subramanian et al., 2010). However, the mechanisms of action of HDAC inhibitors in neurons need to be elucidated. Indeed, there are potential neurological side effects by modulating HDAC functions, and more research is needed to design more specific HDAC inhibitors with the lowest range of neurological side effects (Didonna and Opal, 2015). The neurological sequelae induced by HDAC inhibitors include fatigue, status epilepticus, somnolence, and gait problems (Subramanian et al., 2010). That is, many side effects interfere with arousal and might be linked to altered PPN physiology since this nucleus modulates states of arousal. We addressed the question, does blockade of HDACs affect the manifestation of gamma oscillations generated by N- and/or P/Q-type Ca^{2+} channels?

Histone acetylation

HDACs are classified into four classes based on localization and amino acid sequence similarities, class I HDACs (HDAC1, HDAC2, HDAC3, and HDAC8), class IIa HDACs (HDAC4, HDAC5, HDAC7, and HDAC9), class IIb HDACs (HDAC6 and HDAC10), class II HDACs (HDAC4-7, HDAC9, and HDAC10), class III HDACs (SIRT1-7), and histone deacetylase 11 (HDAC11). The latter shares characteristics with class I and II HDACs but is considered a class IV HDAC based on its distinct physiological roles (Broide et al., 2007; Haberland et al., 2009). While class I HDACs are found in the nucleus, class IIb can travel to the cytoplasm where they can interact with nonhistone proteins.

HDAC1/2 knockdown using lentiviral transfection of cultured hippocampal neurons altered excitatory synaptic transmission and the formation of dendritic spines (Akhtar et al., 2009). These authors showed that HDAC1/2 had different synaptic roles during early developmental stages compared with what was observed in neuronal cultures kept in vitro over 16 days. Synaptic effects of HDAC1/2 knockdown were mimicked by 18–24 h preincubation of cultures with the histone deacetylation inhibitor TSA. Other authors failed to observe similar results after chronic treatment with TSA of hippocampal organotypic cultures, suggesting the existence of molecular compensations associated to the permanent knockdown of HDACs (Calfa et al., 2012). Knockdown of many of the individual HDAC isoforms are lethal, emphasizing the importance of performing acute experiments using HDAC inhibitors in order to assess their role in the modulation intrinsic neuronal properties.

Effects on gamma oscillations

TSA reduced gamma oscillations but not lower-frequency oscillations

Recordings of gamma band oscillations in PPN neurons were performed in slices at 36°C before and in the presence of synaptic blockers (SB) and tetrodotoxin (TTX). Whole-cell current clamp recordings of PPN cells showed that no significant rundown of oscillatory activity was evident during the recording period. The class I and II HDAC inhibitor, TSA (1 μM), induced no membrane potential changes after acute (15–20 min) application. Bath application of TSA significantly increased the input resistance of PPN neurons by 63% ± 23%, while reducing PPN membrane oscillations. The mean reduction of power in the spectra of membrane oscillations before and after bath application of TSA in PPN neurons showed no statistically significant decreases in the power of lower-frequency (10–15, 15–20, or 20–30 Hz) oscillations. However, the gamma band range (30–100 Hz) reduction was significant. This result implies that, normally, the manifestation of gamma band oscillations during waking induces gene transcription.

Preincubation with TSA had similar effects

Class I and II HDACs have been described to act through their association with other proteins at nuclear and/or cytoplasmic sites (Haberland et al., 2009). Therefore TSA might require longer periods of incubation prior to reaching its maximal effects. With this in mind, we preincubated PPN slices with TSA (1 μM) for 90–120 min at 36°C in the presence of SB + TTX. Membrane oscillations were recorded from PPN neurons in control slices and in slices following TSA preincubation, allowing us to compare changes in power spectrum with and without TSA. Recordings from representative PPN cells showed no gamma band frequency oscillations after 120 min preincubation with TSA. On average, low-frequency band power was unaffected, while only gamma band power was significantly reduced by TSA preincubation. Neither membrane potential nor input resistance was altered by TSA preincubation.

TSA reduced calcium currents in PPN cells

We tested whether TSA-mediated reductions in gamma oscillations might be a consequence of its blocking effect on Ca^{2+} currents (I_{Ca}) mediated by high-threshold, voltage-dependent channels. I_{Ca}s were recorded after gaining access to the neuronal intracellular space, and series resistance was compensated and stable. We combined square voltage steps in combination with high Cs^{+}

intracellular pipette solution and synaptic receptor blockers (SB) plus TTX. Recordings of I_{Ca} were carried out for up to 30 min without significant rundown, per previous reports (Kezunovic et al., 2011, 2013; Hyde et al., 2013; D'Onofrio et al., 2015, 2016, 2017; Garcia-Rill et al., 2016; Luster et al., 2015, 2016; Urbano et al., 2016). The holding potential was at first set at −50 mV and then depolarized to 40 mV. Bath application of TSA (1 μM) reduced I_{Ca}. In a group of PPN neurons, the I_{Ca} blocking effects of TSA were observed throughout a wide range of test potentials, with significant reductions in I_{Ca} between −20 and 40 mV (I_{Ca} measured at 2/3 of the slope).

TSA reduced I_{Ca} in cells with P/Q channels

A group of cells that were identified as N-only cells (since I_{Ca}s were blocked by conotoxin [see D'Onofrio et al., 2015, 2016, 2017; Luster et al., 2015, 2016]) showed no significant reduction in I_{Ca}, while a group of cells that were either N + P/Q or P/Q-only showed a significant decrease in I_{Ca}. N-only PPN neurons manifested a 18.5% ± 2.8% TSA-mediated reduction of I_{Ca} compared with before application. Conversely, PPN cells expressing P/Q-type channels (i.e., P/Q only plus N + P/Q types) showed a 47.8% ± 8.8% block after acute TSA application compared with before and a significantly greater effect compared with N-only cells. This result suggests that TSA has a cell-specific effect on those cells with P/Q-type channels but not on those with N-type channels.

Blockade of CaMKII with KN-93 reduced the effects of TSA

Does the TSA-mediated reduction of I_{Ca} require intracellular Ca^{2+}-calmodulin-dependent kinase type II (CaMKII) activation? After preincubating PPN slices with KN-93 (1 μM; >30 min; in SB+TTX at 36°C), we observed that acute TSA (1 μM) was less effective in reducing I_{Ca}. The TSA-induced reduction of I_{Ca} dropped from 56.0% ± 9.6% to 26.3% ± 6.3% with KN-93. In conclusion the acute effect of TSA on I_{Ca} was blunted by KN-93. This finding shows that TSA implements its effects through the P/Q-type channel/CaMKII pathway.

MC1568, an inhibitor of HDAC IIa reduced I_{Ca} in PPN neurons

After acute MC1568 application, no change in membrane potential was observed. MC1568 increased the input resistance of PPN neurons by 12% ± 4% while blocking membrane gamma band oscillations. There was a mean reduction of power in the spectra of membrane oscillations before and following bath application of MC1568 in PPN neurons. No statistically significant decreases were observed in the power of lower-frequency (10–15, 15–20, or 20–30 Hz) membrane oscillations. However, the gamma band range (30–100 Hz) reduction was significant. We also characterized the MC1568-mediated reduction of I_{Ca} before and after preincubation with KN-93 (1 μM). Acute MC1568 (1 μM) reduced I_{Ca}. In addition, the mean reduction of I_{Ca} by MC1568 was blunted by

KN-93. Mean I_{Ca} after MC1568 decreased from 63.7% ± 11.7% in the absence of KN-93 to 23.0% ± 7.0% after KN-93.

Tubastatin A, an inhibitor of class IIb HDAC, and MS275, an inhibitor of class I HDAC, had no effect on PPN gamma oscillations or I_{Ca}

Acute Tubastatin A (150 nM) bath application did not affect gamma oscillations in PPN neurons. No change was observed on either membrane potential or input resistance of PPN neurons. Furthermore, Tubastatin A (500 nM) did not change gamma oscillation mean power spectrum. We also tested the effects of MS275 (500 nM) on oscillations and high-threshold, voltage-dependent I_{Ca}s in PPN neurons. We observed no significant blocking effect on I_{Ca} by MS275, suggesting TSA has effects through HDAC IIa, but not through HDAC I or HDAC IIb.

Summary of intracellular mechanisms

Fig. 13.1 summarizes the effects observed in Urbano et al. (2018). P/Q-type channels (green channel) are modulated by CaMKII (green sextagon) to generate gamma band oscillations. TSA blocked these oscillations and reduced I_{Ca} in PPN cells by inhibiting HDAC I and II (pink). KN-93 blunted the effects of TSA, suggesting that HDAC II action on CaMKII may be required for potential transcription following P/Q-type channel-mediated gamma oscillations (not shown). Inhibition of HDAC I by MS275, or of HDAC IIb by Tubastatin A, had no effects on I_{Ca} or oscillations. However, MC1568 inhibition of HDAC IIa had the same effect as TSA in reducing I_{Ca} and oscillations. These effects were evident in cells with P/Q-type channels (N + P/Q and P/Q-only cells), but there were no effects on N-only cells. The action of P/Q-type channels during waking was proposed to induce high-frequency activity in the cortical EEG that is coherent across distant sites (see Chapter 2 for full description).

Effects on proteins

A significant problem with proteomic and genomic analyses of the brain is the variety of cell types manifesting a myriad of receptors and a host of ionic channels. This makes it difficult to ascribe changes in protein levels and gene expression. We cut sagittal brain stem slices similar to those used for the recording studies described earlier. The preparation we used employed fast synaptic blockers and TTX (same as when recording intrinsic gamma oscillations) to ensure that fast synaptic transmission and circuit activity through action potentials was prevented. The activity present in the system, therefore, was mainly from intrinsic membrane oscillations and limited to high-frequency gamma oscillations mainly in cells manifesting P/Q-type channels. As such, proteomic and genomic changes are more easily ascribed. We then punched the PPN after

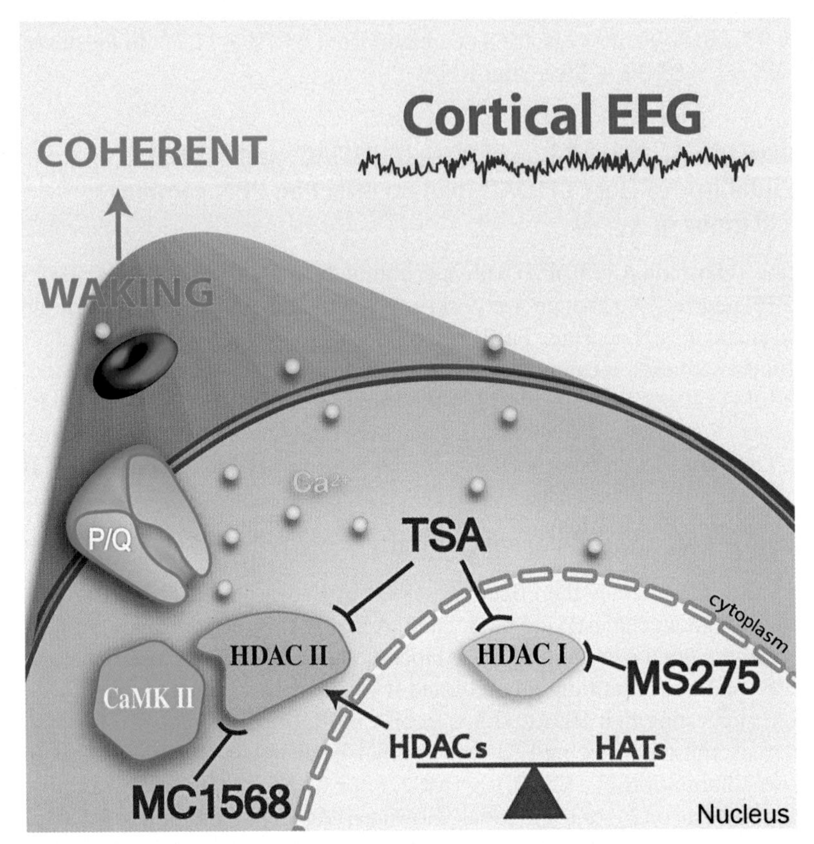

FIG. 13.1 Modulation of gene transcription by waking. P/Q-type channels (green channel) are modulated by CaMKII (green sextagon) to generate gamma band oscillations (low-amplitude, high-frequency activity in the cortical EEG). Such gamma activity during waking manifests coherence across distant sites in the cortex, unlike gamma activity during REM sleep that shows no coherence across distant sites. TSA blocked these oscillations and reduced I_{Ca} in PPN cells by inhibiting HDAC I and II (pink structures). KN-93 blunted the effects of TSA (not shown), suggesting that HDAC II action on CaMKII may be required for potential transcription following P/Q-type channel-mediated gamma oscillations. Inhibition of HDAC I by MS275 (right side), or of HDAC IIb by tubastatin A (not shown), in the nucleus had no effects on I_{Ca} or oscillations. However, MC1568 (left side) inhibition of HDAC IIa had the same effect as TSA in reducing I_{Ca} and oscillations. These effects were evident in cells with P/Q-type channels (N + P/Q and P/Q-only cells), but there were no effects on N-only cells that are modulated by cAMP/PKA (not shown).

30 min exposure to each agent and processed the biopsies for proteomic analysis. Which proteins are affected as a result of intrinsic oscillations presumably participating in arousal? Which genes are up- or downregulated upon activation of oscillations presumed to arise upon waking? These are only some of the potential new directions for future research.

In addition, we performed population recording studies in PPN in which we cut sagittal brain stem slices such as those used for the patch clamp studies described earlier. We did not use synaptic blockers or TTX to allow circuit activity and determine if the population as a whole was affected by exposure to TSA. We used carbachol to stimulate activity in the PPN. Fig. 13.2 shows population recordings from the PPN in four different slices exposed to carbachol

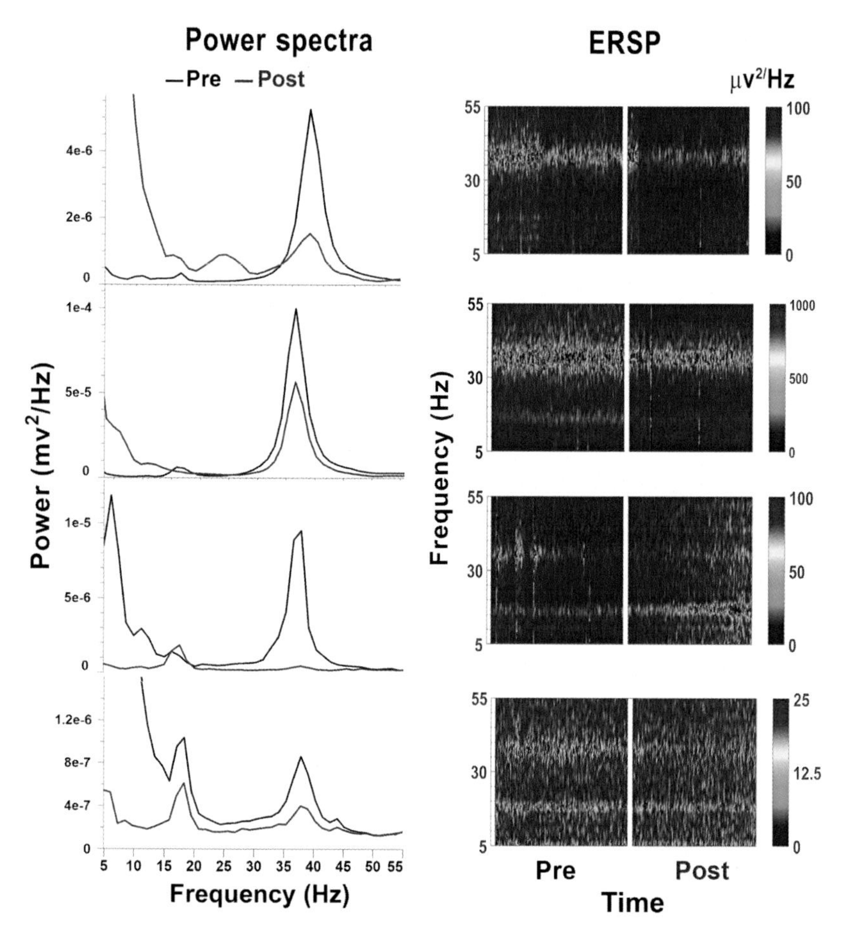

FIG. 13.2 Population responses in PPN before and after TSA. Left side. Power spectra of a single time point midway through the PPN population response with application of carbachol before (pre, black lines) and with application of carbachol after 10min of exposure to TSA (post, red lines). Note that in each of the four slices shown the control response to carbachol before (pre, black lines) showed high-amplitude peaks in the gamma range. On the other hand, TSA was applied for 10min, and then the response to carbachol after TSA (post, red lines) showed markedly reduced peaks in the gamma range. These power spectra are from the minute half way through the carbachol responses before and after TSA. Right side. Event-related spectral perturbations (ERSPs) of the PPN population responses of the same four slices showing gamma band activity throughout the control application of carbachol (pre, left ERSPs). On the other hand, following exposure to TSA for 10min the response to carbachol (post, right ERSPs) showed markedly reduced gamma band activity. These results suggest that TSA reduced gamma band activity in the PPN population as a whole, similar to the reduction of gamma oscillations observed in single patched PPN neurons under synaptic blockers and TTX.

(CAR, 50 μM) for 10 min. On the left are power spectra taken from control CAR recordings (black records, pre) after 2 min of CAR superfusion but before exposure to TSA. All of these slices showed peaks at gamma frequency induced by the administration of CAR. We then performed a 30 min of exposure to TSA (1 μM) while continuing to superfuse with CAR.

Population recordings were again taken after TSA was stopped, but CAR was continued (red records, post). Note the reduction of gamma activity in all of the slices, showing that the response to CAR was reduced after exposure to TSA. On the right side of the figure are event-related spectral perturbations (ERSPs), which are essentially running power spectra, during the first 10 min of CAR exposure, and then, TSA exposure was added (not shown), followed by a second 10 min of CAR exposure after TSA. Note that in all cases, high-frequency activity was reduced, and in some cases, low-frequency activity increased, after TSA. These results suggest that TSA reduced circuit activity in the PPN stimulated with CAR, which normally induces high-frequency activity, but the response is markedly reduced after TSA exposure.

Intracellular calcium regulation

The best description of the evolutionary advance of the role of intracellular calcium processes was provided by Llinas (2001). Llinas describes the pronounced reactivity of the ion and its tendency to crystallize, so that mechanisms had to arise to control its tendency to aggregate. Calcium must be maintained in much lower concentrations in liquid environments than other ions such as sodium or potassium. Free calcium could easily disrupt energy-producing processes that involve phosphorus, that is, oxidative phosphorylation, which are necessary for such functions as muscle contractions and nerve activity. A major component in the harnessing of calcium was the development of calmodulin, which keeps calcium concentrations that are essential for a host of intracellular reactions. Another protein in charge of corralling calcium is NCS-1, which, as we described in Chapters 3 and 4, is overexpressed in schizophrenia and bipolar disorder. We described the potential intracellular disruption in NCS-1 that may lead to some of the changes in gamma band activity in the RAS, along with marked disruption in arousal function.

As we discovered recently, some of the proteins that are activated by the presence of gamma band activity include calcium regulatory proteins such as calcineurin, a calcium- and calmodulin-dependent protein phosphatase, nNOS, which is a neuronal nitric oxide synthase enzyme (interestingly, PPN has one of the highest concentrations of nNOS in the brain), and NCS-1.

Implications

Clinically, we see at least two possible new directions of research to test the involvement of this mechanism in certain disorders. First, blocking histone deacetylation with TSA may have deleterious effects on intrinsic gamma oscillations in the PPN, presumably inhibiting arousal, which may account for some of the soporific side effects of this medication. Second, we previously proposed that insomnia represents not too little sleep but too much waking (Garcia-Rill et al., 2015, 2016a, b). We suggested that insomnia, in at least some cases, may be due to overexpression of P/Q-type calcium channels, since those were proposed to be preferentially active during waking. Moreover, we know that gene regulation can be mediated by voltage-dependent calcium channels (Barbado et al., 2009) and that a number of human disorders result when P/Q-type channels mutate or become dysfunctional (Schorge and Rajakulendran, 2012). Therefore our findings point to a large number of potentially fruitful areas of clinical research that needs to be pursued.

Formulation of the self

There may be a more basic purpose to the neuroepigenetic activities triggered by waking up. Upon waking, sensory information onto PPN neurons induces arousal, a continuous flow of afferent input (Garcia-Rill, 2015; Garcia-Rill et al., 2013). The role of gamma band activity in PPN neurons has been proposed to stabilize coherence related to arousal, providing a stable state during waking (Garcia-Rill, 2015; Garcia-Rill et al., 2013, 2016; Urbano et al., 2016). This constant process was suggested to help evaluate the world around us on a continuing basis, that is, to bestow preconscious awareness (Urbano et al., 2014; Garcia-Rill et al., 2016). The sustained flux of afferent information was termed "bottom-up gamma" (Garcia-Rill et al., 2016a, b), which may provide a remodeling of neurons in this part of the RAS, a daily update of sensory experience. The results we described (Urbano et al., 2018) suggest that the continuous flow of information could modulate transcription and protein activation near membrane/cytoplasmic sites in PPN neurons, which could represent one mechanism behind such remodeling. These are, of course, speculations that require much work to address. In the succeeding text, we present an argument that this mechanism may be a primordial element in the formulation of the self.

In Chapter 2, we described Penfield's proposition of the presence of a "centrencephalic integrating system" that fulfills the role of sensorimotor integration necessary for consciousness (Penfield, 1970). At the wellspring of this process is the RAS, a phylogenetically conserved area of the brain inundated by the continuous flow of internal (especially intrinsic oscillations) and external information (impinging on dendrites). This information modulates sleep-wake cycles, the startle response, and fight-or-flight responses, along with changes in muscle tone and locomotion. Accordingly, activation of the RAS during waking induces coherent activity (through electrically coupled cells) and high-frequency oscillations (through P/Q-type calcium channel and subthreshold oscillations) to sustain gamma activity and support a persistent, reliable state for assessing our world. Perhaps gene transcription in the RAS allows the maintenance of the core identity of the being from day to day.

As soon as we awaken, before the cortex has been fully activated, we know who we are. Instantaneously, we are comfortable in our own skin, despite the fact that several times during the night we experienced slow-wave sleep (SWS), during which cortical columns were firing together without lateral inhibition, thus preventing sensation or perception, basically a period of about 90 min during which there was nobody home. Even during rapid eye movement (REM) sleep, our persona may have participated in our dreams, but sometimes, we were someone else. Only upon waking, and immediately upon waking, do we manifest the realization of our self. This primitive process is much more basic than what we understand as the self. At the risk of being misunderstood, a close parallel seems to be the "id" in Freud's theoretical formulation of the psyche but without the baggage of being driven by merely seeking pleasure and avoiding pain.

This construct included the id, the ego, and the superego, which was a foundation for psychoanalysis. At first the mind was divided into the conscious, preconscious, and unconscious but was replaced by the three-part model. The "id" was the only component of personality present from birth and was responsible for basic, instinctual drives. It was supposed to provide the source for all psychic energy, which is not too distant from a process that generates arousal. While the "id" was attributed a host of bodily needs and wants, including desires, impulses, and sexual and aggressive drives, this would seem overly complex for the simple survival element we describe. Instead, we ascribe the process of waking as providing a primitive formulation of the self as an entity that survived the night, an element that persisted despite the inactivation of the brain during sleep. This primordial element is what we refer to as the formulation of the self, not burdened by the complexity of drives, personality, experience, or learned biases. Presumably, these are recalled once the cortex kicks into full gear. The RAS, as the innate system in charge of survival, arousal, and fight-or-flight responses, would seem to be the ideal candidate to furnish an elemental formulation of the self. Moreover, the persistence of the self across periods of inactivation (SWS) implies a memory that persists, a formulation that lingers, enduring across periods of "mindless void." That memory, we speculate, is provided by the activation of gene transcription upon manifestation of intrinsic gamma band oscillations.

This type of memory is more complex than the phylogenetic memory that contributes to basic brain connectivity and that which marries form to function (Llinas, 2001). These memories represent genetic mutations only modified by natural selection, etc. We do not refer to such mechanisms when we ascribe the formulation of the self as one of the outcomes of gene transcription upon manifestation of intrinsic gamma band oscillations in PPN neurons. One of the fathers of sleep research, Michel Jouvet, posited that one reason why we dream during REM sleep is to preserve our personality (Jouvet, 2001). He suggested that dreaming is necessary for genetic reprogramming of our brain.

Our studies on the effects of TSA on gamma oscillations and the triggering of gene transcription following their manifestation appear to occur through P/Q-type calcium channels that may be active during waking. On the other hand, it is N-type calcium channels that appear to be active during REM sleep. We assume that if there is gene transcription triggered by the N-type channel/cAMP/PKA pathway, it is different than that occurring through the P/Q-type channel/CaMKII pathway. Therefore Jouvet's suggestion is not contrary to our hypothesis, simply an unexplored area that needs attention.

In summary, a consequence of the manifestation of intrinsic gamma oscillations through the P/Q-type channel/CaMKII pathway may be the instantaneous formulation and recall of the self upon waking. This process serves as a primordial survival mechanism for the individual, upon which is presumably superimposed more complex information (provided by the cortex and other regions) as waking progresses. The clinical implications of the dysregulation of such a process are considerable for both neurological and psychiatric disorders. This is an unexplored area that promises to yield valuable insight on the basic workings and the pathological consequences of disease.

Acknowledgments

Supported by NIH award P30 GM110702 from the IDeA program at NIGMS. EGR would also like to express profound gratitude to all of the Federal funding agencies, especially NIH and NSF, which have continuously funded his lab for the last 40 years. We also appreciate the assistance of S. Mahaffey with the figures in this and other chapters.

References

Akhtar, M.W., Raingo, J., Nelson, E.D., Montgomery, R.L., Olson, E.N., et al., 2009. Histone deacetylases 1 and 2 form a developmental switch that controls excitatory synapse maturation and function. J. Neurosci. 29, 8288–8297.

Barbado, M., Fablt, K., Ronjat, M., De Waard, M., 2009. Gene regulation by voltage-dependent calcium channels. Biochim. Biophys. Acta 1793, 1096–1104.

Broide, R.S., Redwine, J.M., Aftahi, N., Young, W., Bloom, F.E., et al., 2007. Distribution of histone deacetylases 1–11 in the rat brain. J. Mol. Neurosci. 31, 47–58.

Cadet, J.L., 2016. Epigenetics of stress, addiction, and resilience: therapeutic implications. Mol. Neurobiol. 53, 545–560.

Calfa, G., Chapleau, C.A., Campbell, S., Inoue, T., Morse, S.J., et al., 2012. HDAC activity is required for BDNF to increase quantal neurotransmitter release and dendritic spine density in CA1 pyramidal neurons. Hippocampus 22, 1493–1500.

Chrobak, J.J., Buzsáki, G., 1988. Gamma oscillations in the entorhinal cortex of the freely behaving rat. J. Neurosci. 18, 388–398.

D'Onofrio, S., Hyde, J., Garcia-Rill, E., 2017. Interaction between neuronal calcium sensor protein 1 and lithium in pedunculopontine neurons. Physiol. Rep. 5, e13246.

D'Onofrio, S., Kezunovic, N., Hyde, J.R., Luster, B., Messias, E., et al., 2015. Modulation of gamma oscillations in the pedunculopontine nucleus (PPN) by neuronal calcium sensor protein-1 (NCS-1): relevance to schizophrenia and bipolar disorder. J. Neurophysiol. 113, 709–719.

D'Onofrio, S., Urbano, F.J., Messias, E., Garcia-Rill, E., 2016. Lithium decreases the effects of neuronal calcium sensor protein 1 in pedunculopontine neurons. Physiol. Rep. 4, e12740.
Didonna, A., Opal, P., 2015. The promise and perils of HDAC inhibitors in neurodegeneration. Ann. Clin. Transl. Neurol. 2, 79–101.
Eckhorn, R., Bauer, R., Jordan, W., Brosch, M., Kruse, W., et al., 1986. Coherent oscillations: a mechanism of feature linking in the visual system? Biol. Cybern. 60, 121–130.
Garcia-Rill, E., 2015. Waking and the Reticular Activating System. Academic Press, New York, p. 330.
Garcia-Rill, E., D'Onofrio, S., Mahaffey, S., 2016b. Bottom-up gamma: the pedunculopontine nucleus and reticular activating system. Transl. Brain Rhythm. 1, 49–53.
Garcia-Rill, E., Kezunovic, N., D'Onofrio, S., Luster, B., Hyde, J., et al., 2014. Gamma band activity in the RAS-intracellular mechanisms. Exp. Brain Res. 232, 1509–1522.
Garcia-Rill, E., Kezunovic, N., Hyde, J., Beck, P., Urbano, F.J., 2013. Coherence and frequency in the reticular activating system (RAS). Sleep Med. Rev. 17, 227–238.
Garcia-Rill, E., Luster, B., D'Onofrio, S., Mahaffey, S., Bisagno, V., Urbano, F.J., 2016. Implications of gamma band activity in the pedunculopontine nucleus. J. Neural Transm. 123, 655–665.
Garcia-Rill, E., Luster, B., Mahaffey, S., Bisagno, V., Urbano, F.J., 2015. Pedunculopontine arousal system physiology-implications for insomnia. Sleep Sci. 8, 92–99.
Garcia-Rill, E., Virmani, T., Hyde, J.R., D'Onofrio, S., Mahaffey, S., 2016a. Arousal and the control of perception and movement. Curr. Trends Neurol. 10, 53–64.
Gray, C.M., Singer, W., 1989. Stimulus-specific neuronal oscillations in orientation columns of cat visual cortex. Proc. Natl. Acad. Sci. U. S. A. 86, 1698–1702.
Haberland, M., Montgomery, R.L., Olson, E.N., 2009. The many roles of histone deacetylases in development and physiology: implications for disease and therapy. Nat. Rev. Genet. 10, 32–42.
Hyde, J., Kezunovic, N., Urbano, F.J., Garcia-Rill, E., 2013. Spatiotemporal properties of high speed calcium oscillations in the pedunculopontine nucleus. J. Appl. Physiol. 115, 1402–1414.
Jouvet, M., 2001. The Paradox of Sleep. MIT Press, Cambridge, MA, p. 227.
Kezunovic, N., Hyde, J., Goitia, B., Bisagno, V., Urbano, F.J., et al., 2013. Muscarinic modulation of high frequency oscillations in pedunculopontine neurons. Front. Neurol. 4 (176(1-13).
Kezunovic, N., Urbano, F.J., Simon, C., Hyde, J., Smith, K., et al., 2011. Mechanism behind gamma band activity in the pedunculopontine nucleus (PPN). Eur. J. Neurosci. 34, 404–415.
Lang, E.J., Sugihara, I., Llinás, R., 2006. Olivocerebellar modulation of motor cortex ability to generate vibrissal movements in rat. J. Physiol. (Lond.) 571, 101–120.
Llinas, R., 2001. I of the Vortex, From Neurons to Self. MIT Press, Cambridge, MA302.
Llinas, R., Grace, A.A., Yarom, Y., 1991. In vitro neurons in mammalian cortical layer 4 exhibit intrinsic oscillatory activity in the 10- to 50-Hz frequency range. Proc. Natl. Acad. Sci. U. S. A. 88, 897–901.
Luster, B., D'Onofrio, S., Urbano, F.J., Garcia-Rill, E., 2015. High-threshold Ca^{2+} channels behind gamma band activity in the pedunculopontine nucleus (PPN). Physiol. Rep. 3, e12431.
Luster, B., Urbano, F.J., Garcia-Rill, E., 2016. Intracellular mechanisms modulating gamma band activity in the pedunculopontine nucleus (PPN). Physiol. Rep. 4, e12787.
Middleton, S.J., Racca, C., Cunningham, M.O., Traub, R.D., Monyer, H., et al., 2008. High-frequency network oscillations in cerebellar cortex. Neuron 58, 763–774.
Palva, S., Monto, S., Palva, J.M., 2009. Graph properties of synchronized cortical networks during visual working memory maintenance. Neuroimage 49, 3257–3268.
Penfield, W., 1970. The Mystery of the Mind. Princeton University Press, Princeton, NJ, p. 157.
Philips, S., Takeda, Y., 2009. Greater frontal-parietal synchrony at low gamma-band frequencies for inefficient then efficient visual search in human EEG. Int. J. Psychophysiol. 73, 350–354.

Schorge, S., Rajakulendran, S., 2012. The P/Q channel in human disease: untangling the genetics and physiology. WIREs Membr. Transport Signaling 1, 311–320.

Simon, C., Hayar, A., Garcia-Rill, E., 2011. Responses of developing pedunculopontine neurons to glutamate receptor agonists. J. Neurophysiol. 105, 1918–1931.

Simon, C., Kezunovic, N., Ye, M., Hyde, J., Hayar, A., et al., 2010. Gamma band unit activity and population responses in the pedunculopontine nucleus (PPN). J. Neurophysiol. 104, 463–474.

Subramanian, S., Bates, S.E., Wright, J.J., Espinoza-Delgado, I., Piekarz, R.L., 2010. Clinical toxicities of histone deacetylase inhibitors. Pharmaceuticals (Basel) 3, 2751–2767.

Urbano, F.J., Bisagno, V., Mahaffey, S., Lee, S.-H., Garcia-Rill, E., 2018. Class I histone deacetylases require P/Q-type Ca^{2+} channels and CaMKII to maintain gamma oscillations in the pedunculopontine nucleus. Nat. Sci. Rep. 8, 13156.

Urbano, F.J., D'Onofrio, S.M., Luster, B.R., Hyde, J.R., Bisagno, V., et al., 2014. Pedunculopontine nucleus gamma band activity-preconscious awareness, waking, and REM sleep. Front. Sleep Chronobiol. 5, 210.

Urbano, F.J., Kezunovic, N., Hyde, J., Simon, C., Beck, P., et al., 2012. Gamma band activity in the reticular activating system (RAS). Front. Neurol. Sleep Chronobiol. 3 (6), 1–16.

Urbano, F.J., Luster, B., D'Onofrio, S., Mahaffey, S., Garcia-Rill, E., 2016. Recording gamma band oscillations in pedunculopontine nucleus neurons. J. Vis. Exp. (115)e546685.

Voss, U., Holzmann, R., Tuin, I., Hobson, J.A., 2009. Lucid dreaming: a state of consciousness with features of both waking and non-lucid dreaming. Sleep 32, 1191–1200.

West, A.C., Johnstone, R.W., 2014. New and emerging HDAC inhibitors for cancer treatment. J. Clin. Invest. 124, 30–39.

Chapter 14

The future of neuroscience and its social impact

Edgar Garcia-Rill

Center for Translational Neuroscience, Department of Neurobiology and Developmental Sciences, University of Arkansas for Medical Sciences, Little Rock, AR, United States

Introduction

Armed with the knowledge in the preceding chapters, what is the impact of our findings, treatments, and outcomes not only for the patient but also for society at large? How can we make our science more accepted and understood by policy makers, politicians, and the public in general? How can what we know help inform the policy makers? This chapter will attempt to describe how we got here and where we should go from here in terms of neurological and psychiatric disorders. To understand that, we need to know how scientists and doctors work and how our hard-won successes originated. We also need to know how our thinking has changed over the years, so that we can assess the progress we have made for the better. Later, we will address the law, specifically the issue of mental health testimony in the courts, highlighting the uninformed and archaic opinions held by judges and legal authorities on how the brain works in health and disease.

Scientists are people too

Lawyers, judges, and the public at large are all awed by the discoveries in the brain sciences. They see the thousands of journals published that each contains hundreds of articles addressing diseases, health care, and novel therapies. This is an impressive and weighty argument in favor of brain research. Neuroscientists themselves are similarly overwhelmed with the massive growth in their own literature and the number of "breakthroughs" published in what seems a weekly basis. But we must always remember that most theories are eventually proved wrong and that is business as usual in science. Patience is required to ensure that new "breakthroughs" are vetted properly by replication, validation, and acceptance.

Arousal in Neurological and Psychiatric Diseases. https://doi.org/10.1016/B978-0-12-817992-5.00014-3

The preeminent philosopher of the 20th century, Sir Karl Popper, believed that science is the search for better answers, not for absolute truth. That is, the aim of science is to achieve better and better explanations. Sir Karl Popper emphasized that a hypothesis can only be empirically tested but can never be proved to be absolutely true and can only be advanced *after* it has been tested (Popper, 1983). Also it is all right for a scientist to be wrong, as long as he is honestly wrong. That is, if you designed and performed the experiment honestly, the wrong answer was arrived at honestly. In addition, a true Popperian neuroscientist will apply the concept of falsifiability (repeatedly questioning your own theories and results) to his or her own theories and research findings, probing weaknesses so that, by surviving, the onslaught of criticism from the one scientist with the greatest familiarity with the experiment comes closer to the truth. But few scientists actually run their own research through such a gauntlet. Many scientists defend their work with a vengeance and criticize opposing theories viciously. Some even censor the work of opponents by rejecting their manuscripts or grant applications. You would think that a scientist would prefer to prove his/her own work false before someone else does, but human nature is strongly swayed by ego.

Moreover, the system exerts pressure on progress. The fact is that proving a theory correct is difficult during the typical period of a grant award, in the 3–5-year range. That is, the funding that was granted for an idea requires strong supporting evidence, expansive progress, and marked success in order for the grant to be renewed for another 3–5 years. However, few "breakthrough" theories can be proved right (or wrong) in such a short period of time; therefore more studies and a further funding period are often called for, thus the need for a renewal application. It is the role of review committees to assess progress and to keep applicants from overselling their findings. Most of the time, multiple reviewers actually agree on the quality of an application; their scores are generally close to each other. The natural tendency of a review committee is to shred an application that has weaknesses, making it difficult to secure funding, but there are instances in which unworthy grants do get funded anyway. However, the opposite is more often the case, that is, due to the shortage of funds, many worthy projects instead go unfunded. Unfortunately, many, many great ideas go unfunded.

On occasion, a novel technique has excellent "wow" value and can actually hide weaknesses. These weaknesses may take time to be recognized, especially when those jumping on the bandwagon defend it out of self-service. Some techniques have been adopted wholesale by an entire field without due consideration for proper controls, one of those involves the use of "knockout" animals (more on this later). In some cases, the particular individual is so well respected; mere reputation can hide small, although usually not glaring, weaknesses. Then there is the "halo" effect from being an applicant from a top 20 medical school, which can provide enough of a nudge to get an award funded, although it may not be better than one from the "backwaters" of science. The question is, will any of those awards lead to a major breakthrough, a new cure, or a novel effective therapy? The answer is that we do not know, only a few will provide actual breakthroughs, but if we do not fund the research, we relegate our lives to the now with no options for the future. This is reckless. For example, say a deadly virus that can be absorbed through a single layer of tissue before entering the brain (e.g., the olfactory system, in which inspired air is one cell layer away from the brain and could be a highway to infection) infects a large segment of the world's population. Such an attack could have been blunted by new technologies or alleviated by novel findings.

So, how can we tell which science to fund? How do we know which discovery is closer to the truth? How can we identify the finding that leads to the new cure? We can design ways to do all these things better but never with absolute certainty. First, we should recognize that we can be "snowed," at least for a while.

Famous wrong neuroscientific theories

The "blank slate" theory proposed by thinkers from Aristotle, to St. Thomas Aquinas, to Locke suggested that everything we know came from experience and education, while nothing came from instinct or natural predisposition. Many of the proponents can be granted some slack because genetics was not in their lexicon at the time. That is, they had incomplete knowledge. An avalanche of data have shown that many traits are inherited, along with many instincts. Therefore we are creatures of both nature and nurture.

At the beginning of the 19th century, "phrenology" posited that many traits could be localized to specific regions of the skull overlying the brain, specifically the cortex, creating detailed maps of these functions. The proponents extrapolated well beyond the available data and in many cases used the process for ulterior motives, including racism. By the 20th century, such pinpoint assignations of skull regions were discredited.

"Humans use only 10% of their brains" came from a misunderstanding of studies on evoked responses in which "primary" afferent input (e.g., vision, touch, and hearing) only activated ~10% of the cortex. This ignored the fact that most cortex is devoted to association areas that continuously process such "primary" information in both serial and parallel manners and may be "quiet" immediately after the sensory stimulus. The fact is as follows: Neurons need to fire; otherwise, their influence on their targets is lessened. Without reinforcement, synapses weaken, almost as if the input was "forgotten." "Use it or lose it" is the principle of brain activity. We use our entire brain, less of it for learned activities, more of it for novel functions.

Many studies testing drugs in animals show limited effectiveness in humans (Heywood, 1990). In fact the sensitivity for most drugs tested on animals has the probability of only a coin toss (~50%) that it will be effective in man. On the contrary, one drug, thalidomide, tested in >10 species hardly ever produced birth defects, except in humans (Shanks et al., 2009). This was a hard lesson to learn.

Most of these and less famous theories were not wrong because of scientific fraud or faulty experiments. They were mainly the result of incomplete knowledge. That includes the common problems of study size and limited technology. We offer that it is also the inadequate application of falsifiability by the proponents that could have prevented at least some of these spectacular failures. This places the responsibility for failures firmly in the neuroscientist's lap.

Interspersed with these famous failures are a host of elegant and surprising discoveries about the brain. At the turn of the 20th century, Ramon y Cajal described the fact that the nervous system is made up of individual cells. Before that the nerves throughout the body were thought to comprise a continuous network. Instead, Cajal's work led to the description of the synapse and later chemical transmission across the gaps between neurons and then to the myriad of transmitters, some of which were found to alter behavior, ultimately leading to the development of psychoactive drugs that modulate mood, movement, and other functions. Pharmacological intervention soon eliminated the need for padded rooms and "lunatic asylums," allowing many patients to live outside an institution. The advent of some of these drugs stopped the use of frontal lobotomy for the treatment of certain psychiatric disorders, mainly on women. Invented by a Portuguese physician, Moniz, who was awarded the Noble prize in 1949, the procedure of frontal lobotomy alleviated some of the symptoms it aimed at treating, but the devastating cost of the side effects (the loss of spontaneity, personality, and memory, etc.) was growingly recognized over the years.

Thirty years ago, we were born with all the cells we were going to have. It turns out we lose cells from birth and throughout puberty, but now, we know there is neurogenesis in the adult, the creation of new brain cells, a totally foreign concept until recently. How to trigger such regeneration is the topic of study in a host of neurodegenerative disorders.

There is little question that brain research has led to remarkable improvements in health and quality of life, so that the rather modest funding targeting the brain has paid off exponentially. While the National Institutes of Health are funded to the tune of ~$35 billion yearly, for all kinds of research from cancer, to the heart, to the brain, spending for research in defense is >10 times that amount. While scientific review committees discuss, dissect, and agonize over a $1 million dollar grant application for almost 1 h, Congress makes billion dollar defense funding decisions in minutes. Probably due to such stringent review the successes in brain research far outweigh the failures. But only some of those successes will result in a novel treatment. In addition, a basic discovery could easily take 20 years to be brought to the clinic.

Moreover, the yearly cost of most brain diseases in terms of medical costs, lost income, and care is in the billions of dollars, each and every year. One new treatment for a disease that was derived from a research program costing, say $5–$10 million, will save those billions every year, year after year after year. For every dollar spent on research, we stand to save thousands every year; for every dollar we do not spend on research, we stand to pay thousands every year, from now on.

Famous techniques

One of the most attractive techniques in medicine is magnetic resonance imaging (MRI) that uses strong magnetic fields stimulated with radio waves to produce field gradients that are then computed into images of the brain. Functional MRI (fMRI) uses blood oxygenation levels to compute the secondary results of neural activity. The standard displays are in black and white and allow the clinician to detect and measure tumors, infarcts, and even infection, as well as bone, fat, and blood. This tool has been a life-saving device for a host of problems in which clear, detailed, and accurate anatomical images are required. With the advent of more sophisticated MRI computation and fMRI, the displays became color coded, so that changes in blood oxygenation could be shown in beautiful false color images. And this is when proponents of the technique began overselling their product.

For example, fMRI is being used in research to make conclusions about the workings of the brain on a minute-to-minute basis. Studies from voluntary movement, to sensory perception, to more complex tasks are being undertaken. Some workers concluded that they could detect truth telling from lying, and so was born the push for fMRI as a lie detector. The issues related to the technology will not be repeated here, merely to emphasize that the field has been remiss in standardizing the generation of images. This has created a field in which the same experiment carried out by different labs produces different images and conclusions. The method suffers from widely varying coding across labs and a complexity that requires several individual decisions regarding the weights of factors that are applied differentially by researchers at multiple stages in the processing of that image. But perhaps the most serious problem is that the technique measures blood flow, not neural activity. It measures the aftermath of brain processes that could have involved both excitation and inhibition. The fMRI is, after all, a static image of an ongoing complex event that already took place. It is like standing up in the middle of a concert and taking a picture of the orchestra. From the frozen positions of the players, some conclusions are offered as to the identity of the musical piece being played.

Although this may be an exaggeration, the fact remains that the mesmerizing effect of the pretty pictures hides the fact that they are based on movable standards founded on unsupported assumptions about how the brain works. Unfortunately, overselling the technology has amassed an undeserved portion of the funding pie. Many have flocked to the technology bandwagon without developing hypotheses and controllable experiments. The BRAIN Initiative has been hijacked by imaging labs to the detriment of other avenues. Hopefully, acceptable limits of the method will be established so that those using more esoteric variables will have to better justify their decisions. At present the value of the technology to the clinical enterprise is without question; it is when complex neural processes are studied using a measure of blood flow that the temptation to extrapolate is too great.

One question that emerges is as follows: what is the harm in throwing money at the problem? Why not overfund an area of research and all the problems will be worked out? Because that is not what happens. We saw this very situation arise when funding agencies began pouring money into AIDS research. Suddenly, funding levels, which historically had funded between 5% and 15% of submitted grant applications, rose to allow funding of 20%–25% of applicants in AIDS research. The pressure was great to make breakthroughs, and the "discoveries" came hard and fast, with seemingly rapid progress toward systematically resolving the problems of such a complicated infectious process. The most responsible labs were soon confronted with the phenomenon that these "discoveries" had not been properly controlled. These workers began to spend resources and time on validating false results and outlandish theories, results that should never have been funded. They were forced to replicate many such findings before they could move the field ahead. The field actually suffered from overfunding. Who knew?

Another technique with which the public at large, including lawyers, is enamored is genetics. This very powerful array of methods has tremendous promise. As future clinical tools, they will provide the answers to a host of medical questions and may even give us cures for some of them. We need to understand the issue of genes and determinism, how genes are not deterministic, but very malleable, likely to produce different proteins under slight changes in condition. Moreover, genes are codependent, such that the expression of certain genes is not only dependent on that of some nearby (in terms of chromosome location) but also certainly dependent on some distant genes.

The field of neurogenetics addresses the links between genes, the brain, behavior, and neurological and psychiatric diseases. As such, it holds great promise for the future of clinical science, but it has created a gap. This promise and research emphasis has attracted the bulk of funding for genetic studies, pushing treatments and cures further away, that is, basic genetic studies tend to push applied research further into the future than most scientists think. This has left a gap between patients who need to be treated now with better methods and those who may be successfully treated with a genetic intervention in 20 or 30 years.

Research grants started migrating away from clinical studies into genetic studies. Because of the complexity of the genome, short-term answers are unlikely. Premature genetic interventions could be catastrophic, yet the power of the technology has moved money away from proximal interventions in the clinic, that is, away from translational neuroscience. Translational neuroscience is designed to bring basic science findings promptly to the clinic (Garcia-Rill, 2012). Translational research is a response to an Institute of Medicine report from 2003, calling for more emphasis in this area (Kohn, 2004). The reason for the concern was the gradual decrease in research grants awarded to MDs (presumably doing research on patients) compared with PhDs (presumably doing research on animals). Over a 10-year period in the 1990s the percentage of MDs with awards had decreased from 20% to 4%, and the percentage may be lower today. While some lip service has been paid to increasing translational research funding, the fact is that most grant reviewers are basic scientists and not very familiar with clinical conditions and human subject research. Animal studies are more easily controlled than human subject studies, so that there is an inherent difference that makes for lower funding levels for human studies. It is incumbent on the research community to correct the discrepancy because they stand to lose public trust. We now live in a world of immediate gratification, and cures far off in the future will not be kindly considered.

Is the emphasis on genetics and molecular biology warranted? Definitely, but not at the expense of advances that could improve the quality of life of patients now rather than later. In addition, some self-evaluation is called for from the molecular biology community. For example, a firm realization of the limits of their own technology is a must. One area is the knockout mouse, in which a genetically modified mouse has undergone a process whereby a gene is kept from expressing or deleted from the mouse's genome. Knocking out the activity of a gene provides knowledge about what that gene does, making a marvelous model for the study of disease. The process is complicated and imaginative, and very effective if properly employed. Because of the variety of genes, the technology has created an opportunity for everyone to have their own knockout mouse. There are a myriad of new genetically modified mouse lines that researchers can make, buy, study, and manipulate.

The three scientists who developed the technology won the Nobel Prize for Physiology or Medicine in 2007. The knockout technology has perhaps been most successful in identifying genes related to cancer biology. These animals allow the study of genes in vivo, along with their responses to drugs, especially since humans and mice share ~80% of genes. The problem has been the inability to generate animals that faithfully recapitulate the disease in man. The uses of this method in neuroscience are numerous, but there are problems. Aside from the typical issues that ~15% of knockouts are lethal and some fail to produce observable changes, there is the fact that knocking out a gene will upregulate many, many other genes and downregulate another large group of genes (Iacobas et al., 2004). In nature, single-gene mutations that survive are very rare, so the knockout is not simply a study of such mutations; it is an attempt at learning all that a single gene does. The problem is that, without knowing which OTHER genes are up- or downregulated, the knockout animal represents an uncontrolled experiment on a creature that never would have existed in nature.

Some problems can be overcome by using conditional mutations, in which an agent added to the diet could make a gene express or cease expressing temporarily. The problem with this approach is that it does not control the up- or downregulation of linked genes whose identities are unknown. In addition, none of these methods measure compensation. Very few researchers verify how the absence of the gene creates compensations in changes in expression of related genes. For example, knocking out a gap junction protein, connexin 36, creates an animal that has no connexin 36, but it appears to lead to over expression of similar other connexins, such as connexin 32 (Spray and Iacobas, 2007). The field of knockout mouse lines is rampant, growingly uncontrolled, and funded way beyond its current scientific affirmation.

Another concern with knockout animals is the comorbidities, many of which go unreported. By this, I mean that animal "models" of human diseases may be similar to one aspect of the human condition, say, memory dysregulation, and workers emphasize that aspect of the "model" to secure funding. However, many mutations have serious conditions that could impact memory formation, for example, metabolic disorders, immune disorders, sensorimotor problems, and morbid obesity. In many cases, these easily observable (in the lab that generated the animals) problems are never published. Until funding agencies insist on full disclosure, the literature will be peppered with faulty or inaccurate leads in the search of true mechanisms.

A final issue is that, as far as the brain is concerned, protein transcription is a long-term process. That is, the workings of the brain are in the millisecond range. Over the last several minutes during the reading of this chapter, transcription was irrelevant. None of the perceptive, attentional, or comprehensive elements of the information on these pages required gene transcription of proteins. Certainly, the long-term storage of the information into memory requires gene transcription, but not before. Our brain takes about 200 milliseconds to consciously perceive a stimulus. Gene action, in the order of minutes to hours, is not in the same galaxy in terms of time. Assessing thought and movement in real time is too fast for genetic methods, but not for two technologies, the electroencephalogram (EEG) and the magnetoencephalogram (MEG).

The EEG amplifies electric signals from the underlying cortex (just from the surface of the brain, not from deep structures), but these signals are distorted by skull and scalp. The MEG measures the magnetic fields of these electric signals but requires isolation by recording rooms, massive computational power, and superconductors that live in very expensive liquid helium. The development of helium-free MEGs is imminent, and it would make the technology less expensive to operate. The MEG is also very useful in producing exquisite localization of epileptic tissue, especially the initial ictal (seizure activity) event. As such, it is reimbursable for diagnostic and surgical uses. As the only real-time localizable measure of brain activity, it is likely to make inroads into the rapid events in the brain.

Transparency and rigor

The National Institute for Neurological Disorders and Stroke convened a meeting of stakeholders on how to improve the methodological reporting of animal studies (Landis et al., 2012). The workshop recommended that at minimum studies should report on sample size estimation, whether and how animals were randomized, whether investigators were blind to the treatment, and the handling of the data. These conclusions stemmed from evidence that many studies could not be replicated due to incomplete reporting and inappropriate implementation of the methods. Only a minority of articles reported randomization but how this was done usually was not described. In very few articles were investigators blind to treatment. The workshop laid out a set of reporting standards for rigorous study design. The article cited numerous instances of erroneous conclusions due to small sample size. For example, with a sample of 10 animals, the results could yield significance at $P<0.05$, or 1 in 20 by pure chance, but with a sample of 24, results would yield a $P<0.001$, or 1 in 100.

An additional factor, publication bias, now enters the equation. If 10 labs performed the study with low sample size and only one obtained significance, purely out of chance, only that lab would publish the results, and the others shelve the study. The need to publish negative results, and for reviewers to accept them if there was proper study design, is becoming growingly important. Errors from incomplete methods also introduces new issues, for example, calling for more extensive methods sections in articles will produce pushback from the publication industry to limit page number. If all the labs had used greater sample size and obtain significance, most would publish, and the finding would be well established. In any case, there needs to be a change in the culture to promote more transparency and rigor into science, despite the added pressures to publish to secure grant support and academic advancement. Despite being enamored with our own theories, we need to strive for a more ethical approach. As scientists, we should be guided by a principle substantiated in a lecture by W. K. Clifford of University College, London, in 1876, entitled "The Ethics of Belief" in which he proposed, "it is wrong always, everywhere, and for anyone, to believe anything upon insufficient evidence" (Clifford, 1877).

The revolution

Before we deal with the future of neuroscience and how it impacts the law, we have to understand the battle within the brain sciences that has led to the current state. Subsequent to Newton's deterministic theories of how the world worked there arose the idea that the brain worked the same way. That is, all brain function could be reduced to the smallest physical components, that is, the ultimate in microdeterminism. This approach was manifested in the brain sciences in the form of "behaviorism", that all actions and thoughts were due to the physicochemical nature of the brain. A major proponent was B. F. Skinner, who considered free will an illusion, and that everything you did depended on previous actions. The fervor for this view was fanned by the advances in molecular biology and the structure of DNA. There was no room for the consideration of consciousness or subjective states. This was the world of the reductive microdeterministic view of the person and the world. These views influenced education and policy, suggesting that the issue was not to free man but to improve the way he is controlled. This was the "behaviorist" one-way reductionist view of the world in general and the brain in particular (Sperry, 1976).

The implication of "behaviorism" for thought and action was that consciousness was an epiphenomenon of brain activity, and that this approach provided a complete explanation of the material world. This deterministic view of the world began to crumble under the weight of the discoveries of quantum mechanics. The old deterministic idea of behavior and the absence of free will were undermined by the advances in quantum mechanics. Behaviorism thus was replaced by a "cognitive revolution" that espoused mental states as dynamic emergent properties of brain activity. That is, there was a two-way street between consciousness and the brain, fused to the brain activity of which it is an emergent property. This is not to imply a dualism, two independent realms, but rather mental states are fused with the brain processes that generate them.

The "cognitive revolution" implied a causal control of brain states downward and upward determinism. This two-way approach offers a solution of the free will versus determinism paradox. This cognitive approach retains both free will and determinism, integrated in a manner that provides moral responsibility (Sperry, 1976). Volition remains causally determined but no longer subject to strict physicochemical laws. This is the current scientific mainstream opinion. It is one that no longer considers there to be a mind versus brain paradox, but a singular, functionally interactive process. Instead of placing the "mind" within physicochemical processes, thought became an emergent property of brain processes.

To use a simplistic parallel, the brain is to thought, and action as the orchestra is to music. Thought and action are emergent properties of the brain just as music is an emergent property of the orchestra. Music cannot exist without the orchestra, just as thought and action cannot exist without the brain. This is a solitary relationship, one in which brain states influence thought and action (downward) and the external world modulates the activity of the brain (upward). The "mind" or consciousness can be viewed as downward control of a system changing due to continuously impinging external inputs.

This new viewpoint combines bottom-up determinism with top-down mental causation, the best of both worlds. The world of reductionism is not entirely rejected, merely considered not to contain all the answers, and an entirely new outlook on nature is manifest. The revolution in science has provided new values, whereby the world is driven not only by mindless physical forces but also by mental human values. This revolution places a premium on values and consistency, principles that apply to all human endeavors including science and the law.

Relevance to the law

The emphasis that the Supreme Court placed on Popper's philosophy in the formulation of the Daubert v. Merrell Dow Pharmaceuticals, Inc, 1993, decision reflects the best elements of the cognitive revolution. The 1993 *Daubert* decision basically involved a mandate by the Supreme Court that required judges to evaluate the scientific validity of "fit" of expert testimony. Until that time, the validity of expert testimony was up to the jury, leading to a host of abuses by "professional" legal experts who created a lucrative career rendering opinions that juries believed, whether correct or not. The Supreme Court based its opinion on Popper's philosophy. Popper redefined an ideological revolution in his book, *Realism and the Aim of Science* (Popper, 1983), and the court understood its value. The principles of falsifiability, diversity of explanation, rationalism (arguing with others), along with error rate, peer review, publication, and general consensus, all reflect the values of Popper's philosophy and apply them to the law. These were integrated into the Superior Court's decision, making judges the gatekeepers of expert testimony, a role not relished by many. Judges are now required to understanding that the probabilistic thinking used by scientists is the key to sound scientific validity determinations. Appellate judges who must decide whether trial judges met their gatekeeping duties need to know how to reason about science. They also need to know that mental state is not an either-or proposition, but a probabilistic statement about the likelihood of being in control of one's faculties at any given time.

Why is this important? In some cases, courts have upheld the constitutionality of proving sanity by the testimony of lay witnesses (Moore v. Duckworth, 1979) or allowing testimony about the likelihood of someone to commit a crime in the future, that is, "future dangerousness" (Barefoot v. Estelle, 1983), neither of which could pass the requirements of *Daubert*. The trend continues in courts of not allowing mental health testimony. A teenager who had been in and out of mental institutions since childhood, diagnosed with paranoid schizophrenia, acting bizarrely and convinced that aliens have invaded, drove around his neighborhood playing loud music until a policeman appeared in answer to neighbors' 911 calls. The young man shot him, believing the officer to be an alien. The murder statute under which the young man was charged requires the knowing killing of a police officer. The court deemed evidence about paranoid schizophrenia proffered to negate intent was irrelevant. This expert testimony was also excluded as irrelevant to the insanity defense, because the statute limited that defense to the accused's ability to know that his act was wrong (Clark v. Arizona, 2006). A young mother, treated and hospitalized repeatedly for paranoid schizophrenia, drowned her young son in the bathtub, because, as she told police, the social interactions and obligations of motherhood were too overwhelming (Commonwealth v. Tempest, 1981). She thought about her actions for days, and planned the murder in detail. That was sufficient, according to the court, to make any evidence about her mental illness irrelevant to negating willful intent. Another young man killed his father, step-mother, and health aide because, he said, God commanded him to aid in their transition to reincarnation as younger, better versions of themselves (State v. Bethel, 2003). This young man had also been diagnosed as a paranoid schizophrenic. Mental state evidence of his mental illness was similarly excluded as irrelevant to the question of intent. In each of these cases, the requisite element of intent was found and the defendants convicted. In each case, expert testimony that might have explained why the defendants could not have knowingly or purposefully acted as they did was excluded as irrelevant.

The current system has filled our so-called "correctional" institutions to the brim with mentally ill persons (Hoke, 2015). These institutions are incapable of dealing with so many of the mentally ill (half of all prisoners suffer from mental illness), and little appropriate care is provided to these persons by jails and prisons (40% of these prisoners recidivate) (Hoke, 2015). We need a new approach to dealing with the mentally ill, especially those who have been incarcerated. Throughout this book, we provide evidence that many of the mentally ill cannot control their thoughts and actions. Therefore, proper diagnoses and treatments are essential to "correctional" success. We must do this not only because it is the right thing to do but also because the costs of keeping many of our mentally ill in jails are becoming prohibitive. What we need now is a new paradigm of human behavior based on empirical studies rather than blind assertions. This begins with information and education.

In Chapter 2, we discussed the idea of preconscious awareness as an essential consequence of arousal. In a recent law review, we provided the example of someone walking along a sidewalk while in conversation yet being "preconscious" of traffic and other pedestrians so as to allow for safe navigation (Garcia-Rill, 2015). We concluded that we are "conscious" while performing the movement and that the activity preceding the intent is a manifestation of "our preconscious awareness." That is, we are preconscious to the performance of our movements and therefore responsible for all of our actions. And yes, free will is alive and well. Studies by Libet in the early 1980s using the readiness potential had suggested that our voluntary movements begin "unconsciously" because brain waves are manifested in advance of the subjective "will" to move. This led to the suggestion that there is no free will. Libet surmised that individuals still have the ability to stop actions or movements of which they are not fully conscious. This became known as "free wont." However, work on brain regions such as the reticular activating system that subserve preconscious awareness, suggest that such activity is present in our brains while conscious (not when we are unconscious), but we are just not paying attention to it, thus the name "preconscious awareness" (Beecher-Monas and Garcia-Rill, 2018).

We argue that the legal concepts of criminal act and criminal intent are outmoded. The legal meaning of choice, intent, and volition originated not from empirical studies about human brains and behavior, but from ungrounded beliefs about human nature. The law still operates on outdated, 19th century assumptions about how human beings function. For example, refusing to admit expert testimony about mental illness is contributing to the complex problem of mental illness among prison inmates and failing to protect society when the convicted are released. Judges should instead admit expert mental health testimony so that the *Daubert* determination and the jury can perform its interpretive and evaluative functions.

References

Barefoot v. Estelle, 463 U. S. 880, 1983.

Beecher-Monas, E., Garcia-Rill, E., 2018. Actus reus, mens rea and brain science: what do volition and intent really mean? Ky. Law J. 166, 265–314.

Clark v. Arizona. 126 S. Ct. 2709, 2006.

Clifford, W.K., 1877. The ethics of belief. Originally published in contemporary review. In: Stephen, L., Pollock, F. (Eds.), Reprinted in Lectures and Essays. Macmillan and Co., London, pp. 1886.

Commonwealth v. Tempest, 496 Pa. 436, 437 A.2d 952, 1981.

Daubert v. Merrell Dow Pharmaceuticals, Inc, 509 U. S. 579, 593, 1993.

Garcia-Rill, E., 2012. Translational Neuroscience: A Guide to a Successful Program. Wiley-Blackwell, p. 168.

Garcia-Rill, E., 2015. Waking and the Reticular Activating System. Academic Press, New York, ISBN: 9780128013854, pp. 330.

Heywood, R., 1990. Clinical toxicity—could it have been predicted? Post-marketing experience. In: Lumley, C.E., Walker, S., Lancaster, Q. (Eds.), Animal Toxicity Studies: Their Relevance for Man, pp. 57–67.

Hoke, S., 2015. Mental illness and prisoners: concerns for communities and healthcare providers. Online J. Issues Nurs. 20, 3.

Iacobas, D.A., Scemes, E., Spray, D.C., 2004. Gene expression alterations in connexin null mice extend beyond gap junctions. Neurochem. Int. 45, 243–250.

Kohn, L.T., 2004. Committee on the role of academic health centers in the 21st century. In: Kohn, L.T. (Ed.), Academic Health Centers; Leading Change in the 21st Century. Institute of Medicine. National Academies Press, Washington, DC.

Landis, S.C., Amara, S.G., Asadullah, K., Austin, C.P., Blumenstein, R., et al., 2012. A call for transparent reporting to optimize the predictive value of preclinical research. Nature 490, 187–191.

Moore v. Duckworth, 443 U. S. 713, 1979.

Popper, K., 1983. Realism and the aim of science. In: Bartley, W.W. III (Eds.), Postscript to the Logic of Scientific Discovery. Harper Torchbooks, New York, pp. 465.

Shanks, N., Greek, R., Greek, J., 2009. Are animal models predictive of humans? Philos. Ethics Humanit. Med. 4, 2.

Sperry, R.W., 1976. Changing concepts of consciousness and free will. Perspect. Biol. Med. 20, 9–19.

Spray, D.C., Iacobas, D.A., 2007. Organizational principles of the connexin-related brain transcriptome. J. Membr. Biol. 218, 39–47.

State v. Bethel, 66 P.3d 840, Kan. 2003.

Index

Note: Page numbers followed by *f* indicate figures and *t* indicate tables.

O

P

Q

R

S